Electrical Transmission System Cascades and Vulnerability

MOS-SIAM Series on Optimization

This series is published jointly by the Mathematical Optimization Society and the Society for Industrial and Applied Mathematics. It includes research monographs, books on applications, textbooks at all levels, and tutorials. Besides being of high scientific quality, books in the series must advance the understanding and practice of optimization. They must also be written clearly and at an appropriate level for the intended audience.

Series Volumes

Bienstock, Daniel, *Electrical Transmission System Cascades and Vulnerability: An Operations Research Viewpoint*
Koch, Thorsten, Hiller, Benjamin, Pfetsch, Marc E., and Schewe, Lars, editors, *Evaluating Gas Network Capacities*
Corberán, Ángel, and Laporte, Gilbert, *Arc Routing: Problems, Methods, and Applications*
Toth, Paolo, and Vigo, Daniele, *Vehicle Routing: Problems, Methods, and Applications, Second Edition*
Beck, Amir, *Introduction to Nonlinear Optimization: Theory, Algorithms, and Applications with MATLAB*
Attouch, Hedy, Buttazzo, Giuseppe, and Michaille, Gérard, *Variational Analysis in Sobolev and BV Spaces: Applications to PDEs and Optimization, Second Edition*
Shapiro, Alexander, Dentcheva, Darinka, and Ruszczyński, Andrzej, *Lectures on Stochastic Programming: Modeling and Theory, Second Edition*
Locatelli, Marco and Schoen, Fabio, *Global Optimization: Theory, Algorithms, and Applications*
De Loera, Jesús A., Hemmecke, Raymond, and Köppe, Matthias, *Algebraic and Geometric Ideas in the Theory of Discrete Optimization*
Blekherman, Grigoriy, Parrilo, Pablo A., and Thomas, Rekha R., editors, *Semidefinite Optimization and Convex Algebraic Geometry*
Delfour, M. C., *Introduction to Optimization and Semidifferential Calculus*
Ulbrich, Michael, *Semismooth Newton Methods for Variational Inequalities and Constrained Optimization Problems in Function Spaces*
Biegler, Lorenz T., *Nonlinear Programming: Concepts, Algorithms, and Applications to Chemical Processes*
Shapiro, Alexander, Dentcheva, Darinka, and Ruszczyński, Andrzej, *Lectures on Stochastic Programming: Modeling and Theory*
Conn, Andrew R., Scheinberg, Katya, and Vicente, Luis N., *Introduction to Derivative-Free Optimization*
Ferris, Michael C., Mangasarian, Olvi L., and Wright, Stephen J., *Linear Programming with MATLAB*
Attouch, Hedy, Buttazzo, Giuseppe, and Michaille, Gérard, *Variational Analysis in Sobolev and BV Spaces: Applications to PDEs and Optimization*
Wallace, Stein W. and Ziemba, William T., editors, *Applications of Stochastic Programming*
Grötschel, Martin, editor, *The Sharpest Cut: The Impact of Manfred Padberg and His Work*
Renegar, James, *A Mathematical View of Interior-Point Methods in Convex Optimization*
Ben-Tal, Aharon and Nemirovski, Arkadi, *Lectures on Modern Convex Optimization: Analysis, Algorithms, and Engineering Applications*
Conn, Andrew R., Gould, Nicholas I. M., and Toint, Phillippe L., *Trust-Region Methods*

Electrical Transmission System Cascades and Vulnerability

An Operations Research Viewpoint

Daniel Bienstock

Columbia University
New York, New York

Society for Industrial and Applied Mathematics
Philadelphia

Mathematical Optimization Society
Philadelphia

10 9 8 7 6 5 4 3 2 1

Publisher	David Marshall
Acquisitions Editor	Elizabeth Greenspan
Developmental Editor	Gina Rinelli
Managing Editor	Kelly Thomas
Production Editor	Louis R. Primus
Copy Editor	Julia Cochrane
Production Manager	Donna Witzleben
Production Coordinator	Cally Shrader
Compositor	Techsetters, Inc.
Graphic Designer	Lois Sellers

Library of Congress Cataloging-in-Publication Data

Names: Bienstock, Daniel.
Title: Electrical transmission system cascades and vulnerability : an operations research viewpoint / Daniel Bienstock.
Description: Philadelphia : Society for Industrial and Applied Mathematics : Mathematical Optimization Society, [2015] | Series: MOS-SIAM series on optimization ; 22 | Includes bibliographical references and index.
Identifiers: LCCN 2015030489 | ISBN 9781611974157
Subjects: LCSH: Electric power transmission–Reliability. | Distributed resources (Electric utilities)–Reliability.
Classification: LCC TK3001 .B47 2015 | DDC 621.319–dc23 LC record available at *http://lccn.loc.gov/2015030489*

Contents

Preface

This monograph explores mathematical issues related to vulnerability analysis of power grids, with a focus on optimization methods. The target audience is primarily made up of operations researchers and applied mathematicians with an interest in electrical power transmission, as well as power engineering specialists who are interested in optimization topics, especially nonconvex optimization and optimization under uncertainty. This monograph will present outlines of appropriate optimization and power engineering fundamentals before embarking on the study of the main topic, which is the rigorous analysis of power grid security. The level of mathematics is intermediate, with occasional developments at greater depth and even some detailed proofs.

The focus of this monograph is the power grid, which can be considered one of the greatest achievements of twentieth-century engineering, and from which populations around the world have attained an improved standard of living. Moreover, modern grids operate around the clock with remarkable constancy and reliability, and the secure delivery of power is taken for granted. These facts can be further explored as follows. The operation of power grids is driven by two core goals: (i) efficient economics, and (ii) secure, i.e., stable, operation. The first goal naturally gives rise to extensive use of optimization at many time scales. The second goal requires a strict adherence to operational criteria designed to keep a grid within a narrowly defined set of available states. This is an overriding principle that trumps even the need for efficiency—it is constantly applied to an extent that a non-power specialist may find surprising. This principle is to a large extent framed through the use of systems of quadratic inequalities designed to reflect the underlying physics of power transport and generation. Often, such nonlinear and nonconvex feasibility problems are recast as optimization problems to account for missing information or infeasibility conditions.

The integration of efficiency and security implies, as a consequence, the need for accurate and efficient optimization as a basic ingredient of grid operation. Altogether the record has been very good. Grids are attaining increasing efficiency and have, overall, proved quite stable.

However, as grids and populations have grown, this success has been punctuated by some high-visibility failures. In fact, during the last fifteen years several significant blackouts have taken place around the world, affecting tens of millions of people and causing significant economic and social disruption. All of these events were caused by a combination of accidental, exogenous causes and human error. The potential for more serious events remains high. Recently, the U.S. National Academies issued a report warning of the possibility of malevolently caused blackouts, carrying an even higher risk.

Blackouts are often complex events because they are caused by a *cascade* of failures that accelerates and outstrips control capabilities. In a cascade, the disabling of a piece of equipment typically causes other pieces of equipment to become overloaded, which in turn increases the likelihood of their failure. Should this process become self-sustaining it will tend to accelerate, overwhelming control capabilities and finally causing a large loss of power delivery—a national-scale blackout.

Since blackouts do happen, it is important to develop methodology for understanding the mechanism of an ongoing cascading failure so that, for example, one can develop fast control algorithms that are provably good, that is to say provably fast and effective. This goal requires appropriate modeling of myriad issues, such as an explanation of how equipment fails and how electrical power "load" is redistributed following a set of outages. As we will discuss, this is an ambitious goal because of the nature of the physics that underlies electrical power, the complexity of the equipment and existing control schemes, and the overabundance of numerical parameters that ideally should be precisely calibrated but in practice are sometimes unknown. A broad part of this monograph targets these issues.

A separate issue concerns the prevention of cascading failures, that is to say the goal of achieving "stability," broadly defined. Modern power grids are clearly stable in the sense that they rarely fail. But since they do occasionally fail, an important question is what structural features constitute a weakness that would be revealed in a failure. The search for such hidden weaknesses becomes an important planning task. The word "hidden" is appropriate in this context since those weaknesses that do remain, despite ongoing improvement efforts, can be considered complex or counterintuitive. The development of an appropriate agnostic methodology for rooting out such weaknesses is therefore an important task. This problem, which is also framed as an optimization problem, is frequently known as the N-K problem and constitutes a second focus of this work.

The need to resolve the above modeling questions in an agnostic manner gives rise to a host of complex and challenging optimization problems, incorporating many facets of modern optimization theory and practice, including linear and nonlinear programming; convex, nonconvex, and discrete optimization; and stochastic and robust optimization. As noted above, this is a natural rather than a forced modeling outcome.

This monograph is organized as follows. Chapter 1 provides an introduction to power engineering which should prove accessible to a reader with a basic applied mathematics background, as well as mathematical statements of several problems addressed in power grid operations. Chapter 2 focuses on mathematical topics related to the computation of power flows in a grid, i.e., the basic computation underlying grid operations today, and introduces the use of semidefinite relaxations, filter methods, and derivative-free methods in this context. Chapter 3 considers the N-K problem, the basic computational task used to assess the reliance of a grid, and describes the use of mixed-integer, nonconvex, and chance-constrained optimization as applied to the N-K problem. Chapter 4 introduces and surveys the modeling of cascading failures of transmission systems, followed by more detailed modeling in Chapter 5. Chapter 6 describes some initial work on rigorous online control schemes that could be used to stop, or slow down, a cascading failure, as well as more detailed models of cascading failures, and appropriate mathematical optimization methods. Throughout, we rely on extensive use of examples designed to reveal counterintuitive behavior of power systems, especially issues related to nonmonotonicity, nonconvexity, and Braess's paradox. Data not given in the text is provided on the book's webpage, `www.siam.org/books/mo22`.

Chapter 1

Power engineering basics

1.1 ▪ Introduction

A power grid, or *interconnection system*, conveys power from generators to demand locations, or *loads*. A power grid comprises two subsystems: a *transmission system* that carries power at high voltage, typically over long distances, and a collection of *distribution systems*, each of which delivers power within a narrow geographical area (such as a city) at lower voltage. Typically, distribution systems are, from a graph-theoretic perspective, either trees or tree-like and are attached to the transmission system at a single point. We can view a transmission system as a graph; in the parlance of the power community the nodes of the graph are called *buses*; the edges of the graph, or *branches*, represent power *lines* used to transmit power between buses. Some of the buses may house one or many generators; and some buses represent points where a distribution system (a *load*) is attached. In what follows we will preferentially use power terminology (e.g., buses and lines) over graph-theoretic terminology. For a thorough study of power systems we refer the reader to any standard textbook (e.g., [8], [25], [106], from which we draw here), but a quick review is appropriate. Also see [176].

From a hierarchical perspective, generators produce power at relatively high voltage (e.g., 20 KV) and are attached to the transmission system by means of *transformers*. Transformers are ubiquitous, and a large transmission system may incorporate them in the hundreds, or more. A transformer is, typically, a two-sided device which takes as input current at a certain voltage and outputs current at a different voltage (while possibly changing phase angles, as well). In the case of transformers used to connect generators to a transmission system, voltage will be raised to a higher level suitable for transmission (e.g., 345 KV and higher); at the opposite end of the hierarchy transformers act to reduce voltage to the lower values used by distribution systems (approximately 13 KV).[1] The reliance on high voltages in transmission systems is motivated by efficiency—power carried at lower voltages will experience thermal losses.

[1]Within a distribution system, transformers are again used to reduce voltages to the much lower levels appropriate for household or commercial consumption.

Basic electromagnetism

The notions of "power," "voltage," and "current" are, likely, informally familiar to the reader. We briefly review some of the underlying physics, which is rooted in the classical theory of electromagnetism. A more thorough, modern view on this topic would involve quantum electrodynamics. We strongly encourage the reader to consult an appropriate physics textbook.

- First, we have the concept of *charge*, which is taken as a primitive (undefined) notion in classical physics. A *point charge* is an idealized concept of a charge occupying a point in space; an electron is an example of a charged particle.
- *Current* represents the movement of charge; more precisely, current is measured as the rate of movement of charge across a surface, per second.
- A collection of charges gives rise to a *force field*; this is a vector field that assigns to each point in space the force vector that would be experienced by a point unit charge located at that point. Likewise, a time-varying *magnetic field* (for example, generated by the motion of a magnet) will give rise to an electric field.
- Another crucial concept is that of *voltage* and the allied concept of *electrical potential energy*. Given an electric field, IF we introduced a new charged particle and moved it ("against" the field), the energy required to effect the movement would be conveyed to the particle in the form of "potential" energy—were we to release the particle, this energy would now be expended as the particle moved back "along" the field (much as raising an object against the force of gravity imbues the object with "potential" energy, which is deployed once we release the object and it falls). Potential electrical energy, per unit charge, is called *voltage*.
- Suppose that a conductor (e.g., a conducting cable) carries a current of i units at voltage v. The product vi has the units of energy/charge×charge/time, or "energy per unit time," that is to say, *power*. A generator produces current at a certain voltage by means of a time-varying magnetic field, which induces the current at that voltage, thereby transferring power from some form (e.g., thermal) to electrical power.
- *Inductance* (arising from magnetism) and *capacitance* are related effects that impart a dynamic relationship between voltage and current. In the presence of inductance, voltage is proportional to the derivative (with respect to time) of current; the reverse relationship is caused by capacitance. *Resistance*, on the other hand, is a purely static effect: the voltage change across the two ends of a conductor is proportional to the current flowing between the ends.

As discussed above, a transmission system can be represented as an undirected graph, where some of the nodes hold generators and some of the nodes represent loads. In order to understand how power flows in a transmission system, we need to understand two additional concepts: AC (alternating current) power, and the Π- or lumped-circuit model of a power line.

Consider first the case of a generator located at a bus k and a line connecting bus k with another bus, m, in an AC grid. The generator will "inject" into the line time-dependent current $i_{km}(t)$ at voltage $v_k(t)$ of the general form

$$v_k(t) = V_k^{\max}\cos(\omega t + \theta_k^V), \quad i_{km}(t) = I_{km}^{\max}\cos(\omega t + \theta_{km}^I). \tag{1.1}$$

Here, $V_k^{\max}$ and $I_{km}^{\max}$ are constants (the *amplitude* of voltage and current, respectively), as are θ_k^V and θ_{km}^I (the *phase* of voltage and current). The parameter ω is also a constant—this is the frequency, which is, under stable conditions, a system-wide parameter. As we will discuss below, it is important in a transmission system that all generators operate at a common frequency (60 Hz or 50 Hz across different countries). From (1.1) we obtain that the (instantaneous) power injected by k into (k,m) at time t is

$$p_{km}(t) = v_k(t)i_{km}(t) = \frac{1}{2}V_k^{\max}I_{km}^{\max}[\cos(\theta_k^V - \theta_{km}^I) + \cos(2\omega t + \theta_k^V + \theta_{km}^I)], \tag{1.2}$$

using relationships from trigonometry. Note that $p_{km}(t)$ may change sign; however, due to the very rapid system frequency, it makes sense to consider the *average* value of $p_{km}(t)$ over one period $T = \frac{2\pi}{\omega}$; thus we obtain

$$p_{km} \doteq \frac{1}{T}\int_0^T p(t) = \frac{1}{2}V_k^{\max}I_{km}^{\max}\cos(\theta_k^V - \theta_{km}^I). \tag{1.3}$$

It is useful to restate the above conditions by means of complex numbers. Note that we can write

$$v_k(t) = V_k^{\max}\,\mathcal{R}\mathrm{e}\,e^{j(\omega t + \theta_k^V)}, \quad i_{km}(t) = I_{km}^{\max}\,\mathcal{R}\mathrm{e}\,e^{j(\omega t + \theta_{km}^I)},$$

where $j \doteq \sqrt{-1}$ and $\mathcal{R}\mathrm{e}$ indicates the real part. Thus, defining

$$V_k \doteq \frac{V_k^{\max}}{\sqrt{2}}e^{j\theta_k^V}, \quad I_{km} \doteq \frac{I_{km}^{\max}}{\sqrt{2}}e^{j\theta_{mk}^I},$$

i.e., the (complex) *voltage at k* and *current injected at k into (k,m)*, we obtain

$$p_{km} = |V_k||I_{km}|\cos(\theta_k^V - \theta_{km}^I) = \mathcal{R}\mathrm{e}(V_k I_{km}^*) \tag{1.4}$$

(here, for a complex number v, $|v|$ denotes its norm and v^* is its conjugate). The quantity p_{km} is known as the *active* (or *real*) power injected by k into line (k,m) (we will discuss this choice for line notation below, in Section 1.2). Equation (1.4) amounts to Ohm's law for AC power. Consequently, it is also of interest to consider the related quantities

$$q_{km} \doteq \mathrm{Im}(V_{km}I_{km}^*) \quad \text{and} \quad S_{km} \doteq p_{km} + jq_{km}. \tag{1.5}$$

The quantity q_{km} is known as the *reactive* power and S_{km} is the *complex* power, injected by k into (k,m). Even though q_{km} seems to arise accidentally from the equations, it does correspond to a quantity of significant importance in power grids. It can be shown that, broadly speaking, operating a transmission system so as to increase the quantities $Q_k = \sum_{km} q_{km}$ at load buses k leads to increased efficiency (and, thus, to increased stability; see below). In fact, the above equations (1.1)–(1.4) apply to any bus k (not just generator buses) and line (k,m), always from the point of view of current or power injected *by* the bus *into* the line.

1.2 ▪ Power flows

To understand such notions of "operation," "efficiency," and "stability," it is important to understand how power and current flow in a grid, and this requires a mathematical model of power lines.

> **Power line notation**
>
> From a mathematical standpoint it is best to model a power grid as a graph with *undirected* edges; principally this is because power can flow in any direction on an edge and a priori it is generally unknown which direction that will be. In what follows, we will denote a line joining buses k and m, from the perspective of bus k, as (k, m). The same line, from the perspective of bus m, will be denoted (m, k). This roughly corresponds to the notion that active or reactive power is injected into a given line from both ends, though possibly in negative amounts. When considering line parameters that do not depend on orientation we will use the notation $\{k, m\}$.

The Π- or lumped-circuit model is an approximation (in fact, an abstraction) of the physical behavior of a line. This model endows a line with several parameters: the (series) *impedance*, $z_{\{k,m\}} = r_{\{k,m\}} + j x_{\{k,m\}}$ (where $r_{\{k,m\}}$ is the *resistance* and $x_{\{k,m\}}$ is the reactance), and the *shunt admittance*, $y^{sh}_{\{k,m\}} = g^{sh}_{\{k,m\}} + j b^{sh}_{\{k,m\}}$ (where $g^{sh}_{\{k,m\}}$ is the *shunt conductance* and $b^{sh}_{\{k,m\}}$ is the *shunt susceptance*). The reader may be intuitively familiar with the concept of resistance; reactance concerns self-inductive effects, and in fact both resistance and reactance quantify opposition to AC power transmission. The shunt parameters concern capacitative effects, in particular related to the interaction of the line with other lines and ground. Typically, $g^{sh}_{\{k,m\}} \ll b^{sh}_{\{k,m\}}$; $g^{sh}_{\{k,m\}}$ is often set to zero. In some calculations, e.g., involving comparatively short lines, $b^{sh}_{\{k,m\}} = 0$. Likewise, resistance is usually an order of magnitude smaller than reactance and is also, sometimes, ignored in computations. The reactance ($x_{\{k,m\}}$) is always positive in a realistic model of a power line.

A parameter related to series impedance is the series admittance,

$$y_{\{k,m\}} \doteq z^{-1}_{\{k,m\}} = g_{\{k,m\}} + j b_{\{k,m\}}, \quad \text{where} \tag{1.6}$$

$$g_{\{k,m\}} = \frac{r_{\{k,m\}}}{r^2_{\{k,m\}} + x^2_{\{k,m\}}} \quad \text{and} \quad b_{\{k,m\}} = -\frac{x_{\{k,m\}}}{r^2_{\{k,m\}} + x^2_{\{k,m\}}}. \tag{1.7}$$

Here, $g_{\{k,m\}}$ ($b_{\{k,m\}}$) is called the conductance (resp., susceptance) of the line. Using these parameters and Ohm's law, one can write relationships involving voltages, power flows, and currents on a line. In particular, when no transformers are present we have (see, e.g. [8, page 8]) that the *complex* current injected into (k, m) at k satisfies

$$I_{km} = y_{\{k,m\}}(V_k - V_m) + \frac{1}{2} y^{sh}_{\{k,m\}} V_k. \tag{1.8}$$

This is, essentially, Ohm's law for AC power. See Figure 1.1, which shows (in gray) the impedance elements associated with the line, together with voltage levels at various points (the V_k, V_m values) and current injections (the I_{km}, I_{mk} values). The two gray vertical boxes represent shunt effects (split into two halves). The case of a line with transformers is somewhat more involved, and the power flow expressions given below become more complex. For completeness, this case will be handled in Appendix B.

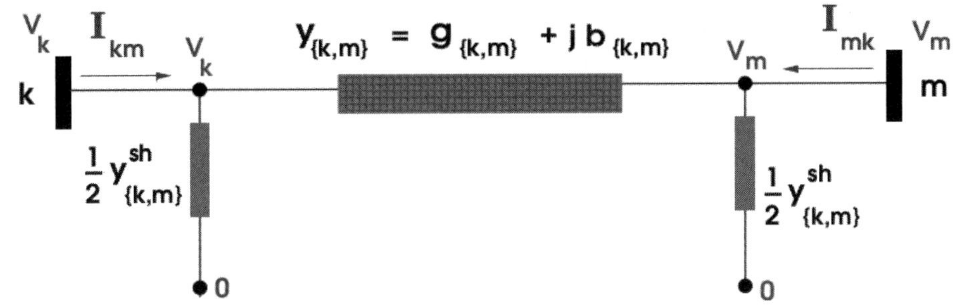

Figure 1.1. Π-*model, no transformer.*

Using (1.8) and the corresponding equation at bus m, we have that if $y^{sh}_{\{k,m\}} = 0$, then $I_{km} = -I_{mk}$ (current conservation), but if $y^{sh}_{\{k,m\}} \neq 0$ then in general current conservation does not hold. Using (1.8) one further obtains equations describing power flows on line (k,m). To do so, we first drop the superscript V to denote by θ_k the voltage phase angle at bus k, and for a line (k,m) we write $\theta_{km} \doteq \theta_k - \theta_m$. And to simplify the equations, we will also assume that $g^{sh}_{\{k,m\}} = 0$. (For complete formulas and derivations see Appendix B.) Then we have

$$S_{km} = V_k I^*_{km} = y^*_{\{k,m\}} |V_k| e^{j\theta_k}(|V_k|e^{-j\theta_k} - |V_m|e^{-j\theta_m}) - j\,\frac{1}{2}\,b^{sh}_{\{k,m\}}|V_k|^2, \tag{1.9}$$

and so

$$p_{km} = |V_k|^2 g_{\{k,m\}} - |V_k||V_m| g_{\{k,m\}} \cos\theta_{km} - |V_k||V_m| b_{\{k,m\}} \sin\theta_{km}, \tag{1.10}$$

$$q_{km} = -|V_k|^2\left(b_{\{k,m\}} + \frac{1}{2}\,b^{sh}_{\{k,m\}}\right) + |V_k||V_m| b_{\{k,m\}} \cos\theta_{km} - |V_k||V_m| g_{\{k,m\}} \sin\theta_{km}. \tag{1.11}$$

Note that, in general, $p_{km} \neq -p_{mk}$; in fact some arithmetic yields

$$p_{km} + p_{mk} = g_{\{k,m\}} |V_k - V_m|^2 \quad (\geq 0). \tag{1.12}$$

The right-hand side of this expression is the amount of active power lost on line (k,m). Thus,

$$\begin{aligned}
\text{active power loss} &= r_{\{k,m\}} \frac{1}{r^2_{\{k,m\}} + x^2_{\{k,m\}}} |V_k - V_m|^2 \\
&= r_{\{k,m\}} \frac{r^2_{\{k,m\}} + x^2_{\{k,m\}}}{(r^2_{\{k,m\}} + x^2_{\{k,m\}})^2} |V_k - V_m|^2 \;=\; r_{\{k,m\}} |y_{\{k,m\}}|^2 |V_k - V_m|^2 \\
&= r_{\{k,m\}} |I_{km}|^2.
\end{aligned} \tag{1.13}$$

(We remind the reader that we are assuming $y^{sh}_{\{k,m\}} = 0$.) This is the familiar RI^2 expression for thermal (resistive) loss on a conductor. Likewise,

$$q_{km} + q_{mk} = -b_{\{k,m\}} |V_k - V_m|^2 - \frac{1}{2} b^{sh}_{\{k,m\}} (|V_k|^2 + |V_m|^2). \tag{1.14}$$

The first term of the right-hand side of (1.14) is nonnegative since $b_{\{k,m\}} \leq 0$ (most commonly it is negative). However, $b^{sh}_{\{k,m\}}$, while generally small, may be positive. Thus, when $b^{sh}_{\{k,m\}} > 0$, it is difficult to categorize the right-hand side of (1.14) as a power loss since it can have arbitrary sign. In fact, it is only the first (and nonnegative) term, $-b_{\{k,m\}}|V_k - V_m|^2$, which should actually be considered as the reactive power loss. We can justify this statement by reviewing (1.8), whose first term is the current across the resistive element (i.e., the middle element labeled $y_{\{k,m\}}$) in Figure 1.1. This element has no shunt component, and thus its corresponding reactive power loss is the first term on the right-hand side of (1.14). In contrast, the second term on the right-hand side of (1.14) corresponds to power injected into the line across the vertical elements in Figure 1.1 (the second term in (1.8) is the current across the left vertical element in Figure 1.1). Using this observation we also note that, through a similar computation as (1.13), we obtain

$$\text{reactive power loss} = x_{\{k,m\}}|I_{km}|^2. \tag{1.15}$$

Comparing with (1.13) and recalling that typically $r_{\{k,m\}}$ is an order of magnitude smaller than $x_{\{k,m\}}$, we see that reactive power losses are much greater than active power losses. Modern transmission grids in normal operation convey active power with small losses (on the order of 2%, say) but reactive power losses are much larger and can pose a stability risk, as will be discussed later.

It is also useful to define, for a bus k, the quantities

$$S_k \doteq \sum_{km} S_{km}, \quad P_k \doteq \sum_{km} p_{km}, \quad Q_k \doteq \sum_{km} q_{km}, \quad \text{and } I_k \doteq \sum_{km} I_{km}, \tag{1.16}$$

the (net) complex power, active power, reactive power, and current (respectively) injected by k into the grid. If bus k houses an operating generator but has no load, $P_k > 0$, whereas if k has a positive load and does not generate power, $P_k < 0$.[2]

Power flows and dynamical systems

Suppose the quantities P_k and Q_k defined in (1.16) are known for each bus k; i.e., for each bus k its net active and reactive power output is given. A specification of voltage magnitudes V_k and phase angles θ_k at each bus k and of power flows p_{km} and q_{km} that satisfies (1.10) and (1.11) for each line (k,m), while delivering the specified output levels P_k and Q_k, corresponds to a physically valid steady-state behavior of the system. In reviewing the above derivations, we can see that this system description originates from (1.1). These equations, in turn, assume a steady-state (e.g., fixed) system behavior, in particular assuming that the frequency ω is time independent. Equations (1.10), (1.11), and (1.16) guarantee (see Section 1.2.5) that all active power generation is matched by active power loads and losses. There is a more comprehensive viewpoint on power systems that does not assume steady-state behavior and where a system may not be "balanced," i.e., generation may not equal loads plus losses. From that perspective a power system can be studied as a *dynamical system* whose equilibrium or fixed points correspond to steady-state (and, thus, balanced) behavior. We will return to this point in Section 6.0.1.

Denoting by I the vector of current injections, and by V the vector of voltages, from (1.8) we obtain

$$I = \mathcal{Y} V, \tag{1.17}$$

[2]The term "house" is somewhat loose; it is more appropriate to say that a generator or load is attached to a bus. Both situations can take place simultaneously.

where the matrix $\mathcal{Y}$ is termed the *bus admittance* matrix and constructed as follows. From (1.8) we see that if $y^{sh}_{\{k,m\}} = 0$ for all lines (k,m), then $\mathcal{Y}$ is a complex-valued weighted graph Laplacian matrix (see [67]), that is to say,

$$\mathcal{Y}_{kk} = \sum_{km} y_{\{k,m\}} \quad \text{for any bus } k, \tag{1.18a}$$

$$\mathcal{Y}_{km} = -y_{\{k,m\}} \quad \text{for any line } km, \tag{1.18b}$$

$$\mathcal{Y}_{km} = 0 \quad \text{otherwise.} \tag{1.18c}$$

When accounting for transformers, (1.8) needs to be modified, and the corresponding matrix $\mathcal{Y}$ is no longer symmetric.

Equations (1.10) and (1.11) are the fundamental relationships between voltage and power in each line of a grid, providing a system-wide model. We stress that these equations reflect a number of modeling simplifications and assumptions:

- They are steady-state equations; in particular the transient dynamics arising from discrete events such as shutting off a generator, or tripping a line, are not captured. The fine-scale behavior of a generator as it changes its frequency or power output (while responding to some exogenous event or control) is also not captured. Very fine time scale modeling of a grid not in steady state would require a system of differential equations that capture, e.g., detailed generator modeling. And if extreme accuracy is desired, one might additionally need an accurate model for line behavior at a fine space scale.

- The Π-model of a line introduces a number of simplifications regarding the physical structure of the line, its *geometry* (i.e., positioning relative to other objects, in particular the spatial arrangement of conductors that make up the line, and support structures), and even the underlying physics (e.g., the model neglects second-order terms).

Additionally, the equations need to be modified to model the behavior of devices such as transformers (though the end result is similar) and other devices, such as FACTS (flexible AC transmission systems) [123].

In the formulas above we have used polar notation to indicate voltages (as well as currents and power flows); the engineering terminology for such quantities denotes them as "phasors." A different perspective is obtained by representing voltages using rectangular coordinates.[3] Write, for each bus k,

$$V_k = e_k + j f_k$$

(where e_k and f_k are real). Then, using (1.8), equation (1.17) has the form

$$\begin{aligned} I_{km} = {} & g_{\{k,m\}}(e_k - e_m) - b_{\{k,m\}}(f_k - f_m) - \frac{1}{2} b^{sh}_{\{k,m\}} f_k \\ & + j \left[b_{\{k,m\}}(e_k - e_m) + g_{\{k,m\}}(f_k - f_m) + \frac{1}{2} b^{sh}_{\{k,m\}} e_k \right]. \end{aligned} \tag{1.19}$$

Using (1.9) and (1.4) it follows that

$$p_{km} = g_{\{k,m\}}(e_k^2 + f_k^2) - g_{\{k,m\}}(e_k e_m + f_k f_m) + b_{\{k,m\}}(e_k f_m - f_k e_m), \tag{1.20}$$

[3] This approach is credited to F. Galiana; see Chapter 5 of [128].

and similarly

$$q_{km} = \\ -\left(b_{\{k,m\}} + \frac{1}{2} b^{sh}_{\{k,m\}}\right)(e_k^2 + f_k^2) + g_{\{k,m\}}(e_k f_m - f_k e_m) + b_{\{k,m\}}(e_k e_m + f_k f_m). \tag{1.21}$$

Let v be the vector $(e_1, e_2, \ldots, e_n, f_1, f_2, \ldots, f_n)^T \in \mathbb{R}^{2n}$ (where $n =$ number of buses). From (1.20) we obtain that

$$v^T M_k v = P_k, \tag{1.22}$$

where M_k is an appropriate $2n \times 2n$ matrix. Thus, to put it simply, P_k is a *quadratic* expression of the values e_m and f_m (for $m = k$ or m adjacent to k), and similarly with Q_k.

Example 1.2.1. *Suppose bus* 1 *is connected to (only) buses* 2 *and* 3. *Then the submatrix* M_1 *restricted to* $e_1, e_2, e_3, f_1, f_2, f_3$ *equals*

	e_1	e_2	e_3	f_1	f_2	f_3
e_1	$g_{\{1,2\}} + g_{\{1,3\}}$	$-\frac{g_{\{1,2\}}}{2}$	$-\frac{g_{\{1,3\}}}{2}$	0	$\frac{b_{\{1,2\}}}{2}$	$\frac{b_{\{1,3\}}}{2}$
e_2	$-\frac{g_{\{1,2\}}}{2}$	0	0	$-\frac{b_{\{1,2\}}}{2}$	0	0
e_3	$-\frac{g_{\{1,3\}}}{2}$	0	0	$-\frac{b_{\{1,3\}}}{2}$	0	0
f_1	0	$-\frac{b_{\{1,2\}}}{2}$	$-\frac{b_{\{1,3\}}}{2}$	$g_{\{1,2\}} + g_{\{1,3\}}$	$-\frac{g_{\{1,2\}}}{2}$	$-\frac{g_{\{1,3\}}}{2}$
f_2	$\frac{b_{\{1,2\}}}{2}$	0	0	$-\frac{g_{\{1,2\}}}{2}$	0	0
f_3	$\frac{b_{\{1,3\}}}{2}$	0	0	$-\frac{g_{\{1,3\}}}{2}$	0	0

The remaining entries of M_1 *equal zero.*

In fact, if bus k has d_k lines attached to it (i.e., if the graph-theoretic degree of k is d_k), then (1.22) only involves $4d_k + 3$ of the entries in v and M_k has at most $8d_k + 2$ nonzeros. The structure in the example can be generalized; we have that M_k always has the structure

$$M_k = \begin{pmatrix} A_k & \bar{A}_k \\ -\bar{A}_k & A_k \end{pmatrix}, \tag{1.23}$$

where A_k is symmetric (and also $\bar{A}_k$ is skew-symmetric—however, this property is not crucial).

Definition 1.2.2. *A* $2n \times 2n$ *matrix* A *is block skew-symmetric if there exist* $n \times n$ *matrices* B *and* C *with* B *symmetric such that*

$$A = \begin{pmatrix} B & C \\ -C & B \end{pmatrix}. \tag{1.24}$$

Thus, M_k is block skew-symmetric, and a similar quadratic expression and considerations apply to reactive power:

$$v^T N_k v = Q_k, \tag{1.25}$$

where N_k is block skew-symmetric. The above derivations apply to a grid without transformers. Detailed transformer modeling can be complex; however, to handle transformers in a transmission system one can use a model reminiscent of that used for power lines, and equations of the form (1.22) and (1.25) will still apply, though the matrices M_k and N_k will not be symmetric.

Laplacian and modified Laplacian matrices occur frequently in the equations above. If no transformers are present and all the $b^{sh}_{\{k,m\}} = 0$, adding up all equations (1.20) for a given bus k will yield an expression for P_k where the coefficient for e_k^2 equals

$$\sum_{km} g_{\{k,m\}}$$

(where the sum is over all lines with one end at k), whereas the sum of all coefficients for terms $e_k e_m$ and $e_k f_m$ equals

$$-\left(\sum_{km} g_{\{k,m\}} - \sum_{km} b_{\{k,m\}}\right),$$

that is to say, the coefficient for e_k^2 is equal to minus the sum of all coefficients for terms $e_k e_m$ and $e_k f_m$, plus the quantity $\sum_{km} b_{\{k,m\}}$—in other words, the matrix M_k is a (modified) weighted Laplacian [67]. As noted above, the $\mathscr{Y}$ matrix in (1.17) is a weighted Laplacian (in the case of zero shunt susceptances). Further, if we combine (1.22) and (1.25) into a single quadratic equation over the complex numbers, then we again obtain a Laplacian matrix.

We stress the substantial complexity introduced by the quadratic equations (1.22)–(1.25). In the operation of power grids a frequent task is to solve computational problems including such equations as constraints, with additional constraints such as bounds on all variables. These are nonconvex constraints that result in NP-hardness of the resulting problems in the worst case. In Chapter 2 we will return to these problems.

A final point here is that instead of relying on rectangular coordinates one may choose to work directly with the polar representation (1.10)–(1.11) of the power flow equations. In fact, this has been the traditional outlook in the power community. This has the added benefit of allowing straightforward modeling of constraints on the phase angle differences, e.g., $|\theta_{km}| \le \theta^{\max}$, which can be important in practice. The disadvantage lies in the increase in complexity. Indeed, recent developments that incorporate semidefinite programming into the analysis of power flows are not easily applicable to formulations in polar coordinates.

1.2.1 ▪ Per-unit system

A common practice in power systems analysis is to scale all physical values (power flows, voltages, and so on) by appropriate constants so that the scaled values take reasonable values (neither too small nor too large, and preferably close to unity). For example, one might scale all power flows (and loads, and generation amounts) by 100 MW. Another example is that of voltage magnitudes. Under stable operation we expect all voltage magnitudes to be approximately equal (see Section 1.3), for example, for all voltage magnitudes equal to 345 KV, scaling by this amount will render all voltage magnitudes approximately equal to 1.0, which makes it easy to quickly identify buses with undesirable voltage values. The scaling approach is known as the *per-unit* (or p.u.) system. It is devised so that the scaling is compatible with the physical laws described above. Popular power flow analysis tools, such as MATPOWER [234], perform calculations using the p.u. system but report some of the results in standard units (in particular, power flows, generation, and losses) because that is meaningful in practical applications—however, voltage magnitudes are reported in the p.u. system because that is most helpful in quick stability assessments.

An additional benefit of the p.u. system is that it makes transformer modeling simple. Transformers change voltage magnitudes, but using the p.u. system throughout makes it possible to rely on "unit" values on both sides of the transformer. See Section 3.3 in [106]. In what follows we will always assume the p.u. system.

1.2.2 ▪ Power flow problems

In the classical **power flow** (PF) or **load flow** problem the overall goal is to determine how power will flow, given a set of generation amounts, demands, and voltage requirements at the buses of a grid. For convenience of notation, we allow buses to (simultaneously) inject power into the grid and also to consume power, and we denote complex power injected into the grid (resp., removed from the grid) at bus k as $P_k^g + jQ_k^g$ ($P_k^d + jQ_k^d$), so that the net power injection at bus k equals

$$(P_k^g - P_k^d) + j(Q_k^g - Q_k^d) \doteq P^k + jQ^k.$$

Each bus is classified as being of type (P,V) (typically, generators) or type (P,Q) (the remaining buses), and the following data are provided as an input:

- for each (P,V) bus k, its active power output P_k^g and load P_k^d, and its voltage magnitude $|V_k|$, and
- at each (P,Q) bus k, P_k^d and Q_k^d.

The objective of the problem is to determine all other operational data:

- at each (P,V) bus k, its voltage angle θ_k and its reactive power output Q_k^g, and
- at every (P,Q) bus k, its complex voltage V_k (e.g., voltage magnitude and phase angle).

In other words, we have two (real) variables per bus. As constraints, we also have two equations per bus: at a (P,V) bus k we have (1.22) as well as

$$e_k^2 + f_k^2 = |V_k|^2, \tag{1.26}$$

while for a (P,Q) bus k we have (1.22) and (1.25). Thus, we have a square system of quadratic (bilinear) *equations*; having solved this system we have all P_k, Q_k, and V_k values and using (1.10) and (1.11) we can then obtain all power flows. For completeness and ease of reference, we now state the entire power flow problem *in rectangular coordinates:*

$$\text{for each } (P,V) \text{ bus } k: \begin{cases} v^T M_k v = P_k, \\ e_k^2 + f_k^2 = |V_k|^2; \end{cases} \tag{1.27}$$

$$\text{for each } (P,Q) \text{ bus } k: \begin{cases} v^T M_k v = P_k, \\ v^T N_k v = Q_k; \end{cases} \tag{1.28}$$

$$v = (e_1, e_2, \ldots, e_n, f_1, f_2, \ldots, f_n)^T \in \mathbb{R}^{2n}. \tag{1.29}$$

Additionally, in power engineering practice each line will have a limit (or rating) on how much power it is allowed to carry. On shorter lines this is a thermal limitation, while on longer lines this rating addresses phase angle (synchronicity) requirements (see Section 1.3). Typically the limit is stated in terms of active power flows; however, it may also be formulated as a current limit or in terms of magnitude of complex power. If present, this constitutes another constraint to be added to the system (1.27)–(1.29). However, most authors refer to (1.27)–(1.29) as the power flow problem proper.

A detail that is frequently introduced in practice is the concept of a "slack bus." This is a particular bus s, usually one that houses a generator, where it is assumed that $V_s = 1$ and P_s and Q_s are unknowns and unconstrained. The purpose of this practice is to facilitate feasibility for the simple reason that (due to active and reactive power losses) it is difficult to know in advance if a particular combination of active power injections and loads will actually turn out to be feasible. Allowing P_s and Q_s to take arbitrary values (even negative) is helpful toward guaranteeing feasibility; moreover, if indeed generator injections and load amounts correspond to a realistic situation, then it should turn out that P_s and Q_s take reasonable values (e.g., values that a generator at s can produce).

Solving power flow problems—a first look

A power flow problem may have zero or multiple solutions, even in extremely simple cases involving a single line and two buses. From a fundamental perspective, power flow problems are intractable: no polynomial-time procedure exists. Nevertheless, from a practical perspective, the situation may not always be quite so dire. Engineers expect that a "reasonable" grid operating under "reasonable" conditions (reasonable values for loads and generation) will always engender a feasible power flow problem. In fact, given that there may be multiple solutions, it is taken as an article of faith that a practicable solution exists and, moreover, that it is unique. Here, practicable means that if implemented the solution will result in stable and desirable operation: high (and nearly equal magnitude) voltages at all buses and small phase angle differences across all lines. Insofar as all real-world grids are (very carefully) operated to remain in a reasonable state at all times, the above statement is essentially tautological—most power flow problems arising in practice have a practicable, feasible solution whose broad parameters can be closely estimated.

In fact, more is true: the preferred algorithmic tool for solving power flow problems is the classical Newton–Raphson method for solving systems of (nonlinear) equations. In modern implementations of Newton–Raphson applied to power flow problems, great care is exercised to leverage sparsity and to maximize numerical stability; however, as the reader can readily appreciate, this does not overcome the fundamental difficulty of the problem. However, engineers have one additional ace up their sleeves (or two). First, most AC power flow problems that need to be solved amount to small perturbations of problems that have already been numerically solved, and such existing solution vectors provide a convenient starting point for Newton–Raphson, which is of course an iterative procedure. With this proviso, most AC power flow problems that need to be solved in practice are actually solved, and extremely efficiently, by Newton–Raphson, typically requiring a remarkably small number of iterations to achieve numerical convergence. An important point is that even though convergence is to *a* feasible solution with desirable properties, whether this solution is truly unique, or even "the best," is highly debatable.[4]

[4]A free implementation of Newton–Raphson for power flow problems is available through the MATPOWER package [234].

And even when a nearly feasible starting point is not known, but we expect that the operation of the grid will be satisfactory, a reasonable starting point is always available: in the p.u. system, $|V_k| = 1$ and $\theta_k = 0$ at each bus k. This is known as the *"flat start"* and in fact does prove successful in many cases.

A different matter altogether concerns a grid under distress—suppose that as part of the analysis of an emergency situation that has not been previously studied, a particular control action is examined: what will be the outcome of the action in terms of voltages, flows, and phase angles? It may be the case that no approximate solution is known for the resulting load flow problem, even if small perturbations of the data are allowed, and that the flat start approach fails for the simple reason that all solutions to the power flow problem, e.g., fail to meet the high-voltage criterion. However, it may still be of interest to find out what will happen as a result of applying the control. In such a setting, which could be common in the midst of a serious emergency, it is likely that Newton–Raphson would fall short. This shortcoming is in fact verified in numerical experimentation.

The basic underlying complication arises when Newton–Raphson fails to converge (or "diverges" as is sometimes stated). This may be due to purely numerical reasons (ill-conditioning, or numerical instability of the underlying implementation) or to fundamental reasons—the problem is truly infeasible. Perhaps surprisingly to non-power researchers, even simple power flow problems can be infeasible (textbook examples consisting of a single line are known; see, for example, [25]). Potentially, an infeasible case could be made feasible via a small change of loads and/or power injections; we might characterize such cases as arising from slight parameter misestimation. Should this happen the computation of a small correction that results in feasibility is of course desirable. There is, however, the possibility that infeasibility of a power flow system results from a much more fundamental reason than simple parameter error and is more closely related to the nature of the underlying transmission network.

Overbye [178], [179] has introduced an iterative algorithm, relying on an elegant method of Iwamoto and Tamura [177] and a technique of Dobson [85], which computes a perturbation to the input parameters to a power flow problem (the P_k and $|V_k|$ quantities at (P, V) buses k and the P_k and Q_k quantities at (P, Q) buses k) to make a power flow problem feasible. In other words, the algorithm computes a vector of power flows with, ideally, minimum 2-norm of the vector of residuals for (1.22), (1.25), and (1.26). In outline the method works as follows: if Newton–Raphson terminates at an infeasible solution to the nominal power flow problem, then the Jacobian of the system equations will be noninvertible. It can then be shown that an element of the null space of the (transpose) Jacobian is orthogonal, in parameter space, to the manifold of right-hand sides to (1.22)–(1.26). Thus a step along this vector will perturb the parameters "toward" feasibility. The algorithm performs a sequence of such steps consisting of a Newton–Raphson computation followed by a perturbation of parameters until a solution to (1.22)–(1.26) is rendered.

We note that the problem of minimally changing the right-hand side vector for (1.22)–(1.26) is itself a nonconvex optimization problem, and therefore the algorithm just described will possibly converge to a local rather than a global optimum. Nevertheless, it is to be expected that if indeed a small perturbation is needed then the algorithm will very likely be successful. In general, however, as demonstrated by Hiskens and Davy [124], the manifold of right-hand side values for (1.22)–(1.26) can be quite nonconvex, in fact, quite intricate, and the problem addressed by [178], [179] may prove extremely challenging. Also see [139]. In Chapter 2 we will return to the topic of power flow problems from the point of view of modern optimization algorithms. We note in passing that some very modern (and provably good) methods for computing solutions to systems of polynomial equations bear a resemblance (perhaps superficial) to the method of Overbye cited above.

1.2.3 ▪ Optimal power flow problems

Optimal power flow (or OPF) problems constitute a much broader family of computational tasks. Here, rather than specific values for any of the operating parameters (the right-hand sides of (1.22)–(1.26), we have ranges. For example, at each bus k we are given parameters P_k^L and P_k^U, as well as the constraint

$$P_k^L \leq P_k \leq P_k^U,$$

with P_k a decision variable defined as per (1.22), and likewise with the Q_k and the $|V_k|$ and the power flow variables. In rectangular coordinates, we therefore have a system of quadratic *inequalities* in the e_k and f_k quantities. In the standard form of OPF used in power engineering practice the goal is to minimize a separable convex quadratic function of active generator power output, i.e., a function of the form

$$\sum_g f_g(P_g), \tag{1.30}$$

where the sum is over the set of generators and each $f_g(P_g)$ is a convex quadratic polynomial of the (real) variable P_g. An objective of form (1.30) is typically used to dispatch (schedule) generators at minimum operating cost. However, many other, typically convex, objective functions have also been considered.

In addition to the constraints described above, several other types arise in power engineering practice. For example, for each line (p,q) we may have a bound on the phase angle difference across the line:

$$|\theta_p - \theta_q| \leq \gamma_{pq}. \tag{1.31}$$

OPF is typically used to dispatch generators. It can also be used to find (near) feasible solutions to power flow problems. In this setting we are given a vector $\tilde{P}$ of (possibly infeasible) active power injections; an optimization problem is then solved with the goal of finding a feasible operating point that is closest (in the L_2-norm sense) to $\tilde{P}$ (this is very similar to the problem considered by Overbye in [178], [179]). From this perspective, OPF can prove a more effective computational tool than the basic power flow—it overcomes simple infeasibility issues and may even point out structural reasons for infeasibility (e.g., some entry of $\tilde{P}$ is unrealistically large or small).

Nevertheless, it is clear that at some level the basic underlying nonconvexity of the power flow equations remains undiminished and the fundamental difficulties underlying OPF are not substantially different than those for the power flow problem (this is obviously clear if the ranges are very narrow). However, the additional flexibility provided by OPF formulations provides quite a bit more firepower to engineers, and a major ingredient in this firepower is the algorithmic methodology that can be deployed.

We will return to the OPF problem in Chapter 2, in particular in Section 2.1.1, where we will examine solution techniques for OPF problems in greater detail.

1.2.4 ▪ DC approximation

Equations (1.22) and (1.25) are quadratic and as such their incorporation into computational tasks (see below) where the V_k vectors are variables introduces significant complexities. Equations (1.10) and (1.11) are arguably no easier. Under normal operating conditions, however, these equations can be approximated by much simpler linear equations. This produces the so-called DC approximation. In its standard form, this approximation is based on three observations:

- Under normal conditions $|V_k| \approx 1$ for each bus k (using the p.u. system).
- Under normal conditions $|\theta_{km}|$ is small for each line (k,m) so that $\sin\theta_{km} \approx \theta_{km}$.
- For any line (k,m), its resistance is much smaller than its reactance, i.e., $r_{\{k,m\}} \ll x_{\{k,m\}}$. Recalling that $g_{\{k,m\}} + j\,b_{\{k,m\}} = (r_{\{k,m\}} + j\,x_{\{k,m\}})^{-1}$, this means $g_{\{k,m\}} \approx 0$ and $b_{km} \approx -1/x_{\{k,m\}}$.

Using these approximations in (1.10), we obtain

$$p_{km} = \frac{\theta_k - \theta_m}{x_{\{k,m\}}}. \tag{1.32}$$

This is a linear equation involving (active) power flow and angles on line (k,m). Rewriting it as $\theta_k - \theta_m = x_{\{k,m\}} p_{km}$, it is reminiscent of Ohm's equation for DC circuits, with p_{km} playing the role of current, phase angles playing the role of voltages, and $x_{\{k,m\}}$ playing the role of resistance (even though it is actually the reactance). The DC approximation is obtained by using (1.32) as the sole descriptor of power flow behavior, with reactive power being ignored. This is the standard approach; though of course one can linearize (1.10) and (1.11) at any operating point and obtain a locally valid linearization. In the DC approximation, the net active power injections P_k in (1.22) or (1.16) are given as inputs, and we may simply abbreviate the entire system of equations as

$$\sum_{km} \frac{\theta_k - \theta_m}{x_{\{k,m\}}} = P_k \quad \text{for each bus } k. \tag{1.33}$$

It is also convenient to write this system in matrix form. Assuming a total of n buses, let $\mathcal{B}$ be the $n \times n$ matrix defined according to the following rules:

$$\mathcal{B}_{kk} = \sum_{km} \frac{1}{x_{\{k,m\}}} \quad \text{for any bus } k, \tag{1.34a}$$

$$\mathcal{B}_{km} = -\frac{1}{x_{\{k,m\}}} \quad \text{for any line } km, \tag{1.34b}$$

$$\mathcal{B}_{km} = 0 \quad \text{otherwise.} \tag{1.34c}$$

The matrix $\mathcal{B}$ is called the *bus susceptance* matrix, so denoted because for each line $\{k,m\}$ its susceptance equals $x^{-1}_{\{k,m\}}$. Recall that the DC approximation assumed that for each line $\{k,m\}$ its resistance $r_{\{k,m\}}$ is zero; in that case the bus admittance matrix $\mathcal{Y}$ (see (1.18)) coincides with $\mathcal{B}$. Using this notation, (1.33) can be restated as

$$\mathcal{B}\,\theta = P, \tag{1.35}$$

where P is the vector indexed by buses, with entries P_k. We assume that $\sum_k P_k = 0$; this is implied by adding all equations (1.35). The bus susceptance matrix can also be written in a more compact manner. Let Y be the diagonal matrix whose entries are the susceptances $y_{\{k,m\}} = x^{-1}_{\{k,m\}}$. Let N denote the node-arc incidence matrix in an arbitrary orientation of the network. Then a direct calculation shows that

$$\mathcal{B} = NYN^T. \tag{1.36}$$

1.2.5 ▪ Flow decomposition theorems

Given the solution to a power flow problem, whether in the AC form or the DC approximation, does the flow have a "global" structure? In classical network flow theory such decompositions are well known [3] and play a very important role in the analysis of models and in the design of efficient algorithms for multiple problems. In this section we show that in both cases such a structure does exist. We begin with the DC case, where we provide a result to be used in Section 3.3.3.

Flow decomposition of DC power flows

Lemma 1.2.3 given below is a standard result concerning solvability of DC power flow systems and more generally of linear systems involving Laplacian matrices; we provide it for completeness and future reference. Prior to proving these results we provide some standard background.

In the following, the *node-arc incidence* matrix of a directed graph with n nodes and m arcs is the $(n \times m)$ matrix N with a row corresponding to each node and a column corresponding to each arc, and such that for each arc (i,j) we have $N_{i,(i,j)} = 1$, $N_{j,(i,j)} = -1$, and all other entries set to zero. This matrix does not have full row rank; it has rank $n-1$ precisely when the underlying graph is connected [3]. Writing, for each line $\{k,m\}$, $y_{\{k,m\}} = x^{-1}_{\{k,m\}}$, a direct calculation shows that $\mathscr{B} = NYN^T$, where $Y = \text{diag}\{y_{\{k,m\}}\}$.

Using this notation, we can rewrite (1.33) (or (1.35)) as the system

$$Np = P, \qquad p_{km} = y_{\{k,m\}}(\theta_k - \theta_m) \quad \forall\,(k,m), \tag{1.37}$$

where p is the (active power) flow vector. The second part of (1.37) is the same as (1.32) and may also be abbreviated as

$$p = YN^T\theta. \tag{1.38}$$

Suppose the graph is connected, and let $\tilde{N}$ be obtained from N by removing any row (i.e., by ignoring (1.33) for a single node). Without loss of generality, assume that this is row n. Then $\tilde{N}$ has full row rank $(= n-1)$ and as a result (using $Y \succ 0$) $\tilde{\mathscr{B}} \doteq \tilde{N}Y\tilde{N}^T$ is invertible. A second implication is that the system $Np = P$ is equivalent to $\tilde{N}p = \tilde{P}$, where $\tilde{P}$ is the vector obtained from P by removing row n.

Lemma 1.2.3. *Under the assumption that* $\sum_k P_k = 0$, *the system* (1.35) *is feasible and there is a unique vector of feasible flows* p, *given by*

$$p = Y\tilde{N}^T(\tilde{N}Y\tilde{N}^T)^{-1}\tilde{P}, \tag{1.39}$$

while the phase angles have one degree of freedom per connected component of the underlying undirected graph. As a corollary, if $P = 0$, *then the unique power flow solution to* (1.37) *is* $p_{km} = 0$ *for all* (k,m).

Proof. Assume without loss of generality that the graph is connected, since system (1.37) decomposes into independent subsystems and $\mathscr{B}$ is decomposed into corresponding square blocks. Then system (1.37) is equivalently written as

$$\tilde{N}YN^T\theta = \tilde{P}. \tag{1.40}$$

However, the sum of columns of N^T is zero. Consequently, if θ solves (1.40), then adding any constant to all entries of θ also yields a solution. We can therefore set, without loss of generality, $\theta_n = 0$. Therefore, denoting by $\tilde{\theta}$ the subvector of θ consisting of the first $n-1$ entries, system (1.40) is equivalent to

$$\tilde{N}Y\tilde{N}^T\tilde{\theta} = \tilde{P}. \tag{1.41}$$

Therefore,

$$\tilde{\theta} = (\tilde{N}Y\tilde{N}^T)^{-1}\tilde{P} = \tilde{\mathscr{B}}^{-1}\tilde{P}$$

(the second equation by definition of $\tilde{B}$) is the *unique* solution to (1.41), and (1.40) has a unique solution up to an arbitrary additive constant. Since $p = YN^T\theta$ and (again) the sum of columns of N is zero, this additive constant is canceled in the computation of p, and thus the unique power flow is given by (1.39) as claimed. ■

Lemma 1.2.3 can be extended by making use of flow decomposition techniques (see Section 3.0.1 for applications in grid security analysis). Below, a *path* will be a graph-theoretic oriented path in the underlying network. For ease of presentation we give a more formal definition.

Definition 1.2.4. *A path is a sequence of buses $\pi = \pi(1), \pi(2), \ldots, \pi(r)$ such that for each $2 \leq i \leq r$, $(\pi(i-1), \pi(i))$ is a line; further, $\pi(i) \neq \pi(j)$ for all $i \neq j$. Finally, $\pi(1)$ is called the source and $\pi(r)$ the sink. A line (k,m) is called forward w.r.t. π if $k = \pi(i)$ and $m = \pi(i+1)$ for some $1 \leq i < r$. If on the other hand $k = \pi(i+1)$ and $m = \pi(i)$ for some $1 \leq i < r$, then the line is called backward w.r.t. π. The source of path π will also be denoted by $s(\pi)$ and the sink by $t(\pi)$.*

The next result is presented in detail for future reference.

Theorem 1.2.5. *Let $(\hat{p}, \hat{\theta})$ be a solution to (1.37). Then there exists a family of paths $\{\pi_h : 1 \leq h \leq H\}$ and reals $\{f_h : 1 \leq h \leq H\}$ with the following attributes:*

(1) *$f_h > 0$ for $1 \leq h \leq H$.*

(2) *For any bus k, we have the following:*

$$\text{if } P_k > 0, \text{ then } \quad P_k = \sum_{h\,:\,k=s(\pi_h)} f_h,$$

and k is not a sink. Further,

$$\text{if } P_k < 0, \text{ then } \quad P_k = -\sum_{h\,:\,k=t(\pi_h)} f_h,$$

and k is not a source. Finally, if $P_k = 0$, then k is neither a source nor a destination for any path.

(3) *If a line (k,m) is forward w.r.t. some path π_h, then (k,m) is not backward w.r.t. any path $\pi_{h'}$, and*

$$0 < \hat{p}_{k,m} = \sum\{f_h : (k,m) \text{ forward w.r.t. } \pi_h\}, \qquad \hat{\theta}_k > \theta_m. \tag{1.42}$$

Conversely, if a line (k,m) is backward w.r.t. some path π_h, then (k,m) is not forward w.r.t. any path $\pi_{h'}$, and

$$0 > \hat{p}_{k,m} = -\sum\{f_h : (k,m) \text{ backward w.r.t. } \pi_h\}, \qquad \hat{\theta}_k < \theta_m. \tag{1.43}$$

Finally, if (k,m) is neither forward nor backward w.r.t. any path π_h, then

$$p_{k,m} = 0 \quad \textit{and} \quad \hat{\theta}_k = \theta_m. \tag{1.44}$$

As a corollary, if we write $\pi_h = \pi_h(1), \pi_h(2), \ldots, \pi_h(r_h)$ for $1 \le h \le H$, then

$$\hat{\theta}_{\pi_h(1)} > \hat{\theta}_{\pi_h(2)} > \cdots > \hat{\theta}_{\pi_h(r_h)}.$$

Proof. Let G denote the underlying directed graph used to represent the grid. We obtain the directed graph $\bar{G}$ by reversing any arc (k,m) with $\hat{p}_{km} < 0$—that is to say, (k,m) is replaced with (m,k), and in that case we write $\hat{p}_{mk} = -\hat{p}_{km} > 0$ (note that we will satisfy $\hat{p}_{mk} = y_{\{m,k\}}(\theta_m - \theta_k)$ since $y_{\{m,k\}} = y_{\{k,m\}}$ by convention). Denoting by $\bar{N}$ the node-arc incidence matrix of $\bar{G}$, we therefore have

$$\bar{N}p = P \quad \text{and} \quad p = Y\bar{N}\theta.$$

Since $\bar{N}p = P$ we can apply standard flow decomposition techniques (see [3]) to the flow $\hat{p}$ in $\bar{G}$. It follows that there is a family of *directed paths* π_h (in $\bar{G}$) and reals f_h (for $1 \le h \le H$) and a family of *directed cycles* γ_j (in $\bar{G}$) and reals g_j (for $0 \le j \le J$) with the following properties, where as before we denote the starting node in π_h by $s(\pi_h)$ and its ending node by $t(\pi_h)$:

(a) $f_h > 0$ for $1 \le h \le H$ and $g_j > 0$ for $0 \le j \le J$.

(b) For any node k, if $P_k > 0$ then $P_k = \sum_h \{f_h : k = s(\pi_h)\}$ and $k \ne t(\pi_h)$ for any $1 \le h \le H$. Similarly, if $P_k < 0$ then $P_k = -\sum_h \{f_h : k = t(\pi_h)\}$ and $k \ne s(\pi_h)$ for any $1 \le h \le H$. Finally, if $P_k = 0$ then $k \ne t(\pi_h)$ and $k \ne t(h)$ for any $1 \le h \le H$.

(c) For any arc (k,m),

$$\hat{p}_{km} = \sum \{f_h : (k,m) \in \pi_h\} + \sum \{g_j : (k,m) \in \gamma_j\}.$$

We claim that $J = 0$, i.e., there are no cycles in the decomposition. To do so, suppose $J \ge 1$, and let cycle γ_1 contain the nodes $v(1), v(2), \ldots, v(k)$ in that order, where $v(1) = v(k)$. Then by (c) for $1 \le i < k$, $\hat{f}_{v(i),v(i+1)} > 0$ and thus $\hat{\theta}_{v(i)} > \hat{\theta}_{v(i+1)}$, leading to the contradiction $\hat{\theta}_{v(1)} > \hat{\theta}_{v(1)}$. Thus indeed $J = 0$ and the second term in the equation in (c) disappears. The rest of the proof of (1)–(3) above is routine. ∎

1.2.6 ▪ A flow decomposition theorem for AC power flows

The reader may wonder if the flow decomposition Theorem 1.2.5 can be extended to AC power flows in some form. A straightforward source-sink flow decomposition for active power flows in general does not exist because of active power losses (see (1.13)). In the case of reactive power flows the situation could be even more complicated because of negative losses, i.e., "gains." However, the corresponding notion of decomposition is related to the theory of *network flows with gains* (in this case, negative gains) or, more generally, *generalized network flows* [3]. It turns out, nevertheless, that appropriate flow decomposition results hold. In particular we will obtain a somewhat simplified mathematical proof of the fact, perhaps obvious to a physicist, that total generation of active power flows equals total loads plus total losses, but we will obtain a detailed version of this result which accounts for the generation/consumption balance by means of flow-carrying paths.

To present the decomposition results we first review some standard concepts. An undirected graph G is denoted by the pair (V,E), where V is the set of vertices and E the set of edges; we may also write $V(G)$ and $E(G)$ for the two sets. Finally, for $v \in V$ we write

$$\delta(v,G) = \{e \in E : e = \{u,v\} \quad \text{for some } v \in V\}. \tag{1.45}$$

Definition 1.2.6. *Let $G = (V,E)$ be an undirected graph. An oriented path in G is a sequence of edges $e_1, e_2, \dots, e_k$ where for $1 \le i \le k$ we can write $e_i = \{u_i, v_i\}$ such that* (a) *for $1 \le i \le k-1$ we have $v_i = u_{i+1}$,* (b) *$u_i \ne u_j$ for all $i \ne j$, and* (c) *$v_k \ne u_i$ for all i. We call u_1 the origin of the path and v_k its destination.*

Comment: Definition 1.2.4 already concerned paths. In fact both definitions are essentially equivalent, but the additional formalism we just introduced will prove useful below. Note that in Definition 1.2.6 the paths are oriented and we are effectively viewing each edge in the path as a directed edge.

Definition 1.2.7. *Let $G = (V,E)$ be an undirected graph. The subdivision of G, denoted G^S, is the graph obtained as follows. The vertex-set of G^S is of the form $V \cup W^S$, where for each edge $e = \{u,v\} \in E$, the set W^S contains a vertex denoted w_e. To obtain the edge-set of G^S we take, for each edge $e = \{u,v\} \in E$, the edges $\{u, w_e\}$ and $\{v, w_e\}$.*

Informally G^S splits each edge of G by introducing a new vertex.

Definition 1.2.8. *Let $G = (V,E)$ be an undirected graph. A flow in G^S is a vector f that assigns to each $e \in E^S$ a real value f_e. For each $r \in V(G^S)$ we write*

$$b_r^f \doteq \begin{cases} \sum_{e \in \delta(r,G^S)} f_e & \text{if } u \in V, \\ -f_{(u,w_e)} - f_{(v,w_e)} & \text{if } r = w_e, \text{ where } e = \{u,v\} \in E. \end{cases} \tag{1.46}$$

We say that v is a source (or sink) for f if $b_v^f > 0$ ($b_v^f < 0$, resp.). Note that by construction

$$\sum_{r \in V(G^S)} b_r^f = 0.$$

Armed with these definitions, we can now return to a concept we already used, that of flow decomposition. To make our final results below easier to follow, we present a formal proof of an auxiliary result.

Lemma 1.2.9. *Let $G = (V,E)$ be an undirected graph, and suppose f is a flow in G^S. Then there exists a family of oriented paths $P^1, P^2, \dots, P^L$ in G^S and positive reals $\mu^1, \mu^2, \dots, \mu^L$ with the following properties:*

(i) *The origin of each path P^i is a source for f; the destination of each path P^i is a sink for f.*

(ii) *For each source v for f we have*

$$b_v^f = \sum_{i=1}^{L} \{\mu_i : v \text{ is the origin for } P^i\},$$

and for each sink v we have

$$b_v^f = -\sum_{i=1}^{L}\{\mu_i \,:\, v \text{ is the destination for } P^i\}.$$

(iii) *Consider any edge $\tilde{e}$ of G^S. Then $\tilde{e} = \{u, w_e\}$, where $u \in V$ and $e = \{u, v\} \in E$. If $f_{\tilde{e}} > 0$ then, for any $1 \le i \le L$, if $\tilde{e} \in P^i$ then $\tilde{e}$ is listed as $\{u, w_e\}$ in the (oriented) description of P^i. On the other hand, if $f_{\tilde{e}} < 0$ then, for any $1 \le i \le L$, if $\tilde{e} \in P^i$ then $\tilde{e}$ is listed as $\{w_e, e\}$ in the (oriented) description of P^i.*

(iv) *Consider any edge $\tilde{e}$ of G^S. Then*

$$|f_{\tilde{e}}| \le \sum_{i\,:\,\tilde{e}\in P^i} \mu^i.$$

Proof. Let D be the directed graph with vertex-set $V(G^S)$ and such that for any edge $\tilde{e} = \{u, w_e\} \in E(G^S)$ (where $u \in V$ and $e \in E$), D contains the arc

$$\hat{e} = \begin{cases} (u, w_e) & \text{if } f_{\tilde{e}} \ge 0, \\ (w_e, u) & \text{otherwise.} \end{cases} \tag{1.47}$$

In either case in the definition of $\hat{e}$ we write $g_{\hat{e}} = |f_{\tilde{e}}|$. Then g is a (standard) flow on D. For any vertex u of D, let $\delta^+(v, D)$ be the set of arcs of D of the form (u, v); likewise let $\delta^-(v, D)$ be the set of arcs of D of the form (v, u). Using this notation let $r \in V(G^S)$, and write

$$B(r) \doteq \sum_{a\in\delta^+(r,D)} g_a - \sum_{a\in\delta^-(r,D)} g_a.$$

If $r \in V$, then, by construction of D and g,

$$B(r) = \sum_{e\in\delta(r,G^S)} f_e = b_r^f$$

(see Definition 1.2.8). On the other hand, if $r \in W^S$, then $r = w_e$ for some $e \in E$, say $e = \{u, v\}$. In that case

$$B(r) = -f_{\{u,w_e\}} - f_{\{v,w_e\}} = b_r^f$$

again. Now we can use the flow decomposition for standard flows. The flow g can be decomposed as a sum of flows on directed paths plus a sum of flows on directed cycles, both of D. Formally, there exists a family of directed paths $\tilde{P}^1, \dots, \tilde{P}^L$ in D and positive reals $\mu^1, \dots, \mu^L$ (possibly $L = 0$) and a family of directed cycles $\tilde{C}^1, \dots, \tilde{C}^K$ in D and positive reals $\gamma^1, \dots, \gamma^K$ (possibly $K = 0$) such that the following hold:

(1) For $1 \le i \le L$ let s^i and t^i be the start and end vertices of $\tilde{P}^i$, respectively. Then $B(s^i) > 0$ and $B(t^i) < 0$.

(2) For each arc $a \in D$,

$$g_a = \sum_{i\,:\,a\in\tilde{P}_i} \mu_i + \sum_{j\,:\,a\in\tilde{C}_j} \gamma_j.$$

As a consequence, for any vertex r, if $B(r) > 0$ then no path $\tilde{P}^i$ ends at r, and

$$\sum_{i:\tilde{P}_i \text{ starts at } i} \mu_i = B(r).$$

And if $B(r) < 0$ then no path $\tilde{P}^i$ starts at r, and

$$\sum_{i:\tilde{P}_i \text{ ends at } i} \mu_i = -B(r).$$

Finally, by (2) we have that for any arc a, $g_a \leq \sum_{i:a\in\tilde{P}_i} \mu_i$. This concludes the proof. ■

We can now use the above results in the context of the AC power flow problem. For simplicity we will have a separate statement for active and reactive power flows. We will use the following terminology.

Definition 1.2.10. *Let $G = (V,E)$ be an undirected graph. Let $P = e_1,\ldots,e_k$ be an oriented path in G^S. Let s be the origin of P and t its destination.*

(a) *We say that P is a full path if $s \in V$ and $t \in V$.*

(b) *We say that P is a left-path or right-path if $s \in V$ and $t \notin V$ ($s \notin V$ and $t \in V$, resp.).*

(c) *We say that P is a pseudopath if $s \notin V$ and $t \notin V$.*

Theorem 1.2.11. *Consider a feasible solution to a feasible AC power flow problem on an undirected graph G, with active and reactive power flow amounts p_{km}, p_{mk} and q_{km}, q_{mk} for each line $\{k,m\}$. Assume that all line conductances are zero, i.e., all $g^{sh}_{\{k,m\}} = 0$. Then we have the following:*

(a) *The sum of active power generation amounts equals the sum of active power loads plus active power losses. Furthermore, there exists a family of full paths (each with a generator bus as origin and a load bus as destination) and a family of left-paths (each with a generator bus as origin and as destination a node in W^S corresponding to a line with positive active power loss) that account for all active power generation, loads, and losses.*

(r) *The sum of reactive power generation amounts plus the sum of reactive power generated by line shunt elements equals the sum of active power loads plus reactive power losses. Furthermore, there exists a family of paths that account for all reactive power generation, loads, and losses. Each path has as origin a generator (with positive reactive power generation) or a shunt element and as destination a bus with positive reactive power load or a node in W^S corresponding to a line with positive reactive power loss. If all shunt susceptances are zero, then each path has a bus with positive reactive power generation as origin.*

Proof. If any bus i hosts both generators and a load we can discount the minimum of the generation and load at i from generation and load, without changing any of the line power flows. We obtain an equivalent system of flows where no bus holds both generation and load.

Next, for any line $e = (i,j)$ let p_{ij} denote the active power injected by i into e.[5] For part (a) we construct a flow f in the graph G^s as follows. Let $\tilde{e} \in E(G^S)$. Then $\tilde{e}$ is of the

[5]With some abuse of notation since we should view lines as undirected.

form $\{i, w_e\}$ for some vertex u and some edge $e \in E(G)$ incident with u, say $e = \{i, j\}$. Then we set

$$f_{\tilde{e}} = p_{ij}.$$

Note that if i is a generation (or load) then $b_i^f > 0$ (resp., $b_i^f < 0$). On the other hand consider any *line* $e = (i, j)$. Since

$$p_{ij} + p_{ji} = \text{loss on } (i, j) \geq 0,$$

it follows that

$$b_{w_e}^f \leq 0,$$

i.e., vertex w_e is not a source, but could be a sink. Now (a) follows by applying Lemma 1.2.9.

To handle part (r) we construct an appropriate flow g in G^s as follows. Let $\tilde{e} = \{i, w_e\} \in E(G^S)$, with notation as above. We set

$$g_{\tilde{e}} = q_{ij} + q_{ij}^{sh},$$

where q_{ij} is the reactive power injected into e at i and q_{ij}^{sh} is the reactive power injected into e by the shunt element next to i. Another application of Lemma 1.2.9 yields statement (r). Additionally, if all shunt susceptances are zero, then all line reactive power losses are nonnegative, and so

$$b_{w_e}^g \leq 0$$

for all lines e. Hence in the flow decomposition all path origins correspond to generator nodes with positive reactive power generation. ■

Remark 1.2.12. *Lemma* 1.2.9(iv) *leaves open the possibility of strict inequality. That would take place for an edge e if some cycle in the family* $\tilde{C}^1, \ldots, \tilde{C}^k$ *contains e. More generally, should cycles appear in the flow decomposition at all? Consider the following example with no active power loads or generation. Here we have a network consisting of a cycle with* 10 *buses, and where each line has zero resistance and unit reactance. Then for each line* (k, m) *we have* $b_{\{k,m\}} = -1/x_{\{k,m\}}$, *and from* (1.10) *we obtain*

$$p_{km} = \frac{|V_k||V_m|}{x_{\{k,m\}}} \sin \theta_{km} = -p_{mk}. \tag{1.48}$$

Hence there is a power flow solution on the cycle, where each bus m has voltage magnitude $|V_m| = 1$, *and the phase angles of the* 10 *buses are* $0, -\frac{\pi}{5}, -\frac{2\pi}{5}, \ldots, -\frac{9\pi}{5}$. *Using* (1.48) *the active power flow on each line equals* $\sin(\pi/5) > 0$.

However, this system of active power flows must be supported by a corresponding set of reactive power flows. The reactive power flow equation corresponding to (1.48) *is*

$$q_{km} = \frac{|V_k|^2 - |V_k||V_m| \cos \theta_{km}}{x_{\{k,m\}}} = q_{mk}. \tag{1.49}$$

With our choice of parameters this implies that at every line (k, m) *of the cycle,* $q_{km} = q_{mk} = 1 - \cos(\pi/5) > 0$. *Hence* (1.48) *requires the operation of generators at every bus, each supplying a common (positive) amount of reactive power and no active power.*

1.3 ▪ Stability

The power flow equations and models given above were derived under the assumption of steady-state operation. Power grids are designed for operation under steady-state conditions and with predictable, constant values of many key parameters. Any development, such as the failure of a line or generator, or a sudden increase in loads, can change parameter values and, worse, introduce time-dependent behavior. Such an unstable situation can prove extremely dangerous, potentially damaging equipment (generators, transformers, lines) and ultimately devolving into long-lasting blackout conditions with large amounts of lost load.

Frequency stability

Paramount among stability requirements is that of *synchronism*. All generators must operate at (or extremely close to) a common frequency; this is the parameter ω in (1.1) and elsewhere. The most common values for ω are 60 Hz (North America) and 50 Hz (Europe). In an abnormal situation we may have generators operating at different frequencies; this condition can be extremely damaging due to high (temporary) current values that can be experienced within generators. In fact, moreover, it is not enough for generators to operate at a common frequency; this should be the prescribed frequency (again, 60 or 50 Hz) because generators (and also some loads, such as traditional electric motors) are calibrated to operate at that frequency.

In the case of a generator frequency deviation, typically a case of frequency *loss*, only an extremely small deviation from the prescribed value is tolerated and whenever possible a control action will be taken to protect the generator. This includes, in the ultimate case, shutting off (tripping) the affected generator for protective reasons. Given that frequency loss is typically caused by a shortage of generation relative to demand (i.e., caused by a generator failure), tripping a generator can actually have a vicious circle effect. The ideal operational range is quite strict; for example a frequency loss of 0.1 Hz is considered very significant and much smaller losses are monitored (in North America, the acceptable range of frequency excursions is ± 0.05 Hz). See [167], [166], [137]. We return to this issue in Section 1.4.

Voltage stability

Transmission systems are designed to operate at high voltage levels. A simple rationale behind this principle arises from Ohm's law—in order to deliver a given amount of power, lower voltages necessitate higher current magnitudes. This translates into higher thermal losses; thus lower voltages (in particular, lower voltages at loads) make it necessary for generators to output more (active) power. Additionally, power lines become more thermally stressed. Either case may cause an unsustainable condition (a generator that reaches its maximum output, or a line that becomes tripped) or more generally may place the system in a precarious situation where for example it is unable to deal with some exogenous contingency. A second line of risk involving low voltages concerns their impact on loads, especially electric motors, whose operation may become compromised. A typical range of acceptable voltage values is $\pm 5\%$ relative to the nominal (ideal) setting.

Thus, ideally, voltage levels should be high, and in fact (e.g., consider (1.12)) they should ideally be *nearly* equal.[6,7] Voltage levels can fall due to some particular condition,

[6]Clearly, phase angle differences should also be small.

[7]There is a practical limit on how high voltage magnitudes can be (due to corona effects on overhead lines).

such as the tripping of a line; in the resulting system, the operating point (i.e., the putatively unique solution to the power flow system) of some of the voltages may be lower than previously. If the drop is manageable the new equilibrium point may be sustainable for an extended period. More commonly, reactive power shortages can cause persistently lower voltages.

Voltage instability ensues when, instead, voltage levels continue to drop. A particularly dangerous situation is that of *voltage collapse*, where perhaps in a system-wide fashion there is a marked decrease of voltages (possibly rapid), which ultimately may result in system failure: the inability of the transmission system to deliver power.

Line stability

Power lines convey power in a transmission system, so their safe operation is of paramount importance. A number of conditions can arise that may cause a line to become disabled. Usually (and almost always) in modern transmission systems such an event does not constitute a "failure" and no physical damage will actually take place; instead the line becomes protectively *tripped*, i.e., nonoperational. We will return to this central topic many times in what follows, so we will just provide a brief introduction here.

A common reason for line tripping is the possibility of *thermal overload*. When current on a line is excessive, the line will become overly hot; in an extreme case this would actually cause the line to mechanically fail, and in a less onerous but still very undesirable case the conducting material on the line will degrade. More commonly, a hot line will sag, which effect may then cause the line to come into physical contact with another object, or to "arc" to an object, in either case causing a short-circuit, characterized by extremely high currents. To protect the line against such extreme events (and events with similar outcomes, such as lightning strikes), equipment will sense such conditions and very quickly trip the line. The *distance relay system* consists of a number of sensors placed on the line which will attempt to determine, in real time, the approximate location of a short-circuit—this approximate determination is then used to trip the line (literally, to open the line) at the point closest to the short-circuit to isolate the condition as narrowly as possible.

Typically, when it has been determined that a line is overloaded, action such as generator rescheduling will be taken to relieve the line. Since under normal conditions thermal overloads will not act instantaneously (e.g., it may take several minutes or longer for the line to reach tripping status), this proves a practical strategy. The determination is based on a numerical rating, or limit, which applies to currents or more often to power flows. A line will have a standard rating and at least one level of emergency ratings. Each rating level is characterized by a maximum period of time that the line is allowed to operate while exceeding the level and before relief action must be taken.

As mentioned above, in the case of very long power lines there is an additional stability criterion: phase angle differences between the ends of the line must be kept small; large phase angle differences are correlated with generator frequency instability, a highly unsafe condition, as discussed above.

1.3.1 ▪ N-1 criterion

The N-1 criterion is a widely applied practice which greatly enhances system security. Broadly stated, the criterion specifies that a power system must be able to tolerate the outage of *one* component (power line, generator, transformer) without degradation. Modern power systems standards, across the world, enforce this criterion. As one might expect, an (N-1)-safe system would likely be able to survive *more* than one single failure, and practical experience indeed suggests that this is the case. Modern power systems have proved remarkably robust and able to endure many forms of abuse without collapsing. Nevertheless, one must contemplate the possibility of more than one simultaneous outage. One way to address this issue involves the N-K criterion, where $K > 1$ is the number of faults to be tolerated. If K is large, a direct implementation of the criterion could prove overly conservative and expensive. On the other hand, a software investigation of N-K robustness of a system would prove useful, for example, as a tool to investigate particular weaknesses in the operation of a system. See Chapter 3.

From a practical perspective, the typical standard specifies that, once *one* outage has taken place, the system operator has a fixed period of time to return the system to (N-1)-safe status (which, following the failure, it is not at). However, what *precise* actions are to be taken is often not specified in the standard, and individual utilities may interpret this mandate in different ways.

1.4 ▪ Power systems control

Modern transmission systems can be very large. For example, the North American Eastern Interconnection covers a land mass comprising the East Coast to the Rockies (with the exception of Texas and Quebec) and from the Gulf of Mexico into Canada (approximately 3.4 million square miles), serving on the order of 230 million customers with a combined load of roughly 650 GW. This system relies on thousands of generators and tens of thousands of transmission lines (spanning tens of millions of miles of wire). A system so large would be very difficult to control as a single, integrated unit. Instead, the system is subdivided into smaller units, called "balancing authority areas" or "control areas"; the Eastern Interconnection includes more than 100 of these [176]. The lines connecting different control areas are known as "tie lines."

The primary function of balancing authority areas is to balance load and generation. Due to real-time demand fluctuation and unforeseeable events (e.g., equipment tripping), it is impossible to attain perfect balance. The resulting imbalance is generally known as the *area control error*, or ACE; from a technical perspective several specific definitions for ACE can be found in the literature.[8] When ACE is observed, the software system known as AGC (or *area generation control*) moves to correct the imbalance by rescheduling generators and by using tie lines to import/export power. Power can thus flow between and even across balancing areas. Under contingency conditions a balancing authority may experience difficulties maintaining low ACE and such power flows can increase. There is also an economic perspective on ACE since different control areas may be financially separate entities.

Balancing generation and load is an extremely complex task whose goal is, on the one hand, the economic operation of a transmission system, and on the other hand and most critically, stable operation. This requires control action across various time scales, and transmission systems implement this in multitier fashion. As pointed out in, e.g., [192], control schemes have evolved in parallel in various countries and even regions (especially since economic deregulation took place), with the result that similar terminologies are employed, but with somewhat different technical interpretations, and the concrete technical details can differ. Here we will provide a broad overview of the control schemes. For additional background see the references provided above; also see [6], [135].

[8] In particular, ACE may include a term to account for frequency imbalance.

1.4.1 ▪ Frequency control

Frequency stability is typically implemented using a three-tier scheme (see, e.g., [166]):

(a) **Primary control**, handling time spans of a few seconds or less, is local and automatic.

(b) **Secondary control** concerns automatic but not necessarily local actions spanning seconds to minutes.

(c) **Tertiary control** encompasses time spans on the order of fifteen minutes to several hours (or even days) and concerns operator-driven actions.

Tertiary control primarily sets the output of generators to match an observed and long-term variation in load. Here we are concerned with a load condition that lasts minutes to possibly hours. Tertiary control is closely related to **OPF** or **optimal power flow**, sometimes also known as **economic dispatch**, a procedure that is periodically run (as often as every five minutes) to determine the setpoints of generators to meet demand estimates for the given time window at minimum cost.

A related operational decision involves the choice of which generators to operate or not over a longer time span (e.g., a 12-hour period); such decisions entail a fixed cost and an appropriate wait period until freshly turned-on generators are able to deliver power. The primary tool in this setting is the **unit commitment** optimization problem, a mixed-integer optimization problem, which has received a large amount of attention in the literature (see, e.g., [203] for an early reference). There is a fixed cost associated with the operation of a generator (in particular so as to model startup/shutdown costs), and the variable cost (a convex function) is approximated with quadratics or piecewise-linear functions. Additionally, the computation sets a fixed (active) power output for operating generators that is modeled as constant throughout the time-span of interest; network constraints (e.g., power line limits) are sometimes included in the computation.[9]

The generator dispatch process is progressively refined, from the several-hours-ahead scope described above, down to a few minutes (using the OPF computation), to adjust to smaller demand fluctuations. An important reason the process starts many hours in advance is that, in the case of large thermal power plants, it can prove extremely uneconomical or even impossible to increase output by a large amount in a short span of time.

At the opposite extreme in the control echelon is **primary control** or **frequency response**. Frequency response includes some of the most sophisticated controls in transmission systems. The purpose of this control, as the name suggests, is to maintain stable (close to nominal) generator frequencies, which, as indicated above, is a critical feature of stable and safe grid operation. As loads fluctuate, real-time imbalances between generation and load can temporarily cause changes in frequency. At a high level (a bit more detail below) this can be outlined as follows. A shortfall in generation causes a loss of frequency (and an excess will cause a speedup). A significant disturbance, such as the loss of a generator, line trips causing islanding (separation of the grid into separate components), or sudden increases in loads, may give rise to larger generation shortfall, with a correspondingly larger frequency drop; frequency response is especially needed in such circumstances. In real time, different generators can experience different transient frequencies, however (very) quickly a system-wide frequency can be said to be attained. Frequency response, a fully automatic mechanism, reacts to system frequency loss by causing appropriate generators

[9]After a long evolution, the use of commercial mixed-integer programs to solve unit commitment problems has largely become standard.

to inject more electrical power to prevent frequency from continuing to drop—the drop is arrested but system frequency is possibly *not* returned to its predisturbance state.

A full understanding of the underlying physics of frequency loss (or gain) as a function of generation imbalances is beyond the scope of this monograph and requires detailed models of the physics of generators; see appropriate chapters in, e.g., [25], [8], [7], [106], [128].

In any case, from an engineering perspective, the relationship between system-wide load change and frequency change is frequently linearized, and appropriate formulas are incorporated into engineering practice. One obtains an equation of the form

$$\Delta L = -R\,\Delta f, \tag{1.50}$$

where $R > 0$ is a constant (not a resistance), ΔL is the change in net load (= total load − total generation), and Δf is the change in frequency. It is important to correctly interpret equation (1.50), which applies postdisturbance and embodies primary control. Consider, for example, the case of a sudden increase in *net* load (which might take place, for example, if a generator unexpectedly trips). In this case $\Delta L > 0$ and (1.50) indicates that $\Delta f < 0$—a proportional decrease in frequency.

The relationship (1.50) enables us to address a common question. The reader may wonder how the grid actually manages to match a sudden increase in loads, in real time. The answer is that the decrease in frequency effectively transfers kinetic energy from the rotors (at the generators) into the transmission system as electrical energy, which is used to match the load increase. Similar considerations apply when we consider a sudden decrease in net loads. The quantity Δf represents a system-wide frequency; R on the other hand is an average representation of the mechanical properties of generators. For a more complete discussion, see Chapter 11 of [25], where R is termed the *regulation constant* or *regulation characteristic*, and $1/R$ the *frequency bias*.

The reader may find (1.50) perplexing; how, exactly, does the equation come about? A key point is that generators are mechanical systems—they impart mechanical power into a generation apparatus, which in turn delivers electrical power to the transmission system; frequency acts as a translating mechanism between the two forms of power. The mathematical relationship between generation and frequency is described by a differential equation called the *swing equation* (see Chapter 14 of [25], Chapter 10 of [8], Chapter 11 of [106]), which is based on a model of a generator that accounts for inertial effects, among others. In its simplest form, the swing equation is used to describe the behavior of a single generator and can be stated as follows, neglecting damping terms:

$$A\frac{d^2\theta}{d^2t} = P_m - P_e. \tag{1.51}$$

Here,

- A is a positive constant related to the inertial properties of the generator and its angular velocity,
- θ is the phase angle of the (rotor of the) generator,
- P_m = mechanical power input into the rotor by the generator, and
- P_e = electrical power output by the generator.

We note that $\frac{d\theta}{dt} = \omega(t)$, the generator frequency, and so (1.51) can be restated as

$$A\frac{d\omega}{dt} = P_m - P_e. \tag{1.52}$$

Equation (1.52) can be interpreted in the context of a sudden generation shortfall in a transmission system and the events that immediately follow. Equation (1.52) precisely quantifies the process. Roughly speaking, we will have that $P_m < P_e$ for the involved generators, and so by (1.52) frequency will decrease. Primary control will then increase the quantities P_m, according to the mechanical characteristics of the generators participating in primary control, to reach a point where $\frac{d\omega}{dt} = 0$, so frequency is constant (but lower than before the contingency).

When considering a transmission system that includes a single generator, the solution of the swing equation will describe transient frequency behavior (converging to the change posited in (1.50)). For more details, see pages 84, 85, 92, 105 in [8] and pages 657–665 in [106]. The swing equation can be extended to describe an entire network including several generators and loads. One obtains a system comprising differential equations, which accounts for dynamic interactions between different generators (and is not simply made up of the individual swing equations) as well as other equations used to describe voltage-current relationships across the network. The solution of such a system can prove challenging, and numerical integration techniques may be employed (see, for example, [106], Section 11.5); such techniques advance time in discrete increments and assume linearized behavior in each time step. Nevertheless, numerical convergence can prove problematic. From a fundamental standpoint, this is not surprising—such a system would be at least as difficult, numerically, as the steady-state system given by, e.g., (1.10)–(1.11). We will return to these issues in Chapters 5 and 6. On the other hand, if a small disturbance is being analyzed, over a small time frame, it is likely that the time-dependent solution will be "near" the predisturbance solution and it is plausible that a (good enough) solution to the dynamical system can be found. If, on the other hand, a larger disturbance is considered, then no near solution may exist and the solvability of the dynamic system becomes more problematic; an optimization goal in particular could become elusive.

A few additional points are in order here. First, even though we speak of system frequency, in the true transient domain each generator may have its own frequency, and system frequency is an average of such generator frequencies (in steady state they will all be very close to the average). Second, not all generators participate in frequency response; i.e., not all generators alter their operation in real time (only those equipped with appropriate governor mechanisms and that have reserve capacity). Finally, it can be difficult to directly measure real-time generation/load mismatch; however, frequency *can* be observed (at generators) and, using the linearized frequency/mismatch formulas one can arrive at a real-time *estimate* of effective load changes. This is the formal technical definition of ACE that is usually employed in the industry; see, e.g., [192], equations (3) and (4). The machinery of frequency response is sophisticated, and much more could be added to the above description. In particular, loads in the form of traditional AC (electrical) motors indirectly contribute to control—in the case of a load increase, motors will slow down, effectively drawing less load; this behavior contributes, positively, to system control. Finally, the overall primary control process is delicate because an improper frequency response mechanism can lead to positive feedback and eventually lead to generator tripping (disconnection) for protective reasons, potentially causing blackouts. See [25, Chapters 6, 7, 14], [167] (and links therein).

We have now outlined tertiary and primary control. There is one remaining layer of the control hierarchy, which takes hold once frequency response is deployed. Post-frequency response, and especially in the case of a large disturbance, a transmission system is left in a vulnerable condition—namely, frequency is stabilized but at a lower value than nominal, and if a second disturbance were to take place the system might be unable to respond and stop additional frequency loss. In fact frequency response cannot be indefinitely maintained. Thus, and operating in a time scale spanning seconds to minutes, secondary control (or **secondary frequency response**) takes hold. Its goal is to reduce ACE to zero and to return system frequency to a nominal value to relieve generators involved in primary response. This is also an automatic procedure, but it is centrally controlled, in response to signaling. Secondary response takes place in the form of specific generators delivering more power, likely in larger quantities than under primary response, and certainly in a more sustained fashion. The term "spinning reserve" is sometimes included under the umbrella of secondary control (see [191] for a discussion on terminology) and describes additional power generation capacity that is available from generators that are already running and connected to the grid. At the end of the secondary response phase, which will typically take minutes, frequency has been restored to the predisturbance target and generators are operating at a different level than after primary control, and, in an ideal scenario, ACEs have been returned to zero.

In cases of heavily loaded systems, or a significant generation shortfall contingency, spinning reserve may prove inadequate. To maintain stable operation of the grid, other generators must be brought online. This may consume a significant amount of time (on the order of thirty minutes or one hour). This additional generation capacity, handled under tertiary control, is then used to restore the spinning reserve capabilities of the grid. A final point is that in the case of a very large generation shortfall, primary and secondary control may still fall short (and bringing other generators online may take too long). Such a situation is critical, and frequency drop and voltage collapse will likely follow (as in the case of the 2003 Italy blackout). A possible remedy is to shed a significant amount of load; this practice is known under the name *emergency load shedding*. This practice can be difficult to implement in a stable manner. Moreover, should load shedding be enacted, it is not clear precisely *how* to do so: proportionally across all loads? Or to favor a particular power flow pattern? To some extent, some of the "smart-grid" initiatives, such as demand response (DR), could address these questions.

We stress that the discussion in the above paragraphs is overly simplified and a bit inaccurate and incomplete, and overall it does not do justice to the overwhelming complexity of the systems currently deployed. It is fair to say that a large part of the complexity in today's grid, and the reason for its success and its seamless operation, lies in the success of the above three-tier structure. We urge the reader to peruse some of the literature we cite; it is eminently readable by a nonexpert (only some mathematical maturity is needed) and quite informative.

1.4.2 ▪ Voltage control

A number of methodologies are possible to deal with falling voltage levels. In particular, it is known (and analysis and simple numerical experimentation verify) that supplying *reactive* power, in particular at or near loads, helps boost voltage levels. At this point, and from a point of view which is not that of a power engineer, it is worth thinking about the meaning of reactive power. In our review this arose in (1.5), as the imaginary part of a complex number whose real part was of interest (i.e., active power). When deciding how to operate a transmission system, it would appear that an engineer should focus on (active) power injections (loads and generator output levels) and complex voltage values (complex, because we need to account for phase angles). These are the relevant quantities from an operational perspective.

However, reactive power injections also have a concrete meaning. Reactive power, which is related to inductance effects, is consumed by electrical motors. Here it may help to think of an electrical motor as a reactor in reverse—electrical motors extract mechanical power from a time-varying magnetic field induced by an AC current; thus inductance effects are important in this context. See [8, Chapter 14].

It may be desirable to have adequate reactive power available near loads; such injections should preferably be done from sources near to the loads because transmission losses of reactive power can be high; i.e., instead of (1.13) we obtain a similar formula but instead using the reactance $x_{\{k,m\}}$, which is as discussed much larger than $r_{\{k,m\}}$ (e.g., ten times as large). See the discussion following (1.14). For example, capacitor banks can be used to inject reactive power. The physical attributes of power lines (e.g., high reactances cause loss of reactive power) also play a role, as noted before.

Finally, from an algorithmic perspective, even if we choose to ignore reactive power loads, reactive power injections can be viewed as physical quantities that are controllable and help parameterize the solution space to obtain feasible solutions.

1.4.3 ▪ Data acquisition

Transmission systems rely on a numerical process known as *state estimation* to guide the real-time application of corrective actions and many other critical tasks. Traditionally, grid operations have relied on SCADA, or supervisory control and data acquisition systems, to obtain relatively coarse grained measurements, which are then corrected and completed (to account for bad and missing data) by the state estimation process.

A relatively new technology is the use of "phasor measurement units," or PMUs, which provide high-frequency data (with signal-processed output provided as often as 30 times per second) which additionally may be GPS-tagged to obtain synchronized measurements with a high degree of accuracy. These measurements include voltage and current (magnitude and angle) and system frequency. As of this writing, PMUs are attaining increasingly widespread use, with the expected goal of improving the state estimation process. See [156].

1.4.4 ▪ Other technical elements

The summary above significantly simplifies how transmission systems operate. Transformers are a key component which we have neglected to analyze; however, for many practical purposes transformer modeling can be incorporated by modifying the standard Π-model of transmission lines. Another ubiquitous feature is the use of three-phase systems. Finally, we have only briefly touched on the operation of SCADA and PMUs.

Yet another central feature of the daily operation of grids is that of *markets* and more generally the economics underlying operation of a transmission system. A description of this topic falls outside the scope of this book; for a recent publication see [173]. We encourage the reader to consult the readily available literature for further background.

Chapter 2
Computations with the AC power flow model

Broadly stated, two kinds of computational problems are of interest in the context of power flows: the basic *power flow* problem, and the *optimal power flow* (OPF) problem. Although both kinds of problems can be considered to be well solved under routine operating conditions, they also present substantial, fundamental mathematical difficulties that become apparent when considering transmission systems under stressed conditions. Traditional power flow computational methodology has its roots in work performed in the 1960s and 1970s; however, the recent work of Lavaei and Low described below has focused on semidefinite programming techniques. This is a modern, if challenging, approach. The incorporation of such techniques into unit commitment problems, and the accurate modeling of new devices and discrete operational controls, such as line switching (i.e., the deliberate turning off of power lines) make the field more amenable to mixed-integer nonlinear programming techniques, an area of optimization that now enjoys great attention among researchers. In this chapter we will survey the current state of the art.

2.1 ▪ Power flow problems revisited

Here we return to the topic of computations of power flows, which we introduced in Section 1.2.2. Traditional methodologies for solving power flow problems (whether basic power flow or OPF) include those that rely on some version of Newton's method, whether directly (in the basic power flow case) or implicitly (logarithmic-barrier methods adapted to OPF), and techniques involving some form of linearization. At this point, it is worth pointing out some basic facts underlying the essential complexity of the problems being tackled, which we do in the box below.

Solving systems of quadratic equations over the complex numbers

Consider the following polynomial equation on variables v and w over the reals:

$$3v^6w - v^4 + 7 = 0,$$

which is equivalent to the following system of quadratic equations on variables v, v_2, v_4, v_6, w, y, and c:

$$\begin{aligned} c^2 &= 1 \\ v^2 - cv_2 &= 0 \\ v_2^2 - cv_4 &= 0 \\ v_2v_4 - cv_6 &= 0 \\ v_6w - cy &= 0 \\ 3cy - cv_4 &= -7 \end{aligned}$$

Here we have constructed the system to emphasize its purely quadratic nature. A consequence of this is that if there is a feasible solution with $c = -1$, then by changing the sign of y, v, w, and every v_i variable, one obtains a solution with $c = 1$. Then one obtains $v_i = v^i$ for $i = 2, 4, 6$, and also $y = v^6w$, and as a result the last equation states $3v^6w - v^4 + 7 = 0$, as desired. This example highlights the fact that any system of polynomial equations can be equivalently represented by a system of quadratic equations. The study of the solution of systems of polynomial equations has long been of interest to mathematicians; a survey (as of 2002) is provided in Sturmfels [218]. In fact, the efficient solution of a system of polynomial equations constitutes the 17th of Smale's "problems for the 21st century" [210], and there is a connection to Hilbert's Nullstellensatz. A rich body of work has focused on algorithms that compute approximate solutions to systems of polynomial equations; see Renegar [195]; the sequence of papers by Shub and Smale (starting with [205] and ending with [206]); the work by Beltrán and Pardo [20], [21]; and, most recently, the work of Cucker and Bürgisser [76], which culminates in an algorithm for solving a system of complex polynomial equations with complexity $N^{O(\log \log N)}$, where N is the size of the system, thus "nearly" solving Smale's 17th problem. Broadly speaking, the above algorithms resort to homotopy (e.g., continuation) methods, Newton's method, and randomization.

Note that these results concern polynomials over the *complex* numbers—over the reals the problem becomes NP-hard. See Section 1.2.3.

The Cucker–Bürgisser results outlined in the box above are the strongest available in a completely general setting. As noted, the result addresses problems over the complex numbers, and not the reals, as needed in the power flow problem. Solving systems of quadratic equations (or inequalities) over the reals is a far more difficult problem; to be precise it is NP-hard even under significant restrictions. We will discuss that problem starting in Section 2.2.1. Furthermore, the Cucker–Bürgisser algorithm finds one solution, not necessarily a solution with a particular structure, as would be the case when considering the power flow problem. In any case, we can succinctly summarize the state of the art by saying that the solution of *general* quadratically constrained problems is a very intractable problem with no theoretically efficient algorithm known or likely to exist.

The inherent difficulties regarding power flow problems constitute a major obstacle to large-volume simulation of emergency conditions and in particular cascading outages, especially on large grids. The need for many simulations arises from the reliance on stochastics used to model noise and uncertainty as well as many initial scenarios that could set off a cascade. In practice, power flow computations over grids in familiar conditions are aided by prior knowledge of the structure of a realistic solution, characterized by high voltages; low phase angle differences; and other attributes of a stable, or "good," solution. It is often argued that there is *only one* such good solution. In such a case, iterative schemes based on the Newton–Raphson method will quickly converge to a numerically accurate solution.

However, when simulating a cascade or a severe emergency condition, equipment trips or faults will produce subgrids and generator/load configurations that are profoundly unfamiliar (indeed, they may never have occurred in practice), which makes it difficult to predict the structure of a "good" solution to the power flow problem. In fact no such good solution may exist; however, in the context of a simulation, it is important to at least try to predict system behavior, especially when under stress.

This difficult dilemma has prompted many practitioners and researchers to rely on DC approximations when performing large-volume simulations. While DC models are indeed simplified and possibly inaccurate, they are on the other hand reliable, simple, fast, and easy to diagnose. For example, putting aside upper bounds on flow magnitudes, the DC approximation is always feasible (in a connected grid) so long as supplies match demands. Diagnosing infeasibility of an AC power flow formulation (indeed of a system of quadratic inequalities) is a problem of fundamental difficulty. Some work (such as that in [178], [179], discussed later) has relied on a procedure used to adjust problem parameters in the case of nonconvergence to attain a feasible instance. A challenge is that this practice could introduce into simulations idiosyncratic behavior which is an artifact of the algorithms (or implementations), and, likely, such idiosyncrasies may be masked by the overall complexity, noise, and incomplete models that are incorporated in the simulations, e.g., simulations of cascading failures. In other words, the reliance on an apparently accurate model of grid operation (the AC power flow model) instead can become a hindrance because of mathematical difficulties: fundamentally, a source of inaccuracy and an impediment to large-scale testing. Needless to say, the DC approximation also suffers from a set of deficiencies; for example, the approximation does not adequately describe phenomena that can contribute to cascades (such as voltage "collapse"). In Section 5.4.4 we will describe experiments comparing both power flow formulations in the context of simulated emergencies.

Recent developments involving the OPF problem, reviewed next, raise the possibility of a new generation of power flow solvers that are both accurate and fast.

2.1.1 ▪ Solving OPF problems

The AC OPF problem was originally introduced in [58] and has been continuously refined by the power engineering community in the ensuing decades. For a comprehensive review of prior work and a literature review, see [55]. A number of recent methodological advances that could prove highly impactful in the near future have generated great interest in the field. In this section we will briefly examine solution methodologies for OPF (and power flow) problems, with a more complete review later.

A modern trend in optimization has been the refinement of nonlinear programming tools originally conceived in the context of convex optimization, such as interior point algorithms (barrier methods) and augmented Lagrangian techniques, with the goal of application to general (i.e., possibly nonconvex) constraints and/or objectives. In the general context these methods may fail at computing a global optimum, but, in principle, advance knowledge of a (globally) near-optimal solution can be used as a starting point to converge to what should be the global optimum. Problems of this sort are quite common in engineering and science and in a sense share some characteristics with OPF problems—namely, in routine cases one approximately knows the numerical values of an optimal solution, or, at least, structural relationships that can be used to estimate a good solution. A computational routine is then used to correct these approximate guesses to a rigorous optimum. A rationale for claiming "optimality" is that near the optimum the overall problem is convex and techniques devised for convex optimization are likely to succeed.

When one does not know an approximately optimal solution, the hazards are more severe; nevertheless, modern search techniques, such as filter methods, (implemented in, e.g., IPOPT [225]), adroitly and efficiently leverage problem structure to try to get out of local optima. (We note that IPOPT is available as a plug-in for MATPOWER [234].) The resulting techniques may only be locally convergent and may only find locally optimal solutions. As far as we know, no commercial packages for power flow computations are able to certify the degree of suboptimality of the computed solution and of the minimality of changes to problem data that were effected at runtime to achieve convergence.

Sequential linearization is another common technique. Roughly speaking, such methods detect which (nonlinear) constraints are active and opportunistically linearize them. The advantage of this approach is that it enables the use of mature linear programming software. See, for example, [57], [56]. A disadvantage is that linearizing nonconvex constraints leads to globally invalid constraints; thus a particular linearization is only "locally" valid and may need to be removed later on; this necessitates adequate measures to avoid cycling. Sequential linearization methods are robust from a practical standpoint and are easily adapted to incorporate detailed models of complex equipment and for that reason are frequently found in commercial software. We note that existing commercial tools for power flow problems (there are many) rely on a broad spectrum of techniques and heuristics to deal with OPF problems, with sequential linearization and the use of logarithmic-barrier methods as preferred tools.

From a rigorous optimization perspective, a recent and exciting development concerns new methodologies for computing global optima for (some) OPF and OPF-like problems. These methodologies directly tackle the quadratic equation nature of expressions (1.22)–(1.26) and as such fit very well within a modern trend in the wider optimization community which we briefly review here and more thoroughly cover in Sections 2.2 and 2.3. Quadratic optimization problems have long been of interest in the optimization community; a standard (and NP-hard) example is that of quadratic programming with box bounds, that is to say, a problem of the form

$$\min\{c^T x + x^T Q x \, : \, l \leq x \leq u, x \in \mathbb{R}^n\} \tag{2.1}$$

for appropriate vectors c, l, u and symmetric but not necessarily positive-semidefinite $n \times n$ matrix Q. Other, related, problems also arise, in particular incorporating quadratic constraints. See [52] for a recent survey. The associated nonconvexities bring such problems close to traditional discrete optimization, and in some cases binary variables may be explicitly present (as in unit commitment or line switching problems; the latter are discussed in Section 6.3). For this reason this problem class is broadly classified within the wider body of MINLP (mixed-integer nonlinear problems). See the survey [53].

2.2 ▪ Mathematical background

Quadratics, and especially quadratic constraints, are an essential feature of power flow problems (optimal or otherwise). This feature endows the problems with a rich structure, and, at the same time, a drastic level of complexity. Arguably, many of the difficulties underlying power flow problems can be viewed from this standpoint. The optimization community has long studied quadratically constrained problems, with many interesting results emerging in recent years. In this section we will survey this field and highlight implications for power flows.

2.2.1 ▪ Quadratically constrained, quadratic programming problems

As noted above, if we put aside the phase angle difference constraints (1.31), OPF problems can be stated as quadratically constrained, quadratic programming problems, or QC-QPs for short:

$$(\mathcal{Q}): \quad Q^* \doteq \min \ q_0(x) \tag{2.2a}$$

$$\text{s.t.} \ \ q_i(x) \le 0, \ i = 1, \ldots, m, \tag{2.2b}$$

$$x \in \mathbb{R}^n. \tag{2.2c}$$

Here, the q_i are possibly indefinite quadratic functions. Problem $\mathcal{Q}$ is very general; if all functions q_i are linear then we obtain a linear program. More generally, if all quadratics are convex, problem $\mathcal{Q}$ is convex and more specifically it can be formulated as a second-order cone program (SOCP) [107], [47]. SOCP solvers have greatly improved in recent years and are now included in most commercial optimization packages. Nevertheless, it is important to point out that SOCP algorithms are not quite as robust (or fast) as linear programming algorithms, especially in the case of large m or even with m small but poor numerical conditioning; as we will discuss, instances arising from power flow applications seem prone to exhibiting poor numerical qualities.

The fundamental complexity of problem $\mathcal{Q}$ changes when nonconvex quadratics are introduced. The problem becomes NP-hard when (a) q_0 is nonconvex even if all the functions q_i for $i \ge 1$ are linear, or (b) when q_0 is linear but one q_i for $i \ge 1$ is nonconvex.

We include a proof for case (a) for completeness. The integer partitioning problem is the following: given integers w_i, $1 \le i \le n$, decide if there are values $x_i = +1$ or -1 such that $\sum_i w_i x_i = 0$. It is well known that this problem is NP-complete. Given an instance of this problem, consider the instance of problem $\mathcal{Q}$ given by

$$\min \ -\sum_i x_i^2 \tag{2.3}$$

$$\text{s.t.} \ \sum_i w_i x_i = 0, \tag{2.4}$$

$$-1 \le x_i \le 1, \ \ 1 \le i \le n. \tag{2.5}$$

Clearly, the instance of the integer partitioning problem is feasible if and only if problem (2.3)–(2.5) has value $-n$. In fact, given any NP-hard, linearly constrained binary integer programming problem, we can equivalently reformulate it in the form $\min\{c^T x : Ax \le b, x_j \in \{-1,+1\} \ \forall j\}$, which is equivalent to $\min\{c^T x - M \sum_j x_j^2 : Ax \le b, -1 \le x_j \le 1 \ \forall j\}$ when $M > 0$ is large enough. A similar analysis yields case (b).

These points highlight that problem $\mathcal{Q}$ can be *quite* hard—as the preceding paragraph shows, via $\mathcal{Q}$ we can encode many very intractable NP-hard problems. Despite this negative outlook, there is a subclass of problems $\mathcal{Q}$ that can be solved efficiently even when q_0 is nonconvex—this is the trust-region subproblem and generalizations, reviewed below.

Returning to problem $\mathcal{Q}$, a classical approximation technique relies on constructing the so-called semidefinite relaxation, obtained as follows. Consider the positive-semidefinite $n \times n$ matrix X whose i,j entry is $x_i x_j$, that is to say

$$X = xx^T. \tag{2.6}$$

For $0 \leq i \leq m$ write $q_i(x) = x^T A_i x + c_i^T x + b_i$ for appropriate symmetric matrix A_i, vector c_i, and scalar b_i. The semidefinite relaxation of $\mathcal{Q}$ is the semidefinite program

$$\begin{aligned}
(\mathcal{R}):\ \min\ & A_0 \bullet X + c_0^T x + b_0 && (2.7a)\\
\text{s.t. } & A_i \bullet X + c_i^T x + b_i \leq 0,\ i = 1, \ldots, m, && (2.7b)\\
& X \succeq xx^T, && (2.7c)\\
& X \in S^n,\ \ x \in \mathbb{R}^n. && (2.7d)
\end{aligned}$$

Here we denote by S^n the set of symmetric $n \times n$ matrices, and for two $n \times n$ matrices A and B,

$$A \bullet B = \sum_{i,j} a_{ij} b_{ij} = \text{trace}(AB).$$

Problem $\mathcal{R}$ coincides with $\mathcal{Q}$ if we can guarantee that constraint (2.7c) holds as an equality, i.e., that the optimal matrix X has rank one. Otherwise $\mathcal{R}$ is simply a relaxation of $\mathcal{Q}$. The study of semidefinite relaxations of QCQPs has proved a fertile research field, with special focus on conditions under which the optimal matrix X has low rank.

Suppose we define

$$\check{Y} = \begin{pmatrix} 1 \\ x \end{pmatrix} \begin{pmatrix} 1 \\ x \end{pmatrix}^T = \begin{pmatrix} 1 & x^T \\ x & X \end{pmatrix}. \tag{2.8}$$

Note that $\check{Y}$ and X contain as entries all products $x_i x_j$, and thus enforcing $\check{Y} \succeq 0$ amounts to a (strong) global constraint on the vector x. Further, constraints (2.7c)–(2.7d) can be equivalently stated in terms of the matrix $\check{Y}$ introduced in (2.8), namely by stating

$$\begin{aligned}
& \check{Y}_{0\cdot} = (1\, x^T), && (2.9a)\\
& \check{Y} \in S^{n+1},\ \ \check{Y} \succeq 0, && (2.9b)
\end{aligned}$$

where $\check{Y}_{0\cdot}$ indicates the zeroth row of $\check{Y}$ (with coordinates indexed $0, 1, \ldots, n$).

The trust-region subproblem and related problems

The trust-region subproblem is a problem of type $\mathcal{Q}$ where $m = 1$ and $q_1(x)$ is a positive-definite quadratic. By changing coordinates if necessary, we can assume that $q_1(x) = ||x||^2 - 1$. In other words, we are minimizing a (general) quadratic subject to a unit-ball constraint. This problem has long been of interest to the optimization community because of applications to general nonlinear programming [73] (also see Sections 3.3.1, 3.3.2) and, in extensions to the problem, to combinatorial optimization. As a result, a rich set of results on this problem is available, both from a practical and from a theoretical standpoint. See [110], [164], [194], [4], [19].

Of particular interest here is the solution of the trust-region subproblem as a semidefinite program. Using notation as above, the trust-region subproblem is

$$\min\{x^T A_0 x + c_0^T x + b_0 : ||x|| \leq 1, x \in \mathbb{R}^n\}. \tag{2.10}$$

Then [194] this problem can be equivalently written using its semidefinite (or SDP) relaxation

$$\min\{A_0 \bullet X + c_0^T x + b_0 : \text{trace}(X) \leq 1, X \succeq xx^T, X \in S^n, x \in \mathbb{R}^n\}. \tag{2.11}$$

The equivalence between (2.11) and (2.10) implies that there is an optimal solution $(\bar{X}, \bar{x})$ to (2.11) such that $\bar{X} = \bar{x}\bar{x}^T$, that is to say, $\bar{X}$ has rank 1. This condition is "automatically" satisfied at optimality, a remarkable fact.

Thus, the trust-region subproblem is a setting where semidefinite relaxations are exact. However, it should be stressed that formulating the trust-region subproblem as the semidefinite program (2.11) does *not* yield the fastest algorithm. Practical algorithms (with strong theoretical underpinnings) rely on sophisticated applications of Newton's method (see [73]). From a complexity theory viewpoint, the problem can be solved to tolerance ϵ in time $O(\log\log \epsilon^{-1})$ and in polynomial time otherwise (whereas the dependence on ϵ incurred by SDP algorithms is $\log \epsilon^{-1}$); see [232].

The computational and methodological success underlying the trust-region subproblem has led researchers to study extensions where side constraints are imposed in addition to $||x|| \leq 1$. Of course, given the NP-completeness result we discussed above, we cannot expect to retain polynomial solvability in the most general setting. Sturm and Zhang [217] showed that if one linear constraint is added to the formulation the problem can be solved in polynomial time; likewise Ye and Zhang [233] show how to handle the case of two linear side constraints. Recently, Burer and Anstreicher [51] showed that the case of two *parallel* side constraints can be formulated as a semidefinite program with conic side constraints; this approach was extended by Burer and Yang [54], who have shown that an arbitrary family of linear side constraints can be added so long as they do not intersect within the unit ball.

The S-lemma

Returning to problem $\mathcal{Q}$, we note that the polynomial-time solvability of its special case arising as the trust-region subproblem hinged on the positive-definiteness of its single constraint. Our interest in problem $\mathcal{Q}$, however, stems from the analysis of power flow problems, where the quadratic constraints will be indefinite.

It turns out, nevertheless, that the case of $\mathcal{Q}$ with $i = 1$ is endowed with a good characterization, which yields a polynomial-time algorithm. This is the celebrated S-lemma of Yakubovich [231].

Lemma 2.2.1 (see [231]). *Let $f, g : \mathcal{R}^n \to \mathcal{R}$ be quadratic functions and suppose there exists $\bar{x} \in \mathbb{R}^n$ such that $g(\bar{x}) > 0$. Then*

$$f(x) \geq 0 \quad \textit{whenever} \quad g(x) \geq 0$$

if and only if there exists $\gamma \geq 0$ such that $(f - \gamma g)(x) \geq 0$ for all x.

For several proofs of this surprising result, see [188]. The S-lemma can be used as an algorithmic framework for minimizing a quadratic subject to a quadratic constraint. Let

p, q be quadratic functions of a variable $x \in \mathbb{R}^n$ and let $\alpha \geq 0$ be a real. Then for given $\alpha \in \mathbb{R}$, and assuming there exists $\bar{x}$ with $g(\bar{x}) > 0$,

$$\min\{p(x) : q(x) \geq 0\} \geq \alpha \tag{2.12}$$

if and only if there exists $\lambda \geq 0$ so that

$$p(x) - \alpha - \lambda q(x) \geq 0 \quad \text{for all } x \in \mathbb{R}^n. \tag{2.13}$$

In other words, the minimization problem in (2.12) can be approached as a simultaneous search for two reals α and $\lambda \geq 0$, with α largest possible such that (2.13) holds; this task can be formulated as the solution to a semidefinite program. The S-lemma has found applications in many domains in optimization and in particular in robust optimization; see [188] and [108].

We note that the problem handled by the S-lemma is strictly more general than the trust-region subproblem. Nevertheless, the problem only addresses the case of one constraint, and it is of interest to study the extension of the S-lemma, or more precisely the optimization problem in (2.12), to include additional side constraints. See [214]. However, the body of knowledge on such extensions is even more limited than that for the trust-region subproblem. Recently, [39] obtained the following result: for any fixed choice of integers $m_0 \geq 1$, $m_1 \geq 0$ there is a polynomial-time algorithm for solving any problem of the form

$$\min \; x^T Q x + c^T x \tag{2.14a}$$
$$\text{s.t. } ||x - v^{0,i}|| \leq r_{0,i}, \; 1 \leq i \leq m_0, \tag{2.14b}$$
$$||x - v^{1,i}|| \geq r_{1,i}, \; 0 \leq i \leq m_1, \tag{2.14c}$$
$$Ax \geq b, \tag{2.14d}$$

where Q is not assumed to be positive semidefinite, provided that the polyhedron defined by $Ax \geq b$ has a polynomial number of faces that intersect the feasible region. This generalizes the result in [217] that handles the case with $k = 2$ and no linear side constraints. The algorithm in [39] does not, however, formulate (2.14) as a convex optimization problem and as such is not directly usable in the formulation of a problem with additional constraints.

2.2.2 ▪ More-general systems of quadratic inequalities

In problem (2.14) all quadratic constraints have the same curvature, up to sign. What can be said about QCQPs over general quadratic constraints? Clearly, given the NP-hardness result given above, we cannot expect polynomial-time solvability if the number of quadratic constraints is not fixed. However, an interesting result due to Barvinok is available when considering a bounded number of quadratics.

Theorem 2.2.2 (see [15]). *For each positive integer K there is a polynomial-time algorithm to decide if a system of the form*

$$x^T M_i x = 0, \quad 1 \leq i \leq K, \tag{2.15a}$$
$$||x||^2 = 1, \tag{2.15b}$$

is feasible, where $x \in \mathbb{R}^n$ *and* M_i *is an* $n \times n$ *matrix for* $1 \leq i \leq K$.

It is worth pointing out several features of this result. First, the algorithm is *nonconstructive*: when the system (2.15) is feasible, a solution is not given. A second and perhaps more subtle feature concerns the *computational model* used in [15]. It turns out that, as stated in [15], the algorithm assumes that all computations are performed in infinite precision over the reals (however, [15] also sketches the extension to the standard, i.e., "bit" or Turing model of computing). It is worth digressing here. This model of computation, called the "Real-RAM" model [44], differs from the standard or bit model of computing that underlies modern complexity theory. The bit model fully accounts for the complexity of numerical operations as a function of the precision attained. In contrast, the Real-RAM model operates over the real numbers (as opposed to integers or rationals in the bit model) and assumes that arithmetic operations can be carried out at infinite precision while only expending unit computational cost.

These speculations are not idle because one can easily come up with examples of systems of the form (2.15) in low dimension, where there is a unique solution and it is irrational. In fact the system (over $\mathbb{R}^1$) $x_1^2 = 2$ has exactly that behavior. In the standard (bit) model of computing we cannot produce a solution even for this simple system in polynomial time—only an approximate solution. Thus, under the bit model, we can only hope for an algorithm that produces explicit approximate solutions to (2.15) and whose complexity depends polynomially on $\log \epsilon^{-1}$, where ϵ is the guaranteed tolerance.

Note that by not guaranteeing an explicit solution, the algorithm in [15] deftly avoids this last difficulty. However, there is an even more subtle hazard. Algorithms for problems of the form (2.15) and, more generally, polynomial optimization problems, are frequently of an enumerative nature. By this we mean algorithms that enumerate a number of candidate "cases" which could give rise to a solution and then analyze those cases. Each case may amount to a convex optimization problem, or at least to a linear algebra problem that can be addressed using a standard numerical recipe. Often, each such case may itself give rise to subcases. As an example, consider a procedure that first estimates the eigenvalues of a symmetric matrix and where each eigenvalue λ, if positive, is used to construct *another* matrix of the form $M(\lambda) = A + \lambda B$ whose eigenvalues will in turn be estimated. Then each eigenvalue of $M(\lambda)$ constitutes a candidate for a solution to a problem. The difficulty in this procedure arises from the "if positive" requirement for λ. Quite possibly a positive but near-zero value λ could be misestimated as a negative value and the procedure as stated would fail by rejecting λ. Under the Real-RAM model this difficulty would not arise, but under the bit model it would. In actual practical computation the difficulty certainly could arise and the overall algorithm might output a substantially incorrect answer. Thus, without a graceful handling of such numerically ambiguous cases, the algorithm could not be considered complete. Note: One can restate the algorithm in [15] so that it does operate under the bit model [18].

A final set of points concerning system (2.15) is that the constraints are all equations, and pure quadratics, and that the system does not address the optimization problem $\mathcal{Q}$. In fact, Grigoriev, Pasechnik, and de Klerk [111] have announced that for each fixed m, there is a polynomial-time algorithm for problem $\mathcal{Q}$. It is not clear whether this result relies on the Real-RAM or on the bit model of computations.

Next we describe a recent result on problem $\mathcal{Q}$. This result requires an assumption which can be viewed as a relaxation of the result of Barvinok's.

Assumption W. For each fixed m there is an algorithm that, given a system of quadratic polynomial inequalities $q_i(x) \leq 0$ $(1 \leq i \leq m)$, where $q_i \succ 0$ for at least one index $i \geq 1$, and $0 < \epsilon < 1$, either proves that the system is infeasible or obtains a vector $y \in \mathbb{R}^n$ with $q_i(y) \leq \epsilon$ for $1 \leq i \leq m$. The complexity of this algorithm is polynomial in the number of bits in the data and in $\log \epsilon^{-1}$.

Clearly, Assumption W holds if Theorem 2.2.2 holds in the bit model of computation, but not vice versa. The assumption addresses a notion of "weak feasibility" that may be more appropriate in the bit model of computation. Based on this assumption, we have the following result.

Theorem 2.2.3 (see [31]). *Assume Assumption W holds. Then for each fixed $m \geq 1$ there is a polynomial-time algorithm with the following properties, such that given an instance of problem $\mathcal{Q}$ where $q_i \succ 0$ for at least one index $i \geq 1$, and $0 < \epsilon < 1$, the algorithm either*

(1) *proves that the problem is infeasible,*

or

(2) *computes a vector $\hat{x}$ such that $q_i(\hat{x}) \leq \epsilon$ for $1 \leq i \leq m$ and there exists no feasible $x \in \mathbb{R}^n$ with $q_0(x) < q_0(\hat{x}) - \epsilon$.*

The complexity of the algorithm is polynomial in the number of bits in the data and in $\log \epsilon^{-1}$.

As of this writing, the results in this section are primarily of theoretical interest, though, quite possibly, they may yield effective relaxation techniques for general QCQPs using the same philosophy as that employed in linear mixed-integer programming problems, where cutting-plane techniques focus on small subsets of constraints. In the next section we return to techniques for general QCQPs.

2.2.3 ▪ On the rank of solutions to semidefinite programs

As pointed out above, semidefinite programming is a key technique that can be brought to bear on quadratically constrained problems. As was the case in formulation (2.11), it is of particular interest to obtain semidefinite programming formulations that are guaranteed to have low-rank solutions. See [51], [54]. It turns out (see below) that this feature is of pivotal interest in the context of OPF problems and is thus worth reviewing in some detail.

Pataki [180] considers a generic semidefinite program:

$$\min C \bullet X$$
$$\text{s.t. } A^k \bullet X = b^k, \qquad 0 \leq k \leq m, \tag{2.16}$$
$$X \succeq 0. \tag{2.17}$$

Here, C and the A^k are symmetric matrices and the b^k are reals. (We note that from a formal perspective this problem statement is *incorrect*—we should use "inf" rather than "min" in the objective. For brevity we will assume that using "min" is correct.) What structural attributes can we expect an optimal solution X^* to possess? Specifically, what can we say about its rank? The following result is proved in [180] (for related results, see [188]).

Theorem 2.2.4. *Let X^* be an extreme point of the set defined by* (2.16)–(2.17), *of rank r, say. Then*

$$r(r+1)/2 \leq m.$$

This implies, in particular, that $r \leq \frac{\sqrt{1+8m}-1}{2}$; a slight improvement was obtained by Barvinok in [17]. It can be shown that the $\sqrt{m}$ bound is the best possible (see [16], [17] for details).

Recall that in the context of semidefinite relaxations of quadratically constrained problems one is interested in finding rank-one solutions. Below we will see that this is in particular the case when considering OPF problems. More generally, finding very low rank solutions is of interest. Thus there is a large gap between this goal and what is guaranteed by Pataki's theorem. This field remains one of great interest, as is the general topic of QCQPs.

Moment relaxations

The semidefinite programming methodology for QCQPs turns out to be a special case of a much more general family of techniques, the *moment* relaxation. This is an approach that has been pioneered by Lasserre [144], [143], but with very deep connections to many other areas in mathematics. Another approach in intellectual proximity is the *sum of squares* or SOS methodology. A complete review of these topics falls outside the scope of this manuscript; for an in-depth review see [145].

Building a moment relaxation of a QCQP

Consider a generic QCQP of the form $\mathcal{Q}$ given in (2.2), and let the integer $k \geq 1$ be given. The *level-k* moment relaxation for $\mathcal{Q}$ is obtained as follows. Let $x^{(k)}$ be the vector consisting of all *monomials* of degree $\leq k$ on x. For example, if $n = 4$ and $k = 3$, the vector $x^{(k)}$ will include the entries

$$1, x_1, x_2, x_3, x_4, x_1x_2, x_1x_3, x_1x_4, x_2x_3, x_2x_4, x_3x_4, x_1^2, x_2^2, x_3^2, x_4^2$$

and

$$x_1^2x_2, x_1x_2^2, x_1^2x_3, x_1x_3^2, x_1^2x_4, x_1x_4^2, x_2^2x_3, x_2x_3^2, x_2^2x_4, x_2x_4^2, x_3^2x_4, x_3x_4^2,$$
$$x_1x_2x_3, x_1x_2x_4, x_1x_3x_4, x_2x_3x_4.$$

Write

$$\check{Y}^{(k)} \doteq x^{(k)T}x^{(k)} \in \mathbb{R}^{\binom{n}{k}} \times \mathbb{R}^{\binom{n}{k}}.$$

Using our previous notation, $\check{Y}^{(1)} = \check{Y}$. Clearly, $\check{Y}^{(k)} \succeq 0$. We can therefore construct a relaxation to problem $\mathcal{Q}$ akin to (2.7), (2.9), but now involving a (much larger) matrix of variables $\bar{Y}$ that approximates $\check{Y}^{(k)}$. The rows and columns of $\bar{Y}$ are indexed by the subsets of $\{1,\ldots,n\}$ of cardinality at most k, and the constraints that we can impose on $\bar{Y}$, in addition to (2.7), are **(a)** $\bar{Y} \succeq 0$, **(b)** the first $n+1$ entries of the zeroth row of $\bar{Y}$ form the vector $(1\,x^T)$, and **(c)** additional structural constraints. These constraints include the statement that $\bar{Y}$ is symmetric, but other constraints as well. For example, in the $n=4$, $k=3$ case, we have

$$\bar{Y}_{\{1,2\},\{3\}} = \bar{Y}_{\{1,3\},\{2\}} = \bar{Y}_{\{2,3\},\{1\}},$$

since each of these matrix elements stands in for the product $x_1x_2x_3$.
For $k > 1$, therefore, the level-k relaxation is stronger than the standard SDP relaxation—but also much larger and correspondingly more difficult to solve.

As the name indicates, the level-k moment relaxation proves a lower bound on the optimal value Q^* for $\mathcal{Q}$. It can be shown that for a given QCQP, there is always a *finite* value of k such that the level-k relaxation is exact, that is to say, the value of the relaxation equals

Q^*. However, the determination of the precise value of the minimum such k for a given problem seems a difficult task. Clearly, as k grows the relaxations become stronger, albeit of mounting computational difficulty. We are not aware of any computational study addressing the application of this methodology to large (or even medium-size) general nature QCQPs; the difficulty of the underlying semidefinite programs remains a stumbling block. In the context of the AC OPF problem, some applications of the moment relaxation to medium-size problems (i.e., thousands, rather than hundreds of thousands, of lines) can be found in [163].

2.3 ▪ Semidefinite programming relaxations to the OPF problem

We now return to the AC OPF problem; we will focus on recent developments relying on semidefinite programming. An early reference is Jabr [130], who departed from traditional approaches to the problem by noting that, e.g., in (1.10) concerning line (k,m) we have expressions $\cos\theta_{km}$ and $\sin\theta_{km}$; thus one can create a variable R_{km} to represent $|V_k||V_m|\cos\theta_{km}$ and another variable $T_{km} = |V_k||V_m|\sin\theta_{km}$, and with this proviso (1.10) becomes

$$p_{km} = |V_k|^2 g_{\{k,m\}} - g_{\{k,m\}} R_{km} - b_{\{k,m\}} T_{km} \tag{2.18}$$

(and likewise with equation (1.11)), with the additional equation

$$|V_k|^2 |V_m|^2 = R_{km}^2 + T_{km}^2. \tag{2.19}$$

This equation is used to approximate the trigonometric identities in the power flow equation; moreover [130] further argues that in some cases (2.19) can be equivalently replaced by the (convex) constraint $|V_k||V_m| \geq R_{km}^2 + T_{km}^2$, and that using this "extended" formulation for the power flow problem enhances its solvability.

A recent and exciting improvement in AC OPF solvability was developed by Lavaei and Low [146]. They consider the standard OPF problem in rectangular coordinates, with objective of the form (1.30) and quadratic constraints (e.g., range constraints on the P_k, Q_k, and $|V_k|^2$ quantities for each bus k). As discussed above, these are indeed quadratic inequalities if rectangular coordinates are used. Lavaei and Low study a number of relaxations of the resulting optimization problem, in particular, a semidefinite relaxation. They furthermore show that under a number of technical conditions, relaxation is exact: it solves OPF, and furthermore this corresponds to obtaining a rank-one optimal solution for a semidefinite program. They argue that the technical conditions that make this outcome possible are "often" satisfied in practice. We will review their work in more detail below.

Following the initial paper [146], a number of researchers have performed investigations on the formulation of Lavaei and Low, both pointing out that there are indeed cases where the relaxation is not exact and presenting many successful cases. This remains an area of active investigation, a welcome development that is likely to yield future positive outcomes. A note of caution regarding these developments is that large-scale semidefinite programming is prohibitively expensive from a computational standpoint; it may even be the case that large-scale SOCPs could also prove impractical. Thus, traditional approaches to OPF may (likely, will) remain competitive at least for routine OPF cases in the near future.

We will now describe the Lavaei–Low approach in somewhat more detail. Suppose rectangular coordinates are used to indicate voltages, e.g., the voltage at bus k is $e_k + jf_k$; let v be (as above) the vector with entries $(e_1, e_2, \ldots, e_n, f_1, f_2, \ldots, f_n)^T$, where n = number of buses. For simplicity of exposition we will assume that all buses that house generators have zero load. Lavaei and Low consider an optimization problem with constraints (see (1.22)–(1.26))

$$\text{for } 1 \le k \le n \begin{cases} P_k^L \le v^T M_k v \le P_k^U, \\ Q_k^L \le v^T N_k v \le Q_k^U, \\ V_k^L \le e_k^2 + f_k^2 \le V_k^U, \end{cases} \tag{2.20}$$

i.e., limits on the active and reactive power injections at as well as on the voltage magnitude at each bus k. The objective is

$$\min \sum_{k=1}^{n} \{ c_k (v^T M_k v)^2 + d_k (v^T M_k v) \},$$

where $c \ge 0$ and $d \ge 0$ are given data, with the property that $c_k = d_k = 0$ whenever bus k does *not* house a generator. Thus, the objective function is a convex quadratic function of the active power generation at each generator, a standard objective in power practice. We can rewrite the OPF problem as

$$\min_{v} \sum_{k=1}^{n} x_k \tag{2.21}$$

$$\text{s.t. } c_k (v^T M_k v)^2 + d_k (v^T M_k v) \le x_k, \quad 1 \le k \le n, \tag{2.22}$$

$$\text{constraints (2.20).} \tag{2.23}$$

This optimization problem is not a QCQP but is easily made into one by replacing constraint (2.22) with the system

$$|v^T M_k v| - \eta_k \le 0, \quad c_k \eta_k^2 + d_k (v^T M_k v) - x_k \le 0$$

for each generator bus k, where η_k is a new variable. Constraint (2.22) can also be restated in matrix form:

$$\begin{pmatrix} x_k - d_k v^T M_k v & -\sqrt{c_k}\, v^T M_k v \\ -\sqrt{c_k}\, v^T M_k v & 1 \end{pmatrix} \succeq 0. \tag{2.24}$$

The fact that the two expressions are equivalent follows using a Schur complement argument. Here, given symmetric matrices A and C, with $C \succ 0$, and a matrix B of appropriate dimensions, the matrix

$$S \doteq \begin{pmatrix} A & B \\ B^T & C \end{pmatrix}$$

satisfies $S \succeq 0$ if and only if $A - BC^{-1}B^T \succeq 0$. This last matrix is known as the Schur complement [47].

Alternatively, and from a simpler perspective, note that (2.22) states that the determinant of the matrix in (2.24) is nonnegative, but because c_k and d_k are nonnegative, that fact also implies that the trace of that matrix is nonnegative. The converse is similar.

We will denote by **OPF** the optimization problem with objective (1.30) and constraints (2.24) (for each bus k) and (2.20).

We can now construct what amounts to the semidefinite relaxation of **OPF**, namely,

$$\textbf{OPF-SDP:} \quad \min_{v} \sum_{k=1}^{n} x_k \tag{2.25}$$

$$\text{s.t. } W \succeq 0 \quad \text{and,} \quad \text{for } 1 \le k \le n: \tag{2.26}$$

$$\begin{pmatrix} x_k - d_k M_k \bullet W & -\sqrt{c_k} M_k \bullet W \\ -\sqrt{c_k} M_k \bullet W & 1 \end{pmatrix} \succeq 0, \tag{2.27}$$

$$P_k^L \le M_k \bullet W \le P_k^U, \tag{2.28}$$

$$Q_k^L \le N_k \bullet W \le Q_k^U, \tag{2.29}$$

$$V_k^L \le T_k \bullet W \le V_k^U. \tag{2.30}$$

In (2.30), T_k is a diagonal matrix with a 1 entry in diagonal positions k and $k+n$ (i.e., in the coordinates of vector v corresponding to e_k and f_k) and zero otherwise. Lavaei and Low argue as follows.

Theorem 2.3.1 (see [146]). *Suppose* $v = 0$ *is not feasible for* **OPF**. *Under appropriate conditions, problem* **OPF-SDP** *has a rank-one optimal solution.*

Thus, provided the "appropriate conditions" hold (we will return to these conditions, and their appropriateness, below), we have a theoretically efficient means of solving problem **OPF**. Completely explaining Theorem 2.3.1 exceeds the scope of this work; however we will outline some of the key ideas in [146].

It turns out that directly analyzing when **SDP-OPF** has a rank-one solution is difficult. Proceeding indirectly, [146] construct what can be shown to be the (Lagrangian) dual of problem **OPF**. For each bus k, and reals λ^U, λ^L, μ^U, μ^L, γ^U, γ^L, let

$$h_k(\lambda^U, \lambda^L, \mu^U, \mu^L, \gamma^U, \gamma^L) \doteq \lambda_k^L P_k^L - \lambda_k^U P_k^U + \mu_k^L Q_k^L - \mu_k^U P_k^U + \gamma_k^L V_k^L - \gamma_k^U V_k^U.$$

Define the problem

$$\textbf{L:} \quad \max \sum_k h_k(\lambda_k^U, \lambda_k^L, \mu_k^U, \mu_k^L, \gamma_k^U, \gamma_k^L) - \sum_k r_k$$

$$\text{s.t. } \lambda_k^U, \lambda_k^L, \mu_k^U, \mu_k^L, \gamma_k^U, \gamma_k^L \ge 0; \; g_k, r_k \in \mathbb{R}, \text{ all } k,$$

$$\sum_k [\lambda_k M_k + \mu_k N_k + \gamma_k T_k] \succeq 0, \tag{2.31}$$

$$\begin{aligned} \text{and } \forall k: \lambda_k &= \lambda_k^U - \lambda_k^L + 2\sqrt{a_k} g_k + b_k, \\ \mu_k &= \mu_k^U - \mu_k^L, \\ \gamma_k &= \gamma_k^U - \gamma_k^L, \\ &\begin{bmatrix} 1 & g_k \\ g_k & r_k \end{bmatrix} \succeq 0. \end{aligned}$$

What is the relationship between problems **OPF**, **OPF-SDP**, and **L**? The following result is proved in [146].

Theorem 2.3.2. *Assume* **OPF** *is feasible.* (a) *Problem* **L** *is the dual of* **OPF**. (b) *Problem* **OPF-SDP** *is the Lagrangian dual of* **L** *with* W *acting as the dual variable for constraint* (2.31). *Further, strong duality holds.*

Thus, problems **OPF-SDP** and **L** have equal value, but of course this may still be strictly smaller than the value of **OPF**. Of course such a gap will not be realized when **OPF-SDP** has a rank-one solution. A key step in [146] is the following assumption.

Assumption C. At optimality for problem **L**, the matrix on the left-hand side of (2.31) has null space of dimension two.

This assumption embodies the "appropriate conditions" in Theorem 2.3.1. Postponing a discussion of this assumption, we can complete a proof of Theorem 2.3.1. Suppose that $\lambda^*, \mu^*, \gamma^*$ are the optimal values for the variables λ, μ, γ in **L**. Similarly, let W^* be optimal for problem **OPF-SDP**. Here, we note that since all matrices M_k, N_k, and T_k are block skew-symmetric (recall Definition 1.2.2), it follows that the matrix

$$A^* \doteq \sum_k \left[\lambda_k^* M_k + \mu_k^* N_k + \gamma_k^* T_k \right] \tag{2.32}$$

is block skew-symmetric as well. But complementary slackness must hold, i.e.,

$$A^* W^* = 0. \tag{2.33}$$

We have that $W^* \succeq 0$; let $W^* = \sum_{i=1}^{2n} \theta_i w_i w_i^T$ be its spectral decomposition, where $\theta_i \geq 0$ for all i. Since $A^* \succeq 0$ (constraint (2.31)), we therefore have that

$$A^* w_i = 0 \quad \text{for all } i \text{ with } \theta_i > 0.$$

By Assumption C, it follows that at most two θ_i can be positive. Given that Theorem 2.3.1 assumes that $v = 0$ is *not* feasible for problem **OPF**, it follows that $W^* \neq 0$, and so at least one θ_i is positive. In summary:

- Exactly one θ_i is positive. Then W^* has rank one, and we are done.
- We have
$$W^* = \theta_1 w_1 w_1^T + \theta_2 w_2 w_2^T,$$
where $\theta_1 > 0$, $\theta_2 > 0$ and w_1, w_2 form an orthonormal pair.

Assuming the second case, write $w_1 = (s_1^T \; t_1^T)^T$, where $s_1, t_1 \in \mathbb{R}^n$, and consider the vector

$$\hat{w}_1 \doteq (-t_1^T \; s_1^T)^T.$$

Then $\hat{w}_1^T w_1 = 0$ and $\hat{w}_1 \neq w_1$. Moreover, for any block skew-symmetric matrix A (with notation as in Definition 1.2.2),

$$\begin{aligned} A \bullet \hat{w}_1 \hat{w}_1^T = \hat{w}_1^T A \hat{w}_1 &= (-t_1^T \; s_1^T) \begin{pmatrix} B & C \\ -C & B \end{pmatrix} \begin{pmatrix} -t_1 \\ s_1 \end{pmatrix} \\ &= t_1^T B t_1 - t_1^T C s_1 + s_1^T C t_1 + s_1^T B s_1 \\ &= (s_1^T \; t_1^T) \begin{pmatrix} B & C \\ -C & B \end{pmatrix} \begin{pmatrix} s_1 \\ t_1 \end{pmatrix} = A \bullet w_1 w_1^T. \end{aligned} \tag{2.34}$$

In particular, $A^* \hat{w}_1 = 0$. Assumption C now implies that

$$\hat{w}_1 = \pm\, w_2.$$

Thus, (using (2.34)) we have that for any block skew-symmetric R,

$$R \bullet W^* = (\theta_1 + \theta_2) R \bullet w_1 w_1^T.$$

It follows that $(\theta_1 + \theta_2) w_1 w_1^T$ is feasible for problem **OPF-SDP** and with equal objective value as W^* and, therefore, optimal, as desired. This concludes the proof of Theorem 2.3.1, under Assumption C.

Now we return to Assumption C, which is clearly the key in arguing that **OPF-SDP** must have a *very low* rank solution, unlike what the Pataki–Barvinok results might threaten. Does Assumption C always hold? This cannot be the case: the OPF problem can be shown to be NP-hard. Reference [162] (also see [148]) presents a number of small, realistic examples that violate Assumption C, and where as a result, solving problem **OPF-SDP** does not produce a rank-one solution. Nevertheless, in [146] it is stated that the IEEE test cases *do* satisfy the condition, after a small perturbation outlined below. Thus, is it the case that Assumption C *nearly* holds for realistic grids? Furthermore, Assumption C is rather technical and requires solving problem **L** for its verification; is there an essentially equivalent statement that is simpler?

With the goal of addressing these questions, [146] proceeds as follows. Assume the following:

(r.1) The subgraph of the given grid obtained by using just those lines with positive resistance is connected.

(r.2) Suppose that at optimality for problem **L1**, the matrix A^* in (2.32) has the following structure:

$$A^* = \begin{bmatrix} H_1^* & 0 \\ 0 & H_1^* \end{bmatrix}. \tag{2.35}$$

In [146] it is argued that (r.1) and (r.2) imply Assumption C. Roughly, the development of this proof is as follows. Using the form (1.23) for M_k (also see Example 1.2.1) and similarly with N_k and T_k, we see that H_1^* is a (generalized) Laplacian matrix, corresponding to a graph that under (r.1) is connected. Thus, classical results show that the minimum eigenvalue for H_1^* has multiplicity 1.

Also, the complementary-slackness condition (2.33) implies that A^* is not positive definite, from which it follows that H_1^* also is not positive definite, and, in fact, the condition that the zero eigenvalue for H_1^* is simple implies that A^* has (up to sign) two unique norm-one eigenvectors for the zero eigenvalue, of the form

$$(u^T \; u^T)^T \quad \text{and} \quad (u^T \; -u^T)^T$$

for some $u \in \mathbb{R}^n$. In other words, Assumption C holds!

Now we will consider assumptions (r.1) and (r.2), beginning with (r.1). Here, we remind the reader that the conductance $g_{\{k,m\}}$ of a line $\{k,m\}$ is effectively proportional to the resistance of that line (see (1.6)–(1.7)) and that this parameter, along with the susceptance, appears in the power flow equations (see, e.g., (1.10)) and thus in the matrices M_k and N_k. A practical concern that this condition brings up is related to the fact that line resistances are typically small—in fact, some or all resistances are frequently set to zero in practical computations. From a theoretical perspective one can always set the resistance of a line to a small parameter ϵ to ensure that the connectivity requirement holds; however this brings up other issues, for example, related to numerical stability. Another way of stating this concern is that (for small but positive resistances) it is possible that the *second*-smallest eigenvalue, while positive, could be very small, so that effectively H_1^* has null space of dimension greater than 1.

Turning to (r.2), [146] presents arguments to posit that we should have

$$A^* \approx \begin{bmatrix} H_1^* & 0 \\ 0 & H_1^* \end{bmatrix}, \tag{2.36}$$

i.e., the off-diagonal blocks should have small entries. Their arguments (and (2.36)) are verified for the IEEE family of problems. As was the case of (r.1), again we have what is essentially a combinatorial condition being approximated; i.e., holding "numerically." This, again, raises the possibility of numerical instabilities.

Putting aside such issues (as well as issues concerning the complexity of solving large semidefinite programs), the central question of interest is how often, in practice, the rank-one condition does hold. In [146] it is reported that most of the IEEE family of problem instances [172] do, in fact, give rise to rank-one solutions of the above SDP, after small perturbations of the reactances to guarantee connectivity of the graph discussed above. Following this additional work, other authors have found evidence of duality gaps, i.e., a nonexistence of rank-one solutions. See, for example, [148]. At any rate, the initial work in [146] has provided a (much-needed) boost for research on efficient, guaranteed solutions to AC OPF problems; work on these topics is now proceeding at a brisk pace.

Moreover, it is clear that there is a large methodological gap between classical results on rank of solutions to semidefinite programs (as reviewed above in Section 2.2) and what is required in formulation **OPF-SDP**. Possibly, future research will attempt to bridge this gap and bring more methodological research regarding SDPs into the realm of AC OPF. Work in [213] explores how the rank-one condition applies to restricted topology cases, such as trees. Also see [46], [157].

A different caveat brought up above concerns the difficulty of solving large-scale semidefinite programs, a task not necessarily viewed as easy in the optimization community. Clearly, however, the Lavaei–Low formulations are very sparse (the total number of nonzeros is proportional to the number of lines). In fact, for each bus k the nonzero pattern of matrices M_k and N_k is that of the neighborhood of bus k, and matrix T_k has two nonzeros. Hence, **OPF-SDP** includes two copies of the nonzero pattern in the adjacency matrix for the underlying graph, plus some diagonal entries. This is significant since typical grids are extremely sparse (average degree approximately 3) and have a simple structure. Some research has targeted the efficient numerical solution of large, sparse SDPs. See, for example, [23]. Reference [162] describes an implementation of the Lavaei–Low approach, which leverages the structure of the M_k and N_k matrices to efficiently solve the SDP with SeDuMi [216] (this relies on techniques motivated by work in [103]). Also see [40]. On the 300-bus IEEE example the code is (relatively) fast; the reported runtime on a modern workstation is approximately twenty seconds. On the Polish grid MATPOWER examples, the code runs in approximately ten minutes. In contrast, commercial OPF codes (which, as described above, are both heuristic and at best guarantee a local optimum) require a few seconds on routine problems. This fact highlights the challenges that the SDP-based solution methods will face in the near future—but it also highlights the main benefit of the SDP methods, which is the theoretical ability to handle any case, even if nonroutine.

This field continues (and will continue, for the foreseeable future) to generate interest. Some recent developments using the moment relaxation method are described in [163], [158]. Also see [185], [83], [63].

We conclude this section by stressing that, at present, there is a debate in the power flow community regarding how often Assumption C holds. It may be the case that for many realistic grids, problem **OPF-SDP** has a *very low* rank solution, but not quite a rank-one solution, and how to obtain a (near-optimal) rank-one solution would then be an interesting challenge. In any case, it is clear that the work in [146] has injected much energy into the power flow computation field, and additional useful results are likely.

2.3.1 ▪ Other convex formulations for power flows

Even though power flow problems can be severely nonconvex, practical experience indicates some degree of near convexity, at least in routine problem instances (see Section 2.4 below). The semidefinite programming family of methods outlined above can be seen as a well-posed attempt to leverage this empirical behavior, though possibly at high computational cost.

Researchers have recently considered exploiting the near-convex behavior of power flows through a variety of means. Here we outline two recent developments.

A logarithmic-barrier method for lossless AC OPF

Equations (1.10) and (1.11) are a general form of the power flow equations using phase angles and voltage magnitudes. Assuming magnitude-one voltages throughout and an (active power) *lossless* system (see (1.12)), we have $g_{\{k,m\}} = 0$ on any line (k,m), and so these equations can be restated as

$$p_{km} = b_{\{k,m\}} \sin(\theta_k - \theta_m), \tag{2.37}$$

$$q_{km} = -(b_{\{k,m\}} + b^{sh}_{\{k,m\}}) + b_{\{k,m\}} \cos(\theta_m - \theta_k). \tag{2.38}$$

Suppose we focus on active power flows, but on the other hand suppose that each line has an active power flow limit. Then we obtain a system of nonlinear equations and inequalities:

$$\sum_{(k,m)} b_{\{k,m\}} \sin(\theta_k - \theta_m) = P_k \quad \forall k, \tag{2.39}$$

$$|b_{\{k,m\}} \sin(\theta_k - \theta_m)| \le \bar{p}_{\{k,m\}}. \tag{2.40}$$

In (2.40) the right-hand side is the power limit on line (k,m). We stress that in (2.39) the sum is over all the lines that have k as a bus. Given values P_k, if a solution to (2.39)–(2.40) could be found (in variables θ), then substituting into (2.38) one obtains a valid solution to a lossless AC power flow system (albeit one with a posteriori computed reactive power injections). The solution of the system is of interest because (2.39) allows us to model phase angles with some accuracy (beyond what the DC approximation can attain).

Here we will describe an approach that uses a method of Boyd and Vandenberghe [48] as extended in [24]. For each line we choose an arbitrary orientation; if a line uses buses k and m and is oriented from k to m, then it is denoted km; otherwise it is denoted mk. An oriented line is termed an *arc*. Consider the optimization problem

$$\min_{\vartheta} \sum_{km} b_{\{k,m\}} \psi(\vartheta_{km}) \tag{2.41}$$

$$\text{s.t.} \sum_{km} b_{\{k,m\}} \vartheta_{km} \; - \; \sum_{mk} b_{\{k,m\}} \vartheta_{mk} = P_k \quad \forall \text{ bus } k, \tag{2.42}$$

$$|\vartheta_{km}| \le \min\{1, \bar{p}_{\{k,m\}}\} \quad \forall \text{ arc } km. \tag{2.43}$$

Here, for $|t| < 1$,

$$\psi(t) \doteq \int_{-1}^{t} \arcsin(y)\,dy.$$

In (2.42) the first sum on the left-hand side is over all lines oriented *out of* bus k; the second sum is over all lines oriented into k. Another point is the following.

Remark 2.3.3. *In the domain $|t| < 1$, the function $\psi(t)$ is convex.*

Proof. This follows because the derivative of arcsin(y) is $(1-y^2)^{1/2}$. ■

It follows that (2.41)–(2.43) is a convex optimization problem. We include a proof of the following result for completeness, since our language differs somewhat from that in [48].

Lemma 2.3.4. *Suppose that ϑ^* is an optimal solution to* (2.41)–(2.43) *with the property that* (2.43) *is slack for every line (k,m). Let θ denote the vector of optimal Lagrange multipliers for constraints* (2.42). *Then for every arc km,*

$$\vartheta^*_{km} = b_{\{k,m\}} \sin(\theta_k - \theta_m). \tag{2.44}$$

Proof. Since all constraints (2.43) are slack, their Lagrange multipliers are zero. First-order optimality conditions then yield that for any arc km,

$$b_{\{k,m\}}(\theta_k - \theta_m) = \arcsin(\vartheta^*_{km}),$$

which implies (2.44). ■

The import of this result is that if its conditions apply, then defining for any line (k,m), $p_{km} = \vartheta^*_{km}$ if the line is oriented away from k, and $p_{km} = -\vartheta^*_{mk}$ otherwise, then we obtain a solution to the nonlinear system (2.39)–(2.40). But there are many impediments to a direct application of this idea in the context of the AC OPF problem.

To begin with, problem (2.41)–(2.43) is not the problem we wish to solve—its objective is artificial. Second, the problem might not even be feasible. And even if it is, some of the inequalities (2.43) may always be binding (or nearly so). The last two issues, however, involve a grid under stress. In general, we *would* expect problem (2.41)–(2.43) to be feasible. And the fact that some inequalities (2.43) must be binding or nearly binding would be a valuable piece of information, and diagnosing this situation is a worthwhile pursuit. In [24], the following approach is applied.

Let P denote the vector of active power injections, let $f(P)$ denote the (convex) objective in the AC OPF problem of interest, for any line (k,m) write $u_{\{k,m\}} \doteq \min\{1, \bar{p}_{\{k,m\}}/b_{\{k,m\}}\}$, and let the *barrier* function $B : \mathbb{R}_+ \to \mathbb{R}_+$ be defined by $B(t) = -\log t$. Where D and ϕ are constants defined below, consider the optimization problem

$$\min_{P,\vartheta,\delta} \; f(P) + D \sum_{km} b_{\{k,m\}} [\psi(\vartheta_{km}) + \phi B(\delta_{km})] \tag{2.45}$$

$$\text{s.t.} \; \sum_{km} b_{\{k,m\}} \vartheta_{km} - \sum_{mk} b_{\{k,m\}} \vartheta_{mk} - P_k = 0 \quad \forall \text{ bus } k, \tag{2.46}$$

$$|\vartheta_{km}| + u_{\{k,m\}} \delta_{km} \le u_{\{k,m\}} \quad \forall \text{ arc } km, \tag{2.47}$$

$$\delta_{km} \ge 0 \quad \forall \text{ arc } km. \tag{2.48}$$

Below we will outline how to choose D and ϕ appropriately small. However, the point of this formulation is that (i) it is convex, and (ii) the use of the barrier function in the objective ensures that in an optimal solution (if one exists) the quantities δ will not be close to zero—in that case the optimal ϑ will satisfy (2.43) with strict inequality. Furthermore, problem (2.45)–(2.48) is convex, and as above one can therefore use Lagrange multiplier arguments. Then the preceding argument regarding slackness of (2.43) can be refined (using that ϕ is large) to argue that, at optimality,

$$\vartheta_{km} \approx b_{\{k,m\}} \sin(\theta_k - \theta_m),$$

where θ_k is the optimal Lagrange multiplier for (2.46); additionally, using the fact that D is *small*, one can further argue that the optimal value of (2.45) equals, approximately, the optimal OPF value as well.

To make these statements precise, suppose we choose a tolerance $0 < \epsilon < 1$ and let $b_{\max} = \max b_{\{k,m\}}$ denote the maximum susceptance, $M =$ number of lines, and $C > 0$ be a *lower bound* on the value of the OPF problem of interest:

$$\min_{P,\theta} f(P) \tag{2.49}$$

$$\text{s.t.} \sum_{(k,m)} b_{\{k,m\}} \sin(\theta_m - \theta_k) = P_k \quad \forall k, \tag{2.50}$$

$$|b_{\{k,m\}} \sin(\theta_m - \theta_k)| < \bar{p}_{\{k,m\}}. \tag{2.51}$$

We then set, in (2.45),

$$D = \frac{\epsilon C}{\pi b_{\max}}, \qquad \phi = (M \log(1/\epsilon))^{-1}.$$

Theorem 2.3.5 (see [24]). (a) *Suppose (P^*, θ^*) is an optimizer for problem* (2.49)–(2.51). *Suppose, further, that for each arc km we have*

$$|\sin(\theta_k^* - \theta_m^*)| \leq (1-\epsilon)\bar{p}_{\{k,m\}}. \tag{2.52}$$

Then, defining $\vartheta_{km}^* = \sin(\theta_k^* - \theta_m^*)$ and $\delta_{km}^* = 1 - |\vartheta_{km}^*|/u_{km}$ for each arc km, we obtain that $(P^*, \vartheta^*, \delta^*)$ is a feasible solution to problem (2.45)–(2.48) *with cost at most $(1+2\epsilon)f(P^*)$.*

(b) *Conversely, let $(\hat{P}, \hat{\vartheta}, \hat{\delta})$ be an optimal solution to problem* (2.45)–(2.48) *with objective value $\hat{K}$, say. Then, with $\hat{\theta}$ obtained by solving problem* (2.41)–(2.43) *we have that $(\hat{p}, \hat{\theta})$ is feasible for the sync-constrained OPF problem and has cost $f(\hat{p}) \leq \hat{K}$.*

(c) *Let $(\hat{P}, \hat{\vartheta}, \hat{\delta})$ be as in* (b), *and suppose that additionally for every arc km we have*

$$|\hat{\vartheta}_{km}| \leq u_{km}(1 - \tilde{\epsilon}_{km}) \tag{2.53}$$

for some value $0 < \tilde{\epsilon}_{km} < 1$. Let $D\hat{\theta}$ be the optimal dual variables for constraints (2.46). *Then for every line km we have*

$$\hat{\vartheta}_{km} = \sin(\hat{\theta}_k - \hat{\theta}_m + \eta_{km}), \tag{2.54}$$

where $|\eta_{km}| \leq (M\, u_{km}\, \tilde{\epsilon}_{km} \log(1/\epsilon))^{-1}$.

The import of this theorem is as follows. Problem (2.49)–(2.51) is the problem we are interested in solving. (a) and (b) state that this problem, and (2.45)–(2.48), are essentially equivalent problems, at least in terms of optimal objective value. However, the solution to (2.45)–(2.48) does not provide optimal phase angles—to obtain those we need to solve a problem such as (2.41)–(2.43), whose solution may produce flows different from the optimal ϑ. (c) addresses this situation—it states that if we compute an optimal solution to (2.45)–(2.48) which is sufficiently slack in the constraints (2.47), then (approximate) phase angles can be obtained. Here note that for given ϵ, if $\tilde{\epsilon}_{km}$ is sufficiently *large*, then $|\eta_{km}|$ is *small* if $M \log 1/\epsilon$ is *large*, i.e., if the barrier parameter ϕ is small.

In other words, if we choose the barrier parameter ϕ large enough, and the conditions in (c) are satisfied, the solution to (2.45)–(2.48) is such that it approximately satisfies the nonlinear equations (2.50) while at the same time attaining near-optimality for problem (2.49)–(2.51). Of course, we cannot guarantee that (2.53) will hold for relatively large $\tilde{\epsilon}$—this is an attribute of the problem instance.

As noted, the solution to problem (2.45)–(2.48) amounts to a step in a logarithmic-barrier method to solve problem (2.49)–(2.51). See [47] for more background. This method would solve a sequence of problems (2.45)–(2.48) for decreasing values of ϵ and ϕ, and starting with an initial solution that satisfies $\delta_{km} > 0$ for all km (which can be computed, if one exists, by solving a linear program). The method can then be used to diagnose whether indeed (2.53) can hold, at optimality, for *some* positive $\tilde{\epsilon}$. Note that as ϕ becomes small, solving the (convex) problem (2.45)–(2.48) may become increasingly challenging from a numerical standpoint, in particular if $\tilde{\epsilon}$ is in fact small.

Linear approximations to AC power flow equations

As discussed above, the DC approximation to power flows is a combination of first- and zeroth-order estimations around the standard operating point. A more sophisticated approach is given in [70] (also see [122]). This method (termed LPAC for obvious reasons) constructs a polyhedral approximation to the cosine terms in the AC power flow equations, and Taylor series approximations to other terms, obtaining a linear programming formulation.

Consider, for example, a line (k,m) where we seek to impose a bound $|\theta_k - \theta_m| \le \theta_{km}^{\max}$ for some quantity $0 < \theta_{km}^{\max} < \pi/2$ (and possibly much smaller than $\pi/2$). Consider equation (1.10), which is used to represent the active power flow on line (k,m) and is repeated here for convenience:

$$p_{km} = |V_k|^2 g_{\{k,m\}} - |V_k||V_m| g_{\{k,m\}} \cos\theta_{km} - |V_k||V_m| b_{\{k,m\}} \sin\theta_{km}. \tag{2.55}$$

In the DC approximation to power flows (a) we would assume known voltage magnitudes all equal to unity, (b) we approximate $\cos\theta_{km} \approx 1$ and thus as a result of (a) cancel the first two terms, and (c) we approximate $\sin\theta_{km} \approx \theta_k - \theta_m$. Even if we do not assume equal voltage magnitudes, estimating $\cos\theta_{km} \approx 1$ replaces the first two terms in (2.55) with

$$|V_k| g_{\{k,m\}} (|V_k| - |V_m|). \tag{2.56}$$

Recall that $g_{km} = \frac{r_{\{k,m\}}}{r_{\{k,m\}}^2 + x_{\{k,m\}}^2}$ and thus is positive when the resistance r_{km} is positive. In such a case, when $|V_k| \approx |V_m|$ (2.56) will in general *underestimate* the first two terms in (2.55) (and, thus, p_{km}).

In other words, we wish to correct the *over*estimate $\cos\theta_{km} \approx 1$. A more accurate approximation (effectively a first-order Taylor approximation) can be obtained by approximating the middle term rather than ignoring it. To this effect, consider any value $\phi \in [-\theta_{km}^{\max}, \theta_{km}^{\max}]$. Then we have that

$$\cos y \leq -(y-\phi)\sin\phi + \cos\phi \quad \forall\, y \in (-\pi/2, \pi/2), \tag{2.57}$$

since the cosine function is concave in the domain $(-\pi/2, \pi/2)$. We can thus approximate $\cos y$ with (2.57), evaluated at a finite set of values $\phi \in (-\pi/2, \pi/2)$ (for example, K evenly spaced values for some large enough K) together with the bound $\cos y \geq \cos\theta_{km}^{\max}$.

A further refinement used in [70] arises from applying the Taylor approximation method to the voltage magnitudes.

This methodology holds great promise, and quite possibly an approach along these lines may replace the reliance on the traditional DC approximation.

DC power flow in rectangular coordinates

As noted above, one of the ingredients in obtaining the DC approximation to the power flow equations is assuming that all voltages have unit magnitude (in the p.u. system). Baldick [12] combines this idea with the flat start concept (Section 1.2.2) and then linearizes the resulting equations in rectangular coordinates.

Consider equation (1.20) describing active power injection at bus k, and repeated here for convenience

$$p_{km} = g_{\{k,m\}} e_k^2 + e_k\left(-g_{\{k,m\}} e_m + b_{\{k,m\}} f_m\right) - f_k\left(b_{\{k,m\}} e_m + g_{\{k,m\}} f_m\right) + g_{\{k,m\}} f_k^2.$$

In a flat start, we would write for any bus p, $e_p = 1 + \tilde{e}_p$, where $\tilde{e}_p \approx 0$, and also assume $f_p \approx 0$. Thus, to first order we can write

$$p_{km} \approx g_{\{k,m\}}(\tilde{e}_k - \tilde{e}_m) - b_{\{k,m\}}(f_k - f_m),$$

and so for any bus k, approximately,

$$P_k = \sum_{km}\left(g_{\{k,m\}}(\tilde{e}_k - \tilde{e}_m) - b_{\{k,m\}}(f_k - f_m)\right). \tag{2.58}$$

A similar approximate relationship can be obtained for reactive power flows:

$$Q_k = \sum_{km}\left(-b_{\{k,m\}}(\tilde{e}_k - \tilde{e}_m) - g_{\{k,m\}}(f_k - f_m)\right). \tag{2.59}$$

As was discussed before, the quantities $g_{\{k,m\}}$ are normally small—setting them to zero we obtain a lossless model (see (1.13)) and in fact the DC approximation assumes as much. Thus, a strategy for approximately solving system (2.58)–(2.59) is the following: first, assuming $g_{\{k,m\}} = 0$ for all lines km, solve (2.58) for the vector f. Then, substitute the solution into (2.59) and solve for $\tilde{e}$. The resulting vector of voltages $1 + \tilde{e}_k + j f_m$ can then be used as an alternative to the flat start for iterative algorithms for AC OPF.

2.4 ▪ Some recent numerical investigations

The research surveyed above spans a wide spectrum of techniques, modern and traditional, incorporating varying degrees of sophistication and computational cost. At the same time, practitioners have typically relied on—at this point—legacy methodology whose conceptual development dates to several decades ago.

Recent work [201], [56] has investigated the numerical properties of typical AC OPF problems as well as the effectiveness of several solvers. The critical issues under consideration are the degree of nonconvexity actually observed in practical problem instances (as opposed to worst-case examples such as those in [124]) and the success of various optimization methods.

The work in [201] considers the 14-, 30-, 57-, 118-, and 300-bus problem instances from the IEEE test family (available with MATPOWER [234]).[10] A first experiment consists of solving, for each case, a number of problem instances where the objective has been randomly perturbed by a very small amount. Then, for each pair of random instances (indexed h and k, say, with optimal solutions v^h and v^k, resp.), a new AC OPF-like problem instance is constructed as follows:

- All AC power flow constraints are used.
- The objective is to find a vector v that minimizes the "distance" to $(v^h + v^k)/2$.

(We recall here that in the AC OPF problem quantities such as generator output levels and voltage magnitudes are variables; our vector v (as well as v^h and v^k) contains entries for such quantities.) The precise definition of "distance" is not important here; however it is a convex function (effectively a sum of scaled square errors) and, moreover, the distance function evaluated at $(v^h + v^k)/2$ takes value zero. In other words, the test simply checks to see if $(v^h + v^k)/2$ is actually a feasible solution to the AC power flow problem, and if not it returns a measure of its infeasibility.

In the experiments reported in [201], *none* of the $(v^h + v^k)/2$ vectors turned out to be feasible, although the infeasibilities were not large. One can interpret this result as indicating a large degree of nonconvexity of the feasible region near the optimal solution. A possible weakness in these experiments is the (potential) local optimality of the computed solutions.

In a second set of experiments in [201] the authors attempt to determine the degree of uniqueness of optimal solutions to AC OPF problems. To this effect, suppose we denote an AC OPF problem by $\min\{f(v) : v \in F\}$, where F denotes the feasible region and $f()$ is a convex quadratic with nonnegative coefficients. Suppose v^* is an optimal solution, and let i be a bus. Then a typical problem considered in [201] is

$$\max |V_i| \tag{2.60}$$

$$\text{s.t. } f(v) \leq (1+\epsilon) f(v^*), \tag{2.61}$$

$$v \in F. \tag{2.62}$$

In (2.61) $\epsilon > 0$ is a small number (0.01 in [201]). Let $|V_i|^{\max}$ denote the value of the optimum. This problem simply aims to find a near-optimal, feasible solution with large voltage magnitude at i—if $|V_i|^{\max} - |V_i^*|$ is large, then we have some evidence of *flatness* of the feasible region near the optimum. As alternatives to the objective (2.60), [201] also considers the objective functions $\min |V_i|$, $\max P_g$, $\min P_g$, $\max Q_g$, and $\min Q_g$, all at each generator g.

To fix ideas, consider the case of the voltage at a bus i, and define the elasticity of voltage at i to be $(|V_i|^{\max} - |V_i|^{\min})/|V_i^*|$ (where $|V_i|^{\min}$ is defined in the obvious way), and likewise define the other elasticity parameters. In the experiments performed in [201] the outcome shows that voltage elasticity, on average, ranges between 3 and 11%, whereas active and reactive power elasticities are much larger (several hundred percent).[11]

[10] All rather small: the 300-bus case, for example, only has 411 lines.

[11] Note that in this case the likely local optimality of computed solutions to problems (2.60)–(2.62) does not invalidate the observation.

Thus, indeed, we see evidence of extensive flatness of the (near) optimal region for the AC OPF problem in the IEEE cases being considered. This is an observation that had already been made, if on grounds less methodologically solid, by the power community. The observation also justifies the use of iterative convex optimization techniques for the AC OPF problem when a starting point sufficiently near the optimum is available, and when solving a problem instance which is a simple variation to one previously solved, then such a starting point would clearly be at hand.

A more complete viewpoint is that for the nonconvexities reported in the first set of experiments in [201], near-optimal solutions are, likely, slightly infeasible as well. At some level this justifies the frequent practice of permitting slight infeasibilities when solving an AC OPF problem (however, it does not justify the claim that such infeasibilities are strictly necessary or minimal). It would be of interest to perform both of the above experiments on much larger and less well behaved problem instances than those used in [201].

Experiments designed to test a range of solver software are reported in [56], using the same test-bed of problem instances as above. The solvers considered were MINOS [211], SNOPT [212], IPOPT [225], CONOPT [74], and KNITRO [174] in the case of nonlinear problems, and GUROBI [115] and CPLEX [125] for linearizations. All of these are high-performance, general purpose optimization packages with a strong record. In all cases recent versions were utilized. In Section 3.3.1 we will examine in greater detail the application of nonlinear programming techniques to general (nonconvex) optimization problems.

The experiments study running time and behavior for three kinds of problems: AC OPF versions (i) in standard form, (ii) with linearized objectives, and (iii) with sequentially linearized constraints. All of these problems require an iterative starting point; the experiments consider the flat start described above, plus several random starts, but not a starting point based on prior knowledge of the problem. A tested feature was the use of so-called slack variables, with high linear objective penalties, which are added to the range constraints for active and reactive power and voltage magnitudes. More properly these would be called artificial variables in the language of linear programming. The DC approximation to the AC OPF problem was also solved (using CPLEX and GUROBI) for comparison.

The purpose of the experiments was to test the speed and robustness of the solvers as well as the quality of the computed solutions; note that none of the tested algorithms guarantee global optimality. A summary of running times is given in Table 2.1, which shows the average performance of the different solvers, on the formulation using artificial variables. Not surprisingly, the 300-bus case provided some of the more challenging tests. Some of the results of the experiments are summarized next; all but the first refer exclusively to the 300-bus case.

(1) Among the nonlinear solvers, CONOPT appeared to be the most robust (always among the fastest and most accurate solvers). In terms of linear solvers, GUROBI was deemed more robust than CPLEX, though it appears that the performance differences were not necessarily one-sided or consistent.

(2) All solvers failed to find a feasible solution unless artificial variables were used. When artificial variables were used, the solvers sometimes failed to remove them at termination.

(3) CONOPT required (on average) approximately 20 seconds to converge. The linearized problems ran approximately twice as fast. By comparison, the DC approximation ran approximately 100 times as fast.

(4) The solutions computed using the linearized problems had value within 3% of the best computed using the nonlinear solvers; however, the algorithms used for the linearized versions failed to converge in the allotted number of iterations.

(5) As a stress test, problems were rerun using increased values for the loads (increased by 20% and 40%). One of the reasons that increasing loads results in harder problem instances is that as power flows increase, line flow constraints become binding. Additionally, reactive power line losses also increase, and voltage constraints become binding or infeasible. In the experiments conducted in [56], the demand-magnified instances resulted in problematic convergence, with failure to converge in some cases, and roughly doubled running times (both for CONOPT). The average CPU time required by the DC approximation changed only slightly.

We stress once again that the 300-bus case was the largest considered in these experiments and is not viewed by practitioners as constituting a significantly challenging example. Modern power grid computations would likely involve examples that are at least 10 or 20 times as large, and possibly 100 time as large. In addition, the IEEE library of test cases is not young—all but one of the examples were developed several decades ago and from a numerical standpoint are well understood (and considered simple).[12] Realistic examples could very well incorporate numerical instability challenges.

Table 2.1. *Performance of nonlinear solvers on AC OPF problems with quadratic costs (from* [56]*) (CPU seconds).*

Case	CONOPT	MINOS	SNOPT	KNITRO	IPOPT
14-bus	0.083	0.052	14.137	0.197	0.432
30-bus	0.175	0.204	0.067	0.626	1.019
57-bus	0.817	2.961	5.749	1.321	2.861
118-bus	8.441	3.424	1.652	129.387	10.906
300-bus	24.500	32.500	14.100	529.300	34.700

A point to be stressed is that the experiments in [56] did not encompass widely used, commercially developed AC OPF solver systems. From a technological standpoint, such systems rely on mathematical methodology that predates, for example, modern semidefinite programming and trust-region methods. Many of the commercial systems are undocumented. From among those that are documented (sometimes indirectly, or partially) our interpretation is that they typically embody sequential search procedures (such as Power World [189], due to Overbye [178], [179]) relying on clever mathematics, linearization, heuristics, and problem data modification when convergence is not attained. Volume testing of such commercial software (and publication of the results) would be a worthwhile research goal. In some cases the financial cost of the software stands as an impediment to this task.

An important point to be noted is that some of the commercial software vendors have reacted to the new methodological developments and are now developing new versions of their products, in some cases after a hiatus spanning decades. We speculate that the new releases will incorporate some of the new methodologies and thus, in combination with industry experience, may attain an entirely new level of performance.

[12] The latest examples available through MATPOWER involve the Polish national grid and have a few thousands buses but were not considered in [56].

In any case, our interpretation of the above performance summary is that it is rather sobering. Using the solver options considered in [56], only the DC approximation emerges as one that can be carried out in large volumes in practicable time.

2.5 ▪ Summary

The AC basic and optimal power flow problems are central in power engineering practice. Typical formulations incorporate quadratic equations or inequalities and as a result are theoretically intractable (NP-hard). Nevertheless, routine problem instances, that is to say instances arising from normal and familiar operating conditions, can be readily solved using standard methodologies for solving systems of equations, such as the Newton–Raphson method, or techniques from convex optimization, such as trust-region methods and logarithmic-barrier algorithms.

When dealing with transmission systems under distress or in unfamiliar configurations, such standard techniques can fail and the problems can become significantly more challenging. In such instances, the widespread use of penalized "artificial" or "slack" variables to allow for small deviations from desired operating points becomes problematic, because algorithm termination with nontrivial values of artificial variables makes it difficult to distinguish between local optimality of the computed penalized solution and infeasibility of the original formulation. This constitutes a significant methodological shortcoming in two different settings. First, the ability to rapidly compute an adequate response to an ongoing critical emergency is an operational imperative. Second, the rapid and agnostic evaluation of many what-if scenarios is a central ingredient in long-term planning of a grid.

New solution methodologies that stem from conic and semidefinite optimization appear quite promising and are very likely to see increased use in the near future; however, it is to be expected that the computational overhead imposed by these techniques will be significant. We estimate that linear (and conic) approximations such as the LPAC family of formulations, and DC approximations, will remain viable competitors in many settings. In the context of this monograph, a central goal is the large-volume analysis of cascading failures of transmission systems. In our experiments (below) we will investigate the use of both AC and linearized formulations of power flow problems. Nevertheless, it seems clear at the time of this writing that the use of AC models causes a visible slowdown in the computations and may even introduce artifacts as a result of specific steps taken to avoid infeasibility (see Section 5.4.4); while on the other hand the DC approximation may lead to incomplete models. Hybrid models incorporating linearization techniques and ad hoc, opportunistic modeling of specific nonlinearities, such as the behavior of a specific subset of power lines or generators, would be a logical extension of currently employed methods for hard mixed-integer programming problems.

Chapter 3

N-K problem

Despite widely publicized failures, national transmission systems have proved quite robust in the past. A reader not familiar with power engineering may wonder how such large, complex systems can manage to operate almost flawlessly. Is it some version of overengineering, on a massive scale indeed, that allows for graceful recovery from possibly frequent equipment outages?

Part of this success lies in the classical N-1 reliability criterion, which is used, in some form, throughout the world, that we introduced in Chapter 1. Typically there are additional criteria in place, for example indicating that after a single-element failure, the remaining system must attain an (N-1)-reliability state within a given span of time (e.g., 30 minutes), though how this second goal is to be attained is not always precisely stated. Another example of a more elaborate requirement is the "N-1-1" criterion, which states, roughly, that any *two sequential* failures, separated by a time interval where specific control actions are allowed, must be tolerated [97]. Additionally, the N-1 criterion can be imposed in a formal planning setting, such as the unit commitment problem [26]. For a more complete overview of practices in North America, see [168], Section TPL (Transmission Planning), and [169].

The implementation of such power grid reliability criteria, in the form of concrete operational directives, has clearly been a success. Yet, problems have recently been experienced throughout the world. Part of the issue here may concern conditions that impede the effective operation of a grid but fall short of outright equipment failures. For example, one might consider the case of a grid which is overloaded due to high demand during poor weather (e.g., high temperatures) and where an element failure takes place. Even if the grid is nominally (N-1)-reliable, the particular combination of factors (some exogenous) under which the failure takes place may not have been captured in tests for the N-1 criterion. Perhaps a stricter criterion might be needed. And of course in a severe event more than one element might fail at once. To address such shortcomings, an alternative that has been suggested to the N-1 criterion consists of the N-K criteria for $K = 2, 3, \ldots$(small). Clearly, enforcing an N-2 criterion places a far more stringent condition on a grid than the N-1 criterion. Additionally, a (software) test to verify that a grid satisfies the N-2 criterion may prove valuable in exposing poor practices or operating details that were previously undetected.

One point hinted at in the previous paragraph concerns precisely *how* the N-K criteria are tested—even for $K = 1$. Since the criteria must be tested under a possibly large number of potential operating environments (time of day, ambient weather, load levels, available mix of generation, and so on), this is a test that will have to be performed using fast algorithms. It is clear that this may not be a trivial task, especially if using an AC power flow model, but even under the DC approximation, the number of cases to examine will be very large. For $K > 1$ an enumerative approach appears quite problematic. For $K \geq 3$, in the case of a grid with tens of thousands of lines and thousands of generators, it appears completely infeasible.

In fact, as we argue later in this section, the realistic evaluation of any particular contingency effectively involves the solution of a potentially difficult optimization problem to incorporate the evaluation of possible remedial actions. Thus, when investigating whether a given system would tolerate all contingencies involving up to K elements, the enumerative approach appears increasingly hazardous. Hence we would ideally *not* want to resort to an enumerative approach: if N is in the tens of thousands (for example, the number of lines and generators in a small- or medium-size national grid) and K is small, the number of cases to be enumerated, $\binom{N}{k}$, obviously grows very rapidly! Table 3.1 somewhat gratuitously depicts this fact.

Table 3.1. $\binom{N}{K}$.

N	$K=2$	$K=3$	$K=4$
1000	499500	166167000	41417124750
2000	1999000	1331334000	664668499500
4000	7998000	10658668000	10650673999000
8000	31996000	85301336000	170538695998000
10000	49995000	166616670000	416416712497500

Given the quantities shown in Table 3.1, we would prefer to substitute overt enumeration with an indirect procedure. One such venue would be provided by stating the task as an appropriate optimization problem. For that purpose we will use the following definition.

Definition 3.0.1. *A ($\leq K$)-contingency is an event where K or fewer elements are faulted. We will use the same terminology to refer to the set of faulted elements.*

We will leave a precise definition of the term "faulted" to the specific problem context. For the time being the reader may view "faulted" as equivalent to "removed from operation," though on occasion "degraded" may be more appropriate. Using this definition, we can test whether a system is (N-K)-reliable by computing a ($\leq K$)-contingency that is most severe—if this contingency is not particularly serious, the system is deemed safe. Alternatively, we could search for a minimum-cardinality contingency that attains a desired level of severity. A clear advantage of such an optimization-driven approach is the ability to *certify*, unambiguously, that a given contingency is most severe from among those of a given cardinality or less, or that a given grid tolerates all contingencies involving up to K faults, for a given K.

However, this is still not a well-defined problem. In particular one needs to specify a particular definition of severity, and different definitions could potentially yield different diagnoses of system safety. A comprehensive definition should account for the fact that a serious contingency will most likely lead to dynamical instability of the grid, but incorporating such dynamics into the search (e.g., into the optimization) will produce a far more difficult problem.

The existing literature, as a result, has usually taken a conservative outlook: a contingency is severe if it leaves the system in a state that is problematic, though not necessarily critical. A typical case would be one where postcontingency, the system operates on a power flow that exceeds the rating of at least one line. Viewing such a contingency as severe is conservative because if a line rating is violated, control actions could possibly succeed in removing the risk that the line will trip. In fact the same principle applies to other types of risk that can be modeled. The prevailing approach in the literature, and in practice, is that such a contingency will be examined, offline, in a separate and more accurate manner (e.g., a numerical integration that may account for system dynamics) to more accurately simulate its actual impact. We will return to these points below. In summary we have a combination of algorithms:

(a) A procedure that uses a simplified model of a grid, for example the DC approximation of power flows, to search for a most severe ($\leq K$)-contingency or a minimum-cardinality contingency of a given severity by solving, typically, an optimization problem. Usually this will be a binary mixed-integer program where the binary variables are used to model faults (yes or no, for each element at risk) and whose objective is to find a most severe contingency according to a simplified, and conservative, risk criterion. The output of this procedure is a list of potentially severe contingencies.

(b) A more accurate and, likely, far slower procedure that is used to evaluate a given contingency to estimate its actual impact.

It is worth analyzing the potential impact of the conservative approach mentioned above in greater detail. In general, the removal of K lines may disconnect the network, even for K relatively small: for example the average degree of national transmission systems is less than four, or even smaller. Typically in a freshly created island (i.e., component) there will be an excess of load over generation or vice versa. Consider first the case of an island where load exceeds generation. Postcontingency, either some of the load will have to be lost (shed) or, if possible, generation will have to increase, or both. Power engineering practice will always favor the minimization of shed load; however, the apportionment of load shedding among the different load locations amounts to a policy decision. Further, the process of causing a generator to increase its output, if the increase is significant, can be slow. This fact is important because in the time span during which generators are increasing output, an excess of demand over supply will be experienced, causing system frequency to drop (see Sections 1.3 and 5.5). If the frequency drop is large enough, generators will trip and the entire island will collapse (very rapidly), with all load lost. An important point here is that these phenomena are not captured by the standard AC power flow model, which is steady state, and of course not by the DC model. In principle, they can be captured through complex dynamical systems; however, as we have pointed out before, it is not necessarily the case that such dynamical systems can be quickly solved (when dealing with large transmission systems) or that all data needed for the solution process are available, or even that the data that are available[13] are actually reliable.

In summary, the accurate analysis of a contingency such that a postcontingency island has a significant demand excess can be quite challenging. Likewise, the analysis of an island where there is a large excess of supply over demand is also complex; in this case generators must quickly decrease their output, which may or may not be feasible. In either case, a completely accurate analysis will almost certainly—at least in today's power engineering practice—require some off-line analysis under human supervision. To put it differently, such analyses would be the opposite of "rapid" and might even incorporate

[13]For offline analysis and certainly for online (real-time) analysis as well.

some ambiguity. As a consequence, the literature we cite in the following section and, indeed, the majority of the literature on N-K problems, has devoted some focus to the study of contingencies where no islands are created—the network remains connected, postcontingency. Later in this chapter we will also survey recent work that does not make this assumption.

Power grid security in contrast with survivable network design

The focus we just described stands in marked contrast to what is found in the literature on a subject that, from the perspective of a researcher outside of power engineering, might be viewed as being in close intellectual proximity to the topics we consider in this book (and is more familiar in the traditional operations research context): the design and analysis of fault-tolerant telecommunications networks, or "survivable" networks, as they have been termed in that literature. See [215], [133], and [190] for wider background and (many) citations. Telecommunications network design problems address the minimum-cost development or expansion of a network used to convey multicommodity traffic from sources to destinations; the primary expense arises in the installation of appropriate devices on the links of this network. Survivability in the face of contingencies is essential. In particular, a common requirement is that the network should be able to deliver all demands (or some fraction of the demands) when any K links are removed; the requirement of postcontingency connectivity has been a central research focus. The study of this problem first arose some 30 years ago, as a result of a number of well-publicized failures of fiber optic networks involving very small values of K (e.g., $K = 1$). See, e.g., [220] for an event that affected millions of people.

The network design literature has primarily relied on mixed-integer programming and discrete mathematics techniques. The problems can be quite difficult; indeed some of the more difficult instances in the MIPLIB library of mixed-integer problems [140] are of this nature. However, the versatility of telecommunications equipment yields a feature that is useful in practice and also in solvability of the optimization problems: telecommunications traffic can be *routed*, i.e., made to follow desirable routes, in particular to avoid overcongesting equipment. In contrast, no such capability exists in power transmission networks: power flows according to the laws of physics, and *how* power will flow (given network parameters) requires a complex computation, as we have seen in prior chapters. This is a crucial difference between the problems we study in this book and those arising in traditional network design.

Returning to the power engineering setting, it would be inaccurate to assume that contingencies that maintain connectivity are simple or uninteresting—quite the contrary. In a dangerous contingency where the network remains connected, we have a case where physics and network structure conspire to make it impossible for the system to deliver all demands, even though generation is sufficient. The discovery of such a contingency therefore reveals a subtle flaw in the underlying grid. In contrast, the discovery of a contingency that creates an island with a large mismatch between supply and demand can be formulated as a binary mixed-integer program. Even though this optimization problem will possibly be difficult, the interpretation of the underlying grid weakness is straightforward.

In the remainder of this chapter we survey prior work on the N-K problem and indicate some future research venues. In Section 3.0.1, we review some classical and more recent approaches to efficient enumeration of ($\leq K$)-contingencies in the case of the N-K problem with line outages under the DC approximation to power flows. In Section 3.1 we examine past work where problem (a) is modeled as a linear binary mixed-integer program. As has been observed in the literature, this family of mixed-integer programs can prove quite difficult. An alternative is to approximate task (a) by relying on nonlinear (and nonconvex) optimization problems; we survey those in Section 3.3. Finally, one approach to making the risk assessment in step (a) less conservative is to attempt to incorporate some model of the postcontingency system dynamics that nevertheless proves computationally practicable in an optimization setting. Some authors have relied on multistage stochastic models, and we review some work in that direction in Section 3.4.

3.0.1 ▪ Accelerating enumeration through fast linear algebra

As argued in the previous section, direct enumeration methods, even for K small, are likely to encounter insurmountable challenges, especially for larger values of N. A line of research in the power engineering community has attempted to make enumeration more rapid by speeding up the analysis of each case and perhaps avoiding some cases altogether. See, for example, [153], [113], [79], [221].

This research cited above has focused on the DC approximation, and in particular on contingencies arising where K lines are removed. As described in Section 1.2.4, the DC approximation for a given grid is described by the equations

$$Np = P, \quad p = YN^T\theta, \tag{3.1}$$

where p is the vector of power flows, θ is the vector of phase angles, and Y is the diagonal matrix of line susceptances. The flow vector solving this system is unique, and it equals

$$p = Y\tilde{N}^T \tilde{\mathcal{B}}^{-1}\tilde{P}, \tag{3.2}$$

where $\tilde{N}$ is obtained by removing one row from N, $\tilde{P}$ is obtained by removing the same row from P, and $\tilde{B} = \tilde{N}Y\tilde{N}^T$ (recall that the bus susceptance matrix equals $\mathcal{B} = NYN^T$; see (1.36)). In principle, one would apply (3.2) to a grid after a contingency and thereby determine if the contingency is problematic in the sense that some of the line flows exceed their respective limits. However, this statement is misleading and, in general, incorrect. Equation (3.2) requires that the underlying network be connected and that $\sum_i P_i = 0$. In the remainder of this section we will only consider contingencies that keep the network connected.

To make details precise, assume that the directed graph G has n nodes and m arcs and that it is connected in the undirected sense. We denote the arc-set by A. Thus, $p \in \mathbb{R}^m$, $\theta \in \mathbb{R}^n$, N is $n \times m$, and Y is $m \times m$. We further assume $\sum_k P_k = 0$. See (1.37), (1.38). The solution to the power flow system is then given by (3.2). Now consider a set of arcs L, the contingency. We are interested in the (rapid) evaluation of (3.2) as applied to the residual network $G - L$ obtained by removing L from G, so that we can decide if, postcontingency, $|p_{ij}| > u_{ij}$ (the limit) for any $(i,j) \notin L$. Clearly, the sticking point is how to quickly compute the inverse of the bus susceptance matrix in $G-L$; let us denote this latter matrix by $\mathcal{B}_{\bar{L}}$.

Recall that $\tilde{\mathcal{B}} = \tilde{N}Y\tilde{N}^T$. If follows that, denoting by Y_L the restriction of Y to elements of L, that is to say

$$Y_L = \text{diag}\{y_{\{k,m\}} : \{k,m\} \in L\},$$

and by $\tilde{N}_L$ the restriction of $\tilde{N}$ to columns of L, we then have

$$\tilde{\mathscr{B}}_{\bar{L}} = \tilde{\mathscr{B}} - \tilde{N}_L Y_L \tilde{N}_L^T.$$

Since we are assuming all susceptances to be positive, it follows that $\tilde{\mathscr{B}}_{\bar{L}}$ is a rank-$|L|$ update of $\tilde{\mathscr{B}}$. At this point we can make use of a well-known result.

Theorem (Sherman–Morrison–Woodbury). *Suppose that for some integers p and k, A is a $p \times p$ matrix, X_1 is a $p \times k$ matrix, C is an invertible $k \times k$ matrix, and X_2 is a $p \times k$ matrix. Assume that both A and $C^{-1} + X_2^T A^{-1} X_1$ are invertible. Then*

$$(A + X_1 C X_2^T)^{-1} = A^{-1} - A^{-1} X_1 (C^{-1} + X_2^T A^{-1} X_1)^{-1} X_2^T A. \tag{3.3}$$

See, e.g., [109] for background.

To apply this result in our setting, we would set $p = n - 1$, $k = |L|$, $A = \tilde{\mathscr{B}}$, $X_1 = N_L = -X_2$, and $C = Y_L$. Then (3.3) would seem to yield

$$\tilde{\mathscr{B}}_{\bar{L}}^{-1} = \tilde{\mathscr{B}}^{-1} - \tilde{\mathscr{B}}^{-1} \tilde{N}_L (Y_L^{-1} - \tilde{N}_L^T \tilde{\mathscr{B}}^{-1} \tilde{N}_L)^{-1} \tilde{N}_L^T \tilde{\mathscr{B}}. \tag{3.4}$$

Since $\tilde{\mathscr{B}}^{-1}$ is (a priori) known (and, quite likely, a sparse factorization of this matrix is available) and since we assume L to be small (and as a result N_L is small and in any case very sparse), it should thus be possible to evaluate (3.4) quite rapidly and thus, through an application of (3.2) to $G - L$, to rapidly perform a risk assessment of that network. More precisely, if we assume that we are dealing with a class of networks with small (i.e., $O(1)$) average degree, then the update needed to compute $\tilde{\mathscr{B}}_{\bar{L}}^{-1}$ from $\tilde{\mathscr{B}}^{-1}$ can be implemented to require, on average, a workload of the order of $O(|L|)$, plus the inversion of an $O(|L| \times |L|)$ matrix. This is a constant amount, i.e., independent of n and m. (The reader acquainted with the literature on Laplacian matrices will find these arguments familiar.) In fact the computation of the power flows in $G - L$ would require $O(m)$ work.[14] Therefore, for any fixed value of K (say, $K = 3$), the entire enumeration of all size $\leq K$-contingencies will require total workload $O(m^{K+1})$.

However, for this approach to be valid, we have to ensure that the hypotheses of the Sherman–Morrison–Woodbury theorem apply. Clearly we have that $A = \tilde{\mathscr{B}}$ is invertible; we thus need to make sure that

$$Y_L^{-1} - \tilde{N}_L^T \tilde{\mathscr{B}}^{-1} \tilde{N}_L$$

is also invertible, and since $Y \succ 0$ this is equivalent to checking that $I_{|L|} - Y_L \tilde{N}_L^T \tilde{\mathscr{B}}^{-1} \tilde{N}_L$ is invertible. This is not always the case, and counterexamples are easy to produce.

Example 3.0.2. *Consider the directed graph G with node-set $\{1,2,3\}$ and arc-set $\{(1,2),(2,3)\}$. Hence*

$$N = \begin{bmatrix} 1 & 0 \\ -1 & 1 \\ 0 & -1 \end{bmatrix}.$$

[14]When $|L|$ is small compared to m, this is a lower bound on the required workload, since the list of arcs not in L has size $m - |L| \approx m$.

Assume all susceptances have value 1.0*, i.e.,* $Y = I_3$*. Suppose we construct* $\tilde{N}$ *by removing the last row of* N*. Then*

$$\tilde{\mathscr{B}} = \tilde{N}\tilde{N}^T = \begin{bmatrix} 1 & -1 \\ -1 & 2 \end{bmatrix} \quad and \quad \tilde{\mathscr{B}}^{-1} = \begin{bmatrix} 2 & 1 \\ 1 & 1 \end{bmatrix}.$$

Suppose L *consists of the singleton* $(2,3)$*. Then* $\tilde{N}_L^T = (0,1)$*, and so*

$$I_{|L|} - Y_L \tilde{N}_L^T \tilde{\mathscr{B}}^{-1} \tilde{N}_L = 1 - (0,1) \begin{bmatrix} 2 & 1 \\ 1 & 1 \end{bmatrix} \begin{pmatrix} 0 \\ 1 \end{pmatrix} = 0.$$

In the example the set L forms a cutset of the underlying graph. But as we argued above, we are primarily interested in cases where L does not define a cutset, that is to say $G-L$ is a connected graph in the undirected sense. Under this assumption, the desired result does hold, as we prove in the next result (which may be folklore). In the proof we denote by $\tilde{N}_L$ the restriction of $\tilde{N}$ to arcs in L.

Theorem 3.0.3. *Suppose* L *is not a cutset of* G*. Then the matrix* $I_{|L|} - Y_L \tilde{N}_L^T \tilde{\mathscr{B}}^{-1} \tilde{N}_L$ *is invertible.*

Proof. Consider any vector $f \in \mathbb{R}^L$ such that

$$(I_{|L|} - Y_L \tilde{N}_L^T \tilde{\mathscr{B}}^{-1} \tilde{N}_L) f = 0, \text{ i.e., } Y_L \tilde{N}_L^T \tilde{\mathscr{B}}^{-1} \tilde{N}_L f = f. \tag{3.5}$$

Here we are indexing f by L for convenience. We will show that under such conditions $f = 0$, which completes the proof. Toward this goal, we will first interpret (3.5) as follows. Let

$$P \doteq N_L f \quad \in \mathbb{R}^n.$$

Thus $\sum_k P_k = 0$. Consider the DC power flow system defined on G by using supply/demand vector P. By Lemma 1.2.3 there is a unique power flow $\hat{p}$ solving this system, given by

$$\hat{p} = Y\tilde{N}^T(\tilde{N}Y\tilde{N}^T)^{-1}\tilde{P} = Y\tilde{N}^T \tilde{\mathscr{B}}^{-1} \tilde{P}, \tag{3.6}$$

where $\tilde{P}$ is the restriction of P to its first $n-1$ rows, and thus

$$\tilde{P} = \tilde{N}_L f.$$

Consequently, (3.6) and the second equation in (3.5) imply that

$$\hat{p}_{ij} = f_{ij} \qquad \forall (i,j) \in L. \tag{3.7}$$

Denote by $\hat{\theta}$ a vector of phase angles which, together with $\hat{p}$, solve the DC system on G with supply/demand vector P. Now, denoting by A the set of all arcs, consider the vector $q \in \mathbb{R}^{A-L}$ defined by

$$q_{ij} = \hat{p}_{ij} \quad \forall (i,j) \in A-L.$$

Then, denoting by N_{A-L} the restriction of N to the arcs in $A-L$ (an $n \times |A-L|$ matrix), we have by definition of $\hat{p}$ and (3.7),

$$P = N\hat{p} = N_L f + N_{A-L} q = P + N_{A-L} q,$$

and so

$$N_{A-L} q = 0.$$

Thus, the pair $q, \hat{\theta}$ is a solution to the DC power flow problem defined on $G - L$ with supply/demand vector 0. By Lemma 1.2.3, $q = 0$, and as a consequence

$$\hat{\theta}_i = \hat{\theta}_j \quad \forall (i,j) \notin L.$$

But $G - L$ is assumed to be connected. Hence there is a constant t such that $\hat{\theta}_i = t$ for all nodes i. Thus

$$\hat{p}_{ij} = y_{ij}(\hat{\theta}_i - \hat{\theta}_j) = 0 \quad \forall (i,j) \in A,$$

and in particular $f = 0$, as desired. ■

The technical results above form a backdrop for much of the work cited above. Here we will primarily focus on the results in [221], which studies the case $K = 2$, i.e., contingencies L with $|L| = 2$. We need a few definitions, where for convenience we label lines using a single index:

(1) For a line a, p_a is the precontingency flow on a.

(2) Given a pair of distinct lines a, b, we write

$$L_b^a \doteq \frac{p_b^a - p_b}{p_a},$$

where p_b^a is the flow, on line b, in the grid $G - a$, *assuming that* $G - a$ *is connected.* If $p_a = 0$ then it is straightforward to show that $p_b^a = p_b$ and we can define $L_b^a = 1$.

The L_b^a quantities have been termed *line outage distribution factors* or LODFs in the literature, and they have a clear interpretation. Clearly $p_b^a - p_b = p_a L_b^a$ is a measure of the impact that removing line a has on the growth of overloading of line b. Now consider a set of two lines a and b, assumed not to form a cutset. Given a different line c, define $p_c^{\{a,b\}}$ to be the flow on c in the grid $G - \{a,b\}$. Then a calculation shows that

$$p_c^{\{a,b\}} - p_c = \frac{L_c^a(p_a + p_b L_a^b)}{1 - L_b^a L_a^b} + \frac{L_c^b(p_b + p_a L_b^a)}{1 - L_b^a L_a^b}. \tag{3.8}$$

It can be shown [221] that the assumption that $\{a,b\}$ is not a cutset implies that the denominator in this expression is nonzero; indeed this turns out to be an application of Theorem 3.0.3. Continuing with (3.8), let us denote the limit of a line k by u_k. Then (3.8) implies that for any line c, $p_c^{\{a,b\}} > u_c$ precisely when

$$\frac{L_c^a}{u_c - p_c}\frac{(p_a + p_b L_a^b)}{1 - L_b^a L_a^b} + \frac{L_c^b}{u_c - p_c}\frac{(p_b + p_a L_b^a)}{1 - L_b^a L_a^b} > 1, \tag{3.9}$$

where it is assumed that $u_c > p_c$. This amounts to assuming that precontingency, the system is secure, i.e., $u_c \geq p_c$—if the strict condition does not hold we can slightly perturb u_c. The above condition can be rewritten in the form

$$A_{a,b} B_{a,c} + A_{b,a} B_{b,c} > 1 \tag{3.10}$$

for appropriately defined matrices A and B. A similar analysis applies when we consider the condition $p_c^{\{a,b\}} < -u_c$. Finding a triple a, b, c such that (3.10) holds, by complete enumeration, would of course be undesirable (total effort proportional to m^3). Instead, [221] suggests the following heuristic speedup method. Initially, we set the lists $\mathcal{U}$, $\mathcal{V}$ to be (in both cases) the set of all ordered pairs of lines.

(1) For each line a compute a line c such that $(a,c) \in \mathcal{V}$ and c maximizes (or minimizes) $B_{a,c}$. Denote the respective quantities by $B_a^{\max}$, $B_a^{\min}$.

(2) For an ordered pair of lines $(a,b) \in \mathcal{U}$, check whether

$$\max\{A_{a,b}B_a^{\max}, A_{a,b}B_a^{\min}\} + \max\{A_{b,a}B_b^{\max}, A_{b,a}B_b^{\min}\} > 1.$$

If *not*, then purge (remove) pair (a,b) from the list $\mathcal{U}$.

Clearly, one execution of (1)–(2) requires $O(m^2)$ time and excludes from $\mathcal{U}$ pairs (a,b) that will not satisfy (3.10) for any line c. A similar method can be used to purge pairs from $\mathcal{V}$. Then (1)–(2) is repeated, with the procedure halting when no pairs are purged from $\mathcal{U}$ and from $\mathcal{V}$. At that point we can resort to applying direct enumeration (e.g., the sparse linear algebra approach outlined before Example 3.0.2) to those pairs that remain in $\mathcal{U}$. As shown in [221], this procedure can prove effective, with the final set $\mathcal{U}$ attaining a linear magnitude ($O(m)$ pairs) in few iterations. In other words, the entire $N-2$ analysis takes time $O(m^2)$ rather than $O(m^3)$, which is the estimate we would obtain from standard sparse Laplacian update estimates as outlined above.

In spite of this interesting approach, it seems clear that for higher values of K a workload growing at least proportional to m^K (if not m^{K+1}) will be difficult to avoid. In view of Table 3.1 a more indirect procedure is called for. In the next several sections we review methods based on optimization techniques.

3.1 ▪ Mixed-integer programming models for the N-K problem

The models we describe in what follows have several common features:

- Only line faults are considered. We associate a binary variable z_{km}, which equals 1 if line (k,m) is faulted.[15]

- The DC flow approximation is used, with line limits and generator output ranges. Recall that if P_k is the (net) active power injection at bus k, the DC power flow equations can be written as

$$\sum_{(k,m)} \frac{\theta_k - \theta_m}{x_{\{k,m\}}} = P_k \quad \text{for each bus } k.$$

  In the above expression, the term for line (k,m) is its corresponding flow p_{km}.

- Postcontingency, the system is deemed stable if loads are met without exceeding line limits.

However, these features give rise to a model that is incomplete and unnecessarily conservative. Consider, for example, a contingency where, after some set of lines becomes faulted, the network is split into two "islands" or components, and in one of the islands the total

[15]We will use "faulted" and "tripped" interchangeably in this chapter.

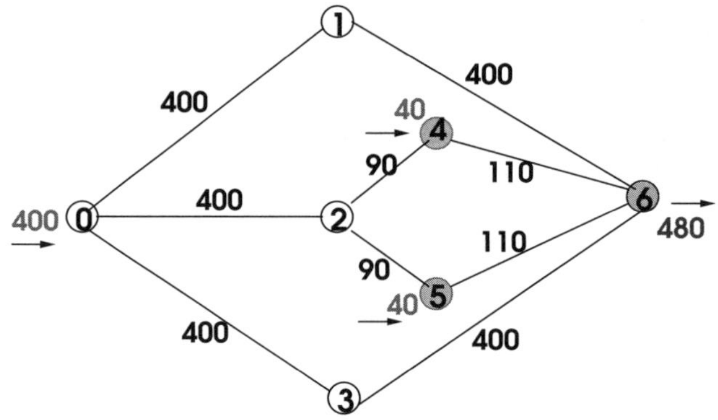

Figure 3.1. *(Example 3.1.1.) A case where controlled line tripping is needed. Line limits are shown.*

demand exceeds the (precontingency) total generation. A simple rule would be to declare the contingency "severe," but that will, in general, prove too conservative, because generators can modify their output (with limitations) in response to precisely such an event. This is an example where in evaluating a contingency we must also take into account a set of feasible recourse actions. If we view the contingency as the action of a fictitious "attacker," or "adversary," we should also view the recourse as the action of another party, which we will term the "controller." This introduces additional complexity.

Besides the simple, generic example we just provided, one can generate subtler examples where more complex recourse actions are available to the controller to illustrate the complexity of the problem.

Example 3.1.1. *Consider the system in Figure 3.1. In this example all lines have unit reactance; the figure shows the line rating next to each line. By convention, generator injections are shown as arrows pointing into buses and loads as arrows pointing out of buses.*

- *Buses 0, 4, and 5 hold generators with zero lower output limit and maximum output, as shown in the figure.*
- *The only load is at bus 6, with demand 480.*
- *The recourse actions available to the controller, postcontingency, are shedding load and tripping at most one line.*

A computation shows that the power flows in this system are 140 *on lines* $(0,1)$, $(1,6)$, $(0,3)$, *and* $(3,6)$; 120 *on line* $(0,2)$; 60 *on lines* $(2,4)$ *and* $(2,5)$; *and* 100 *on lines* $(4,6)$ *and* $(5,6)$. *Consider the contingency where line* $(0,1)$ *is tripped. The power flows, postcontingency, are shown in Figure 3.2.*

Note that lines $(2,4)$, $(2,5)$, $(4,6)$, *and* $(5,6)$ *are all overloaded, the last two by 25 units each. Thus, if the controller does not trip any lines, some load must be shed. To achieve feasibility by proportionally scaling all flows, we will have to scale the load at bus 6 by a factor of* $110/135.0 \approx 0.815$, *which entails a loss of load of approximately* $0.185 * 480 \approx 88.89$ *units. Simple enumeration shows that tripping any single line other than* $(0,2)$ *will require significant load shedding to achieve a stable system. Consider, for example, Figures 3.3 and 3.4.*

In both cases line $(4,6)$ *is highly overloaded and large load shedding will be needed.*

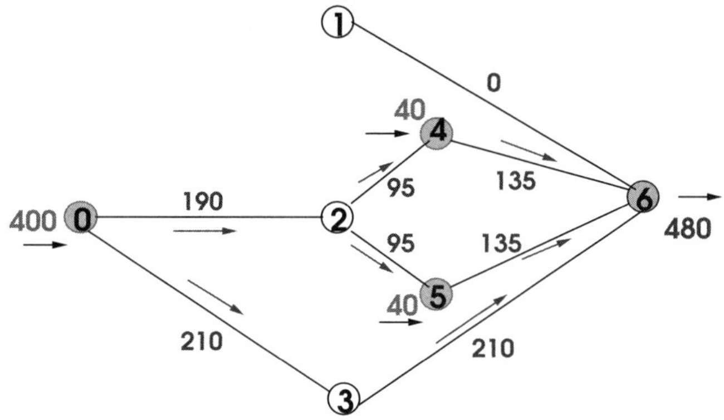

Figure 3.2. *(Example* 3.1.1.*) Flows in the system in Figure* 3.1*, postcontingency.*

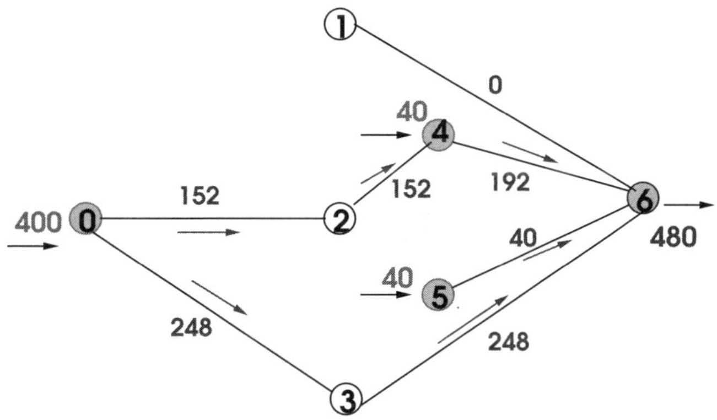

Figure 3.3. *(Example* 3.1.1.*) Unstable solution when line* (2,5) *is tripped.*

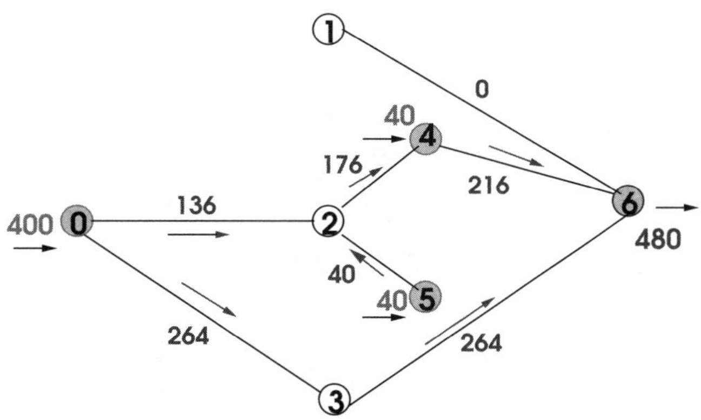

Figure 3.4. *(Example* 3.1.1.*) Unstable solution when line* (5,6) *is tripped.*

On the other hand, if the controller trips line (0,2)*, then the system can still supply the full load with no overloads, as shown in Figure* 3.5*. It is important to note that, postcontingency, line* (0,2) *is not overloaded, whereas* (2,5)*,* (5,6)*,* (2,4)*, and* (4,6) *all are.*

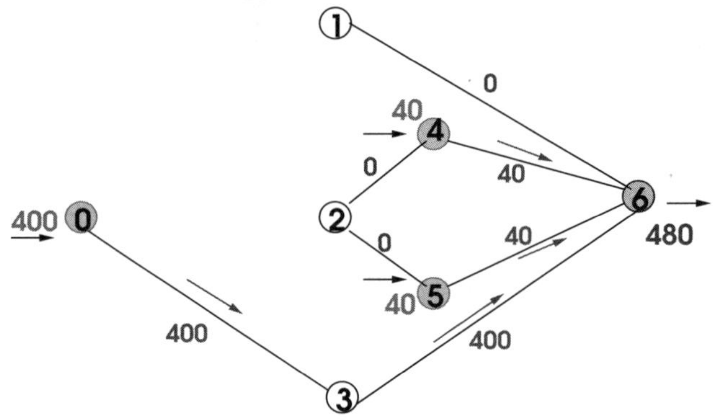

Figure 3.5. *(Example 3.1.1.) Stable solution with no load shedding.*

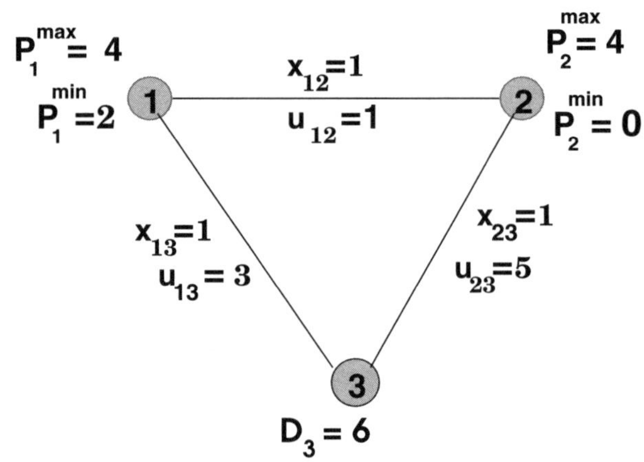

Figure 3.6. *A simple example where the controller must turn off a generator.*

Example 3.1.2. *Consider the 3-bus system given in Figure 3.6, where we have the following:*

1. *Buses 1 and 2 represent generators, with minimum and maximum output limits $P_1^{\min} = 2$, $P_1^{\max} = 4$, $P_2^{\min} = 0$, and $P_2^{\max} = 4$, respectively.*
2. *Bus 3 is a load with demand 6.*
3. *There are three lines; Figure 3.6 shows for each line (k,m) its reactance x_{km} and its rating u_{km}.*
4. *As a recourse action the controller may shed up to 50% of the load and may shut off a generator; however, line limits may not be exceeded.*

Precontingency, the following solution is feasible: $P_1^g = P_2^g = 3$, $P_3^3 = 6$, $p_{12} = 0$, $p_{13} = p_{23} = 3$, $\theta_1 = \theta_2 = 0$, $\theta_3 = -3$. This solution delivers 100% of the demands. On the other hand, consider the attack consisting of the single line $(1,3)$. We note that $P_1^{\min} > u_{12}$, and consequently any recourse action which includes keeping generator 1 in operation is infeasible. However, if the controller shuts off generator 1, the system does admit a feasible solution where 4 units of load are served at bus 3 (load shed $= 2$ units, which is less than 50% of the initial demand 6).

As the above examples show, even in small cases our fictitious controller may need to apply significant analysis to compute a satisfactory response (or even a feasible response). In fact, the controller may need to solve an optimization problem—worse, the problem may be discrete and, possibly, NP-hard. The need for such subtlety on the part of the controller is practical: typically load shedding is undesirable (even in small quantities); line tripping should be analyzed with care so as to avoid undesirable consequences (such as congesting other lines) and should be limited to few instances; and deliberate generator tripping should be reserved for extreme emergencies and, again, with very few units involved.

The requirement to adequately model realistic recourse actions introduces additional complexity into our optimization task of finding the most severe $\leq K$-contingency. That task becomes a (discrete) optimization problem such that the question of whether a vector is feasible is, itself, defined by solving a different (and probably hard) discrete optimization problem. As we will see below (Section 3.1.1), a growing field within the mixed-integer programming community is focused precisely on situations of this type.

It will prove useful to formalize a number of ingredients in the overall problem.

Definition 3.1.3. *Given a transmission system, or "grid," $\mathcal{N}$, we have the following:*

- *An* **attack** $\mathcal{A}$ *specifies a possible action taken by the attacker, for example the removal or partial impairment of a set of elements of the grid. We assume that $\mathcal{A}$ is represented by a vector in the appropriate space.*
- *Given an attack $\mathcal{A}$, the* **surviving grid** $\mathcal{N}(\mathcal{A})$ *is the system resulting from the attack.*
- *A* **configuration** $\mathcal{C}$ *is a valid recourse action taken by the controller. Given an attack $\mathcal{A}$, we denote by* $\mathbf{P}(\mathcal{N}(\mathcal{A}),\mathcal{C})$ *the system of power flow equations in $\mathcal{N}(\mathcal{A})$ under recourse action $\mathcal{C}$. Configuration $\mathcal{C}$ is valid if $P(\mathcal{N}(\mathcal{A}),\mathcal{C})$ is feasible, and we indicate by $\mathcal{R}(\mathcal{N}(\mathcal{A}))$ the set of valid configurations given attack $\mathcal{A}$.*
- *There is a* **performance measure** $\mu(\cdot)$ *indicating the value of a valid recourse action $\mathcal{C}$ taken after an attack $\mathcal{A}$ and indicated by $\mu(\mathcal{N}(\mathcal{A}),\mathcal{C})$. As an example, the* **yield** *attained by $\mathcal{C}$ is the (percentage of) total demand satisfied, postattack, under $\mathcal{C}$. Other examples are weighted sums of served demands, or the sum of served demands larger than a given threshold. If no feasible solution to $\mathcal{N}(\mathcal{A})$ exists, we set $\mu(\mathcal{N}(\mathcal{A}),\mathcal{C}) = -\infty$.*
- *There is a function* $\mathbf{s}(\cdot)$ *such that given an attack $\mathcal{A}$, $s(\mathcal{A})$ provides the cost or "effort" to the attacker of implementing $\mathcal{A}$; in other words $s(\cdot)$ measures the cost of impairing a set of grid elements to various degrees. For example we might have $s(\mathcal{A}) = \|\mathcal{A}\|_1$. Another example is $s(\mathcal{A}) = \|\mathcal{A}\|_0$, i.e., the number of nonzero entries in $\mathcal{A}$. If in addition $\mathcal{A}$ is a 0/1 vector, then we simply count the number of impaired grid elements.*

We differentiate between two problems. In the **minimum effort attack** *problem we are given a threshold value $\mu^{\min}$, and we have the following:*

(mc.1) *We say that an attack $\mathcal{A}$* **defeats** *a configuration $\mathcal{C}$ if either* (a) *$\mu(\mathcal{N}(\mathcal{A}),\mathcal{C})$ is* **at most** $\mu^{\min}$ *or* (b) *no feasible solution to $P(\mathcal{N}(\mathcal{A}),\mathcal{C})$ exists. Otherwise we say that $\mathcal{C}$ defeats $\mathcal{A}$.*

(mc.2) *We say that an attack is* **successful** *if it defeats* **every** *configuration, i.e., $\max_{\mathcal{C}} \mu(\mathcal{N}(\mathcal{A}), \mathcal{C}) \leq \mu^{\min}$.*

(mc.3) *The goal is to find a successful attack $\mathcal{A}$ such that $s(\mathcal{A})$ is minimum. If $\mathcal{A}$ is a 0/1 vector and $s(\mathcal{A}) = \|\mathcal{A}\|_1$, we obtain the* **minimum cardinality** *attack problem informally discussed above.*

In the **maximum severity attack** *problem a value K is given, and we have the following:*

(mi.1) *Only attacks of* $\mathscr{A}$ *with* $s(\mathscr{A}) \leq K$ *are allowed.*

(mi.2) *The goal is to find an allowed attack* $\mathscr{A}$ *such that minimizes* $\max_{\mathscr{C}} \mu(\mathscr{N}(\mathscr{A}), \mathscr{C})$.

In Section 3.1.1 we will formalize the definitions into a family of optimization problems and discuss efficient solution procedures. The minimum effort and maximum severity attack problems are in a sense dual to one another. More precisely, if $\mathscr{A}^*$ is an optimal solution to the minimum effort problem for a given value $\mu^{\min}$, then $\mathscr{A}^*$ is feasible for the maximum severity problem with $K = s(\mathscr{A}^*)$ and thus the value of this problem is at most $\max_{\mathscr{C}} \mu(\mathscr{N}(\mathscr{A}^*), \mathscr{C}) \leq \mu^{\min}$. A similar statement can be made in the reverse direction. In what follows we will primarily focus on the minimum cardinality attack, and we will make a case that it can be made somewhat tractable. We will first remark on some fundamental issues brought up by the above definitions.

First, there is a flaw in the above problem descriptions; namely, both may yield ambiguous answers in borderline cases. For example, in the minimum cardinality attack problem this ambiguity will arise when an attack $\mathscr{A}$ and a configuration $\mathscr{C}$ are such that $\mu(\mathscr{N}(\mathscr{A}), \mathscr{C}) \approx \mu^{\min}$. The root cause of the problem is that the particular choice of the value $\mu^{\min}$ may be either too conservative or too optimistic. Thus, the rule we provided above to decide if $\mathscr{A}$ defeats $\mathscr{C}$ is (mathematically) clear, yet in a practical setting it could be the case that even if $\mu(\mathscr{N}(\mathscr{A}), \mathscr{C})$ is (very slightly) smaller than $\mu^{\min}$ it is still the case that $\mathscr{C}$ could be considered viable—if we have chosen $\mu^{\min}$ large.

Or, even if $\mu(\mathscr{N}(\mathscr{A}), \mathscr{C})$ is (very slightly) larger than $\mu^{\min}$ it could still be the case that attack $\mathscr{A}$ is not really properly handled by $\mathscr{C}$—for example if we have chosen $\mu^{\min}$ small. Between two extremes we have cases that could prove ambiguous. The ambiguity is also revealed and becomes more acute when infeasibility ensues in (mc.1)—the problem cases are those where $P(\mathscr{N}(\mathscr{A}), \mathscr{C})$ is barely feasible or infeasible. The ambiguity, in all cases discussed here, has a numerical instability component, because ambiguous cases are likely to generate inconsistent numerical solutions. In what follows, we will attempt to resolve the ambiguity either by insisting on strict inequalities in the definitions above, or, more commonly, by introducing tolerances in the definitions. Ultimately, unfortunately, one must deal with ambiguity in a per-instance fashion.

To illustrate these issues, consider the following example, where the ambiguity is easily exposed.

Example 3.1.4. *Consider the system shown in Figure* 3.7. *Here we have L load buses, each with demand* 150/*L. We will denote these buses as* $v_1, \ldots, v_L$. *We also have G generators, one located at each bus* $g_1, \ldots, g_G$. *Generator* 1 *has minimum output* 10 *and maximum,* 100. *All others have minimum output* 60 *and maximum output* 100. *The lines joining buses a and b with* $c, d, e,$ *and* f *all have limit* 20. *All other lines have large limits, and all lines have unit reactance.*

Precontingency the system can be feasibly operated under the DC approximation. We operate generators 1 *and* 2 *only, and we set*

$$\begin{aligned}
\theta_{g_1} &= 25/L, \\
\theta_{g_2} &= 62.5, \\
\theta_a = \theta_b = \theta_{g_i} &= 12.5 \quad \text{for } 3 \leq i \leq G, \\
\theta_c = \theta_d = \theta_e = \theta_f &= 0, \\
\theta_{v_i} &= -25/L \quad \text{for } 1 \leq i \leq L.
\end{aligned}$$

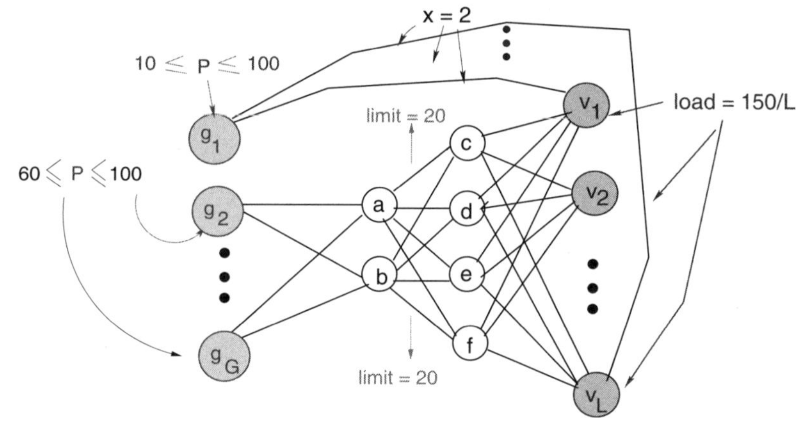

Figure 3.7. *System for Example* 3.1.4; *generators are on the left, loads on the far right.*

We have that the flow on any line joining buses c, d, e, *or* f *and the loads equals* $25/L$; *the flow on each line of the form* $g_1 v_i$ *equals* $50/L$; *thus the total inflow at each load equals* $150/L$. *Similarly, the flow on each line joining buses* a, b *with* c, d, e, f *equals* 12.5 (< 20) *units. Finally, the flow on lines* $g_2 a$ *and* $g_2 b$ *equals* $62.5 - 12.5 = 50$ (*for a total of* 100) *units. Thus generator* g_1 *produces* 50 *units and generator* g_2, 100 *units. The flow on lines* $g_i a$ *and* $g_i b$ *for* $3 \le i \le G$ *equals zero. In summary the solution is indeed feasible and delivers* 100% *of the loads.*

We consider a setting where an attacker may disable up to one generator bus (and all lines incident with it) as well as up to two additional lines, and that the controller is successful, after an attack, if a yield of 80% (*or* 120 *units of demand*) *can be feasibly attained. Consider the contingency in Figure* 3.8, *where generator* g_1 *and all lines incident with it, and lines* ac *and* bc, *are all disabled. To deliver the desired minimum amount of demand we will need to operate exactly two generators (because the generator upper limit is* 100 *and the lower limit is* 60, *but the total load is* 150). *Postcontingency, this is a feasible goal, attained by setting*

$$\theta_{g_2} = \theta_{g_3} = 50,$$
$$\theta_a = \theta_b = \theta_{g_i} = 20 \quad \text{for } 1 \le i \le G \text{ with } i \ne 2,3,$$
$$\theta_d = \theta_e = \theta_f = 0,$$
$$\theta_c = \theta_{v_i} = -40/L \quad \text{for } 1 \le i \le L.$$

Then the flow on any line joining d, e, *or* f *and* v_i (*for* $1 \le i \le L$) *equals* $40/L$; *the flow on any line joining* a *or* b *and* d, e, *or* f *equals* 20; *and the flow on any line joining* g_2 *or* g_3 *and* a *or* b *equals* 30. *Further, the flow on* $g_i a$ *and* $g_i b$ (*for* $1 \le i \le G$ *with* $i \ne 2,3$) *is zero, as is the flow on any line of the form* cv_i (*for* $1 \le i \le L$). *In summary, each load receives an inflow of* $120/L$ *units (for a total of* 120 *units); each bus* d, e, *or* f *has inflow and outflow of* 20 *units; buses* a *and* b *have inflow and outflow equal to* 60 *units; and generators* g_2 *and* g_3 *have outflow of* 60 *units each. All other generators have zero outflow and are not operated.*

The solution is feasible, since total amount of delivered flow is 120 *units, the flow limits are satisfied, and the two operating generators produce an amount of flow within allowable bounds. However, note that a number of feasibility limits are simultaneously attained by this solution: the minimum allowable yield; the line limits on every line joining* a *or* b *and* d, e, *or* f; *and the lower limits on the two operating generators. Moreover, the postcontingency solution we described is one of* $\binom{G-1}{2}$ *isomorphic solutions (each operating two generators) that attain the minimum yield. Finally, there are four similar attacks that attain essentially the*

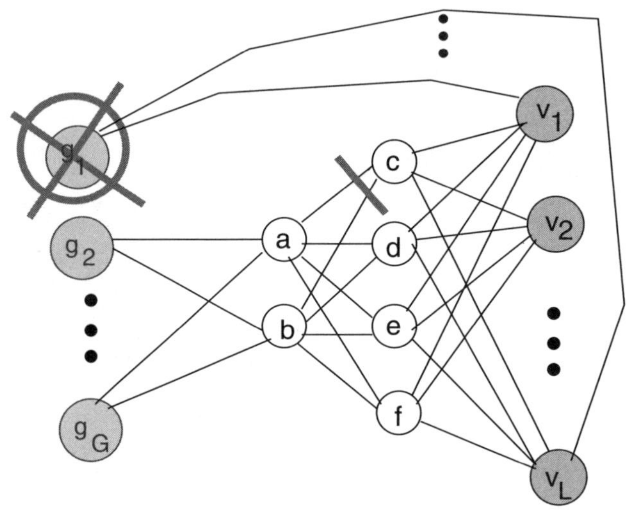

Figure 3.8. *Contingency in Example* 3.1.4.

same outcome (e.g., disable all lines between one of c, d, e, or f and a and b in addition to generator g_1).

Thus, if problem data were to deviate from the assumed values, the purported controller recourse action may not be feasible. The example is therefore ambiguous—whether the controller can defeat one of at least four different attacks depends on precise numerical values.

As the example shows, an algorithm trying to determine the vulnerability of the system would have to examine a possibly large number of ambiguous cases; the ambiguity stems in part from the numerical tolerances incorporated by the algorithm (e.g., a linear system solver) and in any simplification of the underlying models (e.g., the power flow model). A stochastic risk assessment might appear more flexible (as opposed to a simple "yes/no" assessment), but it would still fail to fully account for system reliability, since power systems do exhibit yes/no behavior: for example, equipment will trip and not simply become partly degraded. Perhaps the best that we could hope for is an algorithm that provides a risk assessment and also exposes ambiguous behavior when found.

Postponing these issues, it is clear that, potentially, problems like the minimum cardinality attack problem could be quite difficult, and rather than simply launching ourselves into an immediate solution attempt that relies on some standard methodology we would prefer to first attempt to understand the overall problem behavior. In particular we would like to understand whether the solution is likely to possess some useful structure. An example would be *monotonicity*: is it true that the most severe attack involving exactly K lines is at least as severe as the worst attack involving exactly $K-1$ lines?

Should the monotonicity property hold, we could use it to develop a number of simple heuristics for the minimum cardinality attack problem; for example, we could start from an attack $\mathscr{A}$ of large cardinality that defeats the controller and then remove one element at a time until we reach cardinality K (while hopefully retaining a successful attack). Or, starting from a small cardinality attack, use a heuristic that adds one element at a time, perhaps computed by enumeration over all elements, and choose that element that most increases the severity of the attack. Unfortunately, the following example, adapted from [42], proves that monotonicity does not hold.

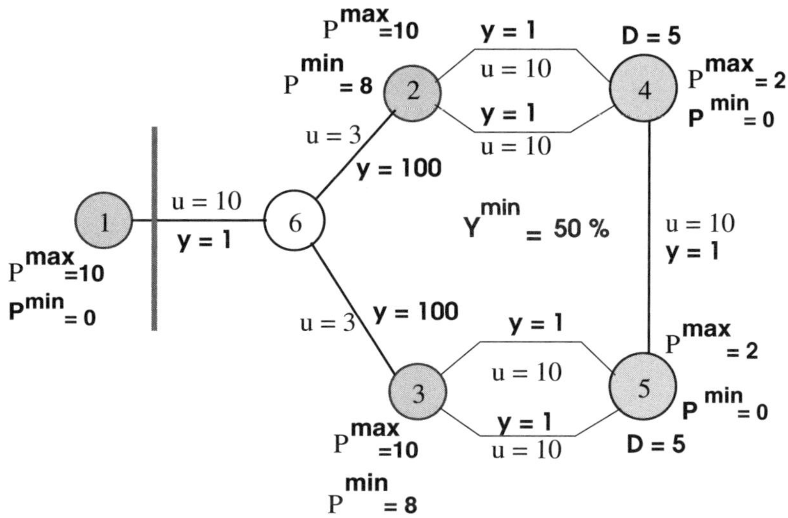

Figure 3.9. *Example 3.1.5, with a successful attack with $K = 1$ but none with $K = 2$.*

Example 3.1.5. *Consider the example in Figure 3.9. Here, buses 4 and 5 have both loads and generation colocated (a common feature in practice). The figure shows, next to each line, its limit (u) and its susceptance (y). Also, there are two parallel copies of lines (2,4) and (3,5), each with capacity 10 and unit susceptance. We assume that the attacker may disable up to two lines. The recourse actions available to the controller are load shedding and shutting off of generators; the performance measure μ is the yield, and the minimum allowable yield is $Y^{\min} = 0.5$. It is easy to see that with no attack the controller can feasibly attain a yield of 100%: we operate generators 4 and 5 to generate 2 units each, do not operate generators 2 and 3, and operate generator 1 to send 3 units of flow along the each of the paths $1-6-2-4$ and $1-6-3-5$ (the flow on, e.g., the two parallel (2,4) lines is evenly split). This is achieved by setting $\theta_1 = 7.53, \theta_6 = 1.53, \theta_2 = \theta_3 = 1.5$, and $\theta_4 = \theta_5 = 0$.*

On the other hand, the attack consisting of the single line (1,6) is successful. To see this, note that under this attack the controller cannot operate both generators 2 and 3, since their combined minimum output exceeds the total demand. On the other hand, at least one of these generators must be operated, since the combined output of generators 4 and 5 does not reach 50% of the total demand. Thus, without loss of generality suppose that only generator 3 is operated, and assume by contradiction that a feasible solution exists—then this solution must route at most *3 units of flow along $3-6-2-4$, and (since $P_3^{\min} = 8$)* at least *5 units of flow on (3,5) (both copies together). In such a case, the phase angle drop from 3 to 5 is at least 2.5, whereas the drop from 3 to 4 is at most $\frac{3}{100} + \frac{3}{100} + 1.5 = 1.56$. In other words, $\theta_4 - \theta_5 \geq 0.94$, and so we will have $p_{45} \geq 0.94$. Thus, the net inflow into node 5 is at least 5.94, which exceeds the demand at that node. Hence the attack is indeed successful.*

However, there is no *successful attack consisting of line (1,6) and another line. To see this, note that if one of (2,6), (3,6), and (4,5) is also removed then the controller can* just *operate one of the two generators 2, 3 and meet 8 units of demand. Suppose that (say) one of the two copies of (3,5) is removed (again, in addition to (1,6)). Then the controller operates generator 2, sending 2.5 units of flow on each of the two parallel (2,4) lines, achieved by setting $\theta_2 - \theta_4 = 2.5$. The controller also routes 3 units of flow along $2-6-3-5$ by setting $\theta_2 - \theta_5 = 3.06$. Consequently, $\theta_4 - \theta_5 = 0.56$ and $p_{45} = 0.56$, resulting in a feasible flow which satisfies 4.44 units of demand at 4 and 3.56 units of demand at 5.*

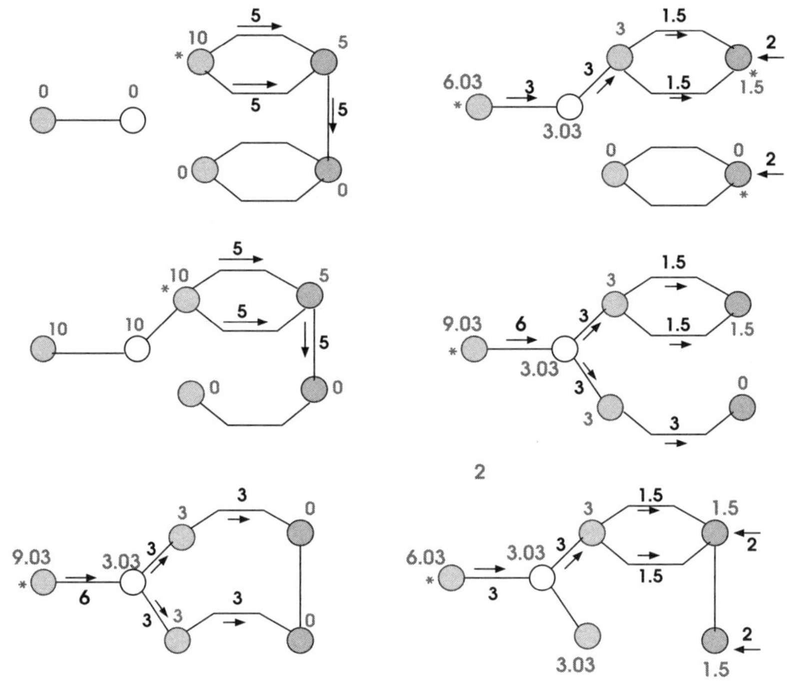

Figure 3.10. *Example* 3.1.5. *Unsuccessful attacks with* $K = 2$ *not including* $(1,6)$.

In fact, we can show that no *successful attack of cardinality* 2 *exists—there is always some way of operating the generators to feasibly meet* 50% *of the demand. This is shown in Figure* 3.10, *which shows flows and phase angles and indicates operating generators by an asterisk.*

Hence, this example displays nonmonotonicity: the version of the attack problem where exactly two lines must be disrupted is not stronger than that where one line can be disrupted.

By elaborating on this construction, one can generate examples for every $K > 1$ and every subset $S \subseteq \{1,\ldots,K\}$ such that for any $1 \leq h \leq K$ there is a contingency of cardinality h that defeats the controller for any $h \in S$, but there does not exist such a contingency if $h \notin S$.

Clearly, the minimum cardinality attack problem is indeed quite complex. The nonmonotonicity noted in the above example is due to the availability of recourse. We can view the recourse as leveraging a form of Braess's paradox [49]. More broadly, the complexity simply reflects the nature of the problem, which falls into the larger class of so-called bilevel programs.

3.1.1 ▪ The N-K problem as a bilevel program

Using the notation in Definition 3.1.3, the minimum cardinality attack problem can be formulated as follows:

$$\min \sum z_{km} \tag{3.11a}$$

$$\text{s.t. } \max\{\mu(\mathscr{C}) : \mathscr{C} \in \mathscr{R}(z)\} \leq \mu^{\min}, \tag{3.11b}$$

$$z_{km} \in \{0,1\} \quad \text{for all lines } km. \tag{3.11c}$$

Here, given a 0/1-vector z, the corresponding attack $\mathscr{A}(z)$ impacts the support of z, and we abbreviate $\mathscr{N}(\mathscr{A}(z))$ as $\mathscr{N}(z)$, $\mathscr{R}(\mathscr{N}(\mathscr{A}))$ as $\mathscr{R}(z)$, and $\mu(\mathscr{N}(z),\mathscr{C})$ as $\mu(cC)$. Note: (3.11b) includes those cases where $\mathscr{R}(z)=\emptyset$, i.e., there is no recourse available to the controller. Problem (3.11) is a classical bilevel programming problem: its feasibility region is defined by an optimization problem.

Example 3.1.6 (Example 3.1.5, continued). *We apply formulation* (3.11). *Given a binary vector z used to indicate those lines that are removed by the adversary, the inner problem, i.e., the maximization on the left-hand side of* (3.11b), *is the following:*

$$\max D_4 + D_5 \tag{3.12a}$$
$$s.t.\ P_1 = p_{16}, \tag{3.12b}$$
$$P_2 = p_{26} + p^u_{24} + p^d_{24}, \tag{3.12c}$$
$$P_3 = p_{36} + p^u_{35} + p^d_{35}, \tag{3.12d}$$
$$P_4 = p^u_{42} + p^d_{42} + p_{45} + D_4, \tag{3.12e}$$
$$P_5 = p^u_{53} + p^d_{53} + p_{54} + D_5, \tag{3.12f}$$
$$0 = p_{16} + p_{26} + p_{36}, \tag{3.12g}$$
$$|p_{26}| \le 3,\ |p_{36}| \le 3,\ |p_{ij}| \le 10 \textit{ for all other lines } ij, \tag{3.12h}$$
$$p_{26} = 100(1-z_{26})(\theta_2-\theta_6), \tag{3.12i}$$
$$p_{36} = 100(1-z_{36})(\theta_3-\theta_6), \tag{3.12j}$$
$$p_{ij} = (1-z_{ij})(\theta_i-\theta_j) \quad \textit{for all other lines } ij, \tag{3.12k}$$
$$0 \le P_1 \le 10\omega_1, \quad 8\omega_2 \le P_2 \le 10\omega_2, \tag{3.12l}$$
$$8\omega_3 \le P_3 \le 10\omega_3, \quad 0 \le P_4 \le 2\omega_4, \quad 0 \le P_5 \le 2\omega_5, \tag{3.12m}$$
$$\omega_i = 0 \textit{ or } 1, \quad 1 \le i \le 5, \tag{3.12n}$$
$$0 \le D_4 \le 6, \quad 0 \le D_5 \le 6. \tag{3.12o}$$

Here, D_i is the demand served at bus i and P_i is the power generated by bus i. We denote by p_{ij} the flow on line ij from i to j (by convention, $p_{ji}=-p_{ij}$). The flow on the upper (resp., lower) copy of line 24 *is indicated by f^u_{24} (f^d_{24}) and similarly with line* 35. *Thus, the constraints before* (3.12h) *describe flow balance. Constraints* (3.12i)–(3.12k) *indicate that the flow on an interdicted line ij (that is, one with $z_{ij}=1$) must equal zero, but if $z_{ij}=0$ the flow and the phase angle difference are related as per the DC approximation (the "*100*" for lines* 36 *and* 26 *is due to the susceptance of value* 100 *on these lines; see Figure* 3.9*). In constraints* (3.12l)–(3.12m), *variable ω_i is set to* 1 *when generator i is not tripped; these constraints are used to limit generator outputs and demand levels.*

Below we will see other formulations for the inner problem that are better suited for analysis.

Bilevel programming problems arise in many contexts. In the spirit of (3.11), a generic bilevel programming problem can be written as

$$F^* \doteq \min F(z) \tag{3.13a}$$
$$\text{s.t. } \max\{G(p) : p \in R(z)\} \le G_0, \tag{3.13b}$$
$$z \in \mathscr{Z}, \tag{3.13c}$$

for an appropriate set $\mathscr{Z} \subseteq \mathbb{R}^n$ and where the *recourse sets* $R(z) \subseteq \mathbb{R}^m$ are appropriately defined. Other, essentially equivalent forms are possible. For example, we may consider a problem of the form

$$\min\{F(z)+H(u) : z \in \mathscr{Z},\ u \in \operatorname{argmax}\{G(p) : p \in R(z)\}\}$$

for an appropriate function H, as in [82]. For example, by setting $H(u)=0$ if $G(u) \le G_0$ and $H(u) = M$ (large) otherwise, we recover problem (3.13) (and vice versa). However, we will focus on formulation (3.13) because it is closer to our problem of interest. In this formulation, $\mathscr{Z} \subseteq \mathbb{R}^n$; for each $z \in \mathscr{Z}$, $R(z) \subseteq \mathbb{R}^m$; $G : \mathbb{R}^m \to \mathbb{R}$; and G_0 is a given value. A case of particular interest is that where for $z \in \mathscr{Z}$ the set $R(z)$ is defined by a convex system, for example

$$R(z)=\{p \in \mathbb{R}^m : ||z||^2 + ||p||^2 \le g\}$$

(for some $g > 0$), or a linear case,

$$R(z)=\{p \in \mathbb{R}^m : Az+Bp \le d\}, \tag{3.14}$$

for appropriate matrices A and B and vector d. Note, however, that this case is already more restricted than those arising in Example 3.1.4 or 3.1.5, where the recourse sets are defined by mixed-integer programs, and where the sets $R(z)$, while convex for any given z, are quadratic in p and z (we have explicitly shown this for Example 3.1.5 in Example 3.1.6).

As is the case in the minimum effort attack problem that we introduced above, a frequent application domain concerns critical infrastructure, with an "upper-level" optimization problem used to model the actions of a (possibly fictitious) attacker, and a "lower-level" problem akin to (3.11b) used to model recourse actions. Frequently, such problems are also termed "interdiction" problems. Many authors have focused on problems of this type, see, e.g., [129], [151], [131], [229]; also see [36]. We will survey some of this work below; a common thread found in this literature is the reliance on problem-specific structure to produce a practicable solution. This strategy is of course motivated by the extreme difficulty caused by the bilevel structure of the problems.

Bilevel programming and, more generally, multilevel programming problems have received extensive attention in the literature. A clear connection with game theory has been identified by many authors. From the perspective of problems interesting in the context of this book, Bard and Moore [13], [14] presented early algorithmic ideas. For a more complete and recent set of references, see the PhD thesis [82], which additionally presents general cutting-plane and branching algorithms. An older source is the online bibliography at `http://www.netlib.org/bibnet/subjects/multilevel-prog.html`. In fact the field of bilevel programming is enormous, with problems in close intellectual proximity (such as equilibrium-constrained optimization problems, or MPECs) attracting much attention [155]. Despite this extensive literature, it is clear that bilevel programming is still a wide-open research topic, with many significant challenges that are yet to be overcome.

Returning to formulation (3.13), we note that subtle issues arise if we require a strict inequality in (3.13b)—if all variables in the problem are continuous, then we intuitively expect that such a substitution may have limited impact (some thought will show that the intuition is wrong), but certainly if we allow discrete variables a difference can arise if we insist on a strict inequality. Nevertheless, in selected cases where $R(z)$ is convex, duality can then be used to replace (3.13b) with an equivalent convex system, thus replacing the bilevel program with an equivalent "single-level" optimization problem in a higher-dimensional space. This will hold in particular in the case that $G(p)$ is linear and $R(z)$ is as in (3.14). If $F(z)$ and $\mathscr{Z}$ are additionally convex then the bilevel program is a convex problem. In the linear $R(z)$ case, with F linear and $\mathscr{Z}$ discrete, we will obtain a problem that may be easy to reformulate as a standard linear integer program.

In the next sections we will investigate difficulties that arise in bilevel programs, in particular those that arise directly from the N-K problem. We will then use these observations to suggest algorithmic ideas, and we will review results from the literature.

Toward the use of Benders decomposition

Consider the case of problem (3.13) where the recourse sets are as in (3.14) and where additionally $G(p) = g^T p$ for some $g \in \mathbb{R}^m$. To facilitate the discussion below, we restate the bilevel program under these specific assumptions:

$$F^* \doteq \min F(z) \tag{3.15a}$$

$$\text{s.t. } \max\{g^T p : Bp \leq d - Az\} \leq G_0, \tag{3.15b}$$

$$z \in \mathscr{Z}. \tag{3.15c}$$

If we "apply" linear programming duality to this formulation we obtain

$$\tilde{F} \doteq \min F(z) \tag{3.16a}$$

$$\text{s.t. } d^T \pi - \pi^T Az \leq G_0, \tag{3.16b}$$

$$B^T \pi = g, \tag{3.16c}$$

$$z \in \mathscr{Z}, \quad \pi \in \mathbb{R}^q_+, \tag{3.16d}$$

where q is the number of rows in the matrices A and B. But there is a caveat here: as we will see below, bilevel programs can exhibit significant subtlety, and there is a need to examine the purported equivalence between problems (3.15) and (3.16) with some care.

In particular suppose that there is *no* vector $\pi \in \mathbb{R}^q_+$ such that $B^T \pi = g$. Then not only is problem (3.16) infeasible, but for any $z \in \mathscr{Z}$, the linear program

$$LP(z) : \max\{g^T p : Bp \leq d - Az\} \tag{3.17}$$

is either infeasible or unbounded. Suppose we further assume that for any $z \in \mathscr{Z}$ the set $R(z)$ (in this case $\{p : Bp \leq d - Az\}$) is bounded, as would happen in N-K problems. Then (3.17) is not unbounded and so it is infeasible for any $z \in \mathscr{Z}$—using the language we derived for the N-K problem in Definition 3.1.3 we would say that any attack defeats the controller. In such a case the bilevel problem becomes the much simpler "one-level" or standard optimization problem $\min\{F(z) : z \in \mathscr{Z}\}$, which would be tackled using standard techniques.

To summarize the above, we present the following result for future reference.

Lemma 3.1.7. *Assume* (3.15) *has an optimal solution* z^* *and that there exists* $\pi_0 \in \mathbb{R}^q_+$ *with* $B^T \pi_0 = g$. *Then problems* (3.15) *and* (3.16) *have the same value.*

Proof. We claim that a vector $\pi^* \in \mathbb{R}^q_+$ exists such that the pair (z^*, π^*) satisfies (3.16b)–(3.16c). To prove this, suppose first that $R(z^*) \neq \emptyset$. Then linear programming duality indeed guarantees that a vector π^* as desired exists. On the other hand, suppose $R(z^*) = \emptyset$. But the dual of $LP(z^*)$, a minimization problem, is feasible (π_0 is feasible for it) and so it is unbounded, and once again a vector π^* as stated exists. So $\tilde{F} \leq F(z^*) = F^*$. Further, consider any pair $(\tilde{z}, \tilde{\pi})$ feasible for (3.16). Then it follows that $\tilde{z}$ is feasible for (3.15) (this includes the case where $LP(\tilde{z})$ is infeasible). Thus $F(\tilde{z}) \geq F(z^*)$. ■

Thus when the hypotheses of this lemma hold, formulation (3.16) is certainly correct. Unfortunately, this formulation is also not linear. Thus, if $\mathscr{X}$ defines a linearly constrained mixed-integer set, and F is linear, then we might seek a cutting-plane strategy to solve problem (3.15), and the (near) success of formulation (3.16) appears to suggest an application of linear programming duality. In particular, given the two-level structure of this formulation one might be tempted to develop a procedure based on Benders decomposition, an algorithmic construct originally presented in [22] and briefly outlined next.

In one of its applications, Benders decomposition considers an optimization problem of the form

$$\min_x \{c^T x + v(x) : x \in X \subseteq \mathbb{R}^n\}, \quad \text{where} \tag{3.18a}$$

$$v(x) \doteq \min_y \{h^T y : Ay \geq b - Bx,\ y \in \mathbb{R}^m\}, \tag{3.18b}$$

where A, B, and b are of appropriate dimensionality. The set X typically will include linear constraints and may also include integrality of a subset of the variables. Thus the problem in (3.18a) describes a two-stage setting where the variables x describe top-level decisions which are followed by a second-stage set of decisions. Note the similarity to problem (3.15). A case of particular importance is the one where the second-stage problem splits into several independent subproblems, that is to say $y = (y^1, \dots, y^k)$, where $y^i \in \mathbb{R}^{m_i}$ with $\sum_{i=1}^k m_i = m$, and A has the structure

$$A = \begin{bmatrix} A_1 & 0 & \dots & 0 \\ 0 & A_2 & \dots & 0 \\ 0 & 0 & \dots & A_k \end{bmatrix}, \tag{3.19}$$

where for $1 \leq i \leq k$, A_i has m_i columns. In that case, for $x \in X$, the problem in (3.18b) decomposes into k independent problems. This setting arises in multiscenario problems, where (for $x \in X$) we pay a "here-and-now" cost $c^T x$ plus a second-stage cost, which is a sum of costs to be paid over the different scenarios, all of which are affected by the choice of x.

In terms of formulation (3.18), Benders decomposition can be sketched as the following iterative algorithm. This algorithm maintains, at each iteration, a "master" formulation:

$$\min_x c^T x + v \tag{3.20a}$$

$$\text{s.t. } x \in X, \tag{3.20b}$$

$$v \geq \alpha_i^T x + \beta_i \quad \forall\, i \in F. \tag{3.20c}$$

Here $\alpha_i \in \mathbb{R}^n$ and $\beta_i \in \mathbb{R}$ for each $i \in F$, and the family F is such that each inequality in (3.20c) is valid, that is to say, for each $x \in X$ we have $v(x) \geq \alpha_i^T x + \beta_i$ for all $i \in F$. Thus (3.20) (denoted the "master" problem) is a relaxation for problem (3.18), i.e., the value of problem (3.20) is a lower bound for that of (3.18).

Given this fact, suppose that $(\tilde{x}, \tilde{v})$ is an optimum solution to (3.20). Hence

$$\tilde{v} = \max_{i \in F} \{\alpha_i^T \tilde{x} + \beta_i\}.$$

Consequently, if

(a) the system $Ay \geq b - B\tilde{x}$, $y \in \mathbb{R}^m$ is feasible, and

(b) $v(\tilde{x}) \leq \tilde{v}$,

then in fact $v(\tilde{x}) = \tilde{v}$ (by the validity of (3.20c)) and so by the relaxation property just mentioned, $\tilde{x}$ is optimal for (3.18). On the other hand assume that (a) holds, but

$$v(\tilde{x}) > \tilde{v}.$$

Then duality implies that there exists a (nonnegative) vector $\tilde{\alpha}$ such that

$$\tilde{\alpha}^T A = h \quad \text{and} \tag{3.21}$$
$$\tilde{\alpha}^T (b - B\tilde{x}) > \tilde{v}. \tag{3.22}$$

But (3.21) and $\tilde{\alpha} \geq 0$ imply that for any $x \in X$ such that the system $Ay \geq b - Bx$, $y \in \mathbb{R}^m$, is feasible we must have

$$v(x) \geq \tilde{\alpha}^T (b - Bx).$$

In short, the inequality

$$v \geq \tilde{\alpha}^T (b - Bx)$$

is valid and cuts off $(\tilde{x}, \tilde{v})$. The iteration is completed by adding this inequality to the master, and repeating. The case where (a) above does not apply (i.e., $Ay \geq b - B\tilde{x}$, $y \in \mathbb{R}^m$, is infeasible) is similarly handled (using the Farkas lemma instead of duality).

This completes our brief sketch of Benders decomposition, which as we have alluded to before has often been used in the literature, sometimes implicitly. This is due to the power and versatility of the basic technique and to the fact that properly implemented, Benders decomposition can prove extremely effective. In this regard, part of the appeal is that the technique can leverage modern implementations of the simplex method that are well suited to cutting-plane algorithms.

Returning to our prototype bilevel program (3.13), an application of Benders optimization would amount to an iterative algorithm that at each step solves two linear programs: first, a "master problem" from the perspective of minimizing $F(z)$, which produces a vector $\hat{z}$ as well as an estimate for F^*, and second, a "subproblem" or "inner" problem that, should $\hat{z}$ prove too optimistic for (3.15b), uses linear programming duality to cut it off. This produces a cut that is added to the master problem.

However, some care is necessary. As an example of a failed algorithm, consider the (wrong) idea in the following box.

Template 3.1.8. Incorrect Benders decomposition in the linear recourse case

Initialization. Set $h = 0$. We start with the formulation consisting of (3.16a), (3.16c), (3.16d).

Step 1. Let $\check{z}^h$ solve the current formulation.

Step 2. If $LP(\check{z}^h)$ has value at most G_0 then **STOP**; $\check{z}^h$ is optimal.

Step 3. Otherwise let $\check{\pi}^h$ be optimal for the dual of $LP(\check{z}^h)$.
Add the cut $-\check{\pi}^{hT} Az \leq G_0 - d^T \check{\pi}^h$ to the formulation.

Step 4. Increment $h \leftarrow h + 1$ and go to Step 1.

This process appears to mimic formulation (3.16) and it follows the rough structure of a Benders decomposition method, but it is incorrect. In the proof of Lemma 3.1.7, the optimality of vector z^* is certified by *some* dual vector π^*. In contrast, as we iterate through Steps 1, 2, 3 of Template 3.1.8 we will impose weak duality through *all* vectors $\breve{\pi}^h$ encountered as we iterate. Thus, at some iteration h, there may exist vectors $\hat{z} \in \mathscr{Z}$ with the property that $LP(\hat{z})$ has value at most G_0 (so $\hat{z}$ is feasible for (3.15)) and yet the cut added in Step 3 is violated by $\hat{z}$. This would happen if $-\breve{\pi}^{hT} A\hat{z} > G_0 - d^T \breve{\pi}^h$ or, in other words, if $\breve{\pi}^h$ is suboptimal for the dual of $LP(\hat{z})$. Thus the cuts added in Step 3 will in general not be valid. What the template attempts to accomplish is to cut off vector $\breve{z}^h$, but a different methodology needs to be used.

In the case of the (incorrect) algorithm in Template 3.1.8, most of the per-iteration computational overhead will be incurred in Step 1. However, the optimization problem solved in this step would be warm started, yielding substantial savings. Of course the algorithm in Template 3.1.8 is incorrect—however, below (Section 3.1.1) we will consider a different formulation for the bilevel programming problem where a version of Benders decomposition is indeed applicable and can attain numerical success, and it would be desirable for such an algorithm to extend beyond cases where the recourse sets $R(z)$ are defined by a linear program. The incorrect algorithm above, and the bilinear term $\pi^T Az$ in (3.16), suggest that such "valid" algorithms may not be simple; this suspicion will prove correct.

In the following sections we will outline three valid algorithmic ideas for addressing bilevel programming problems of the type that arise in the analysis of the N-K problem. Specifically, and motivated in part by Examples 3.1.4 and 3.1.5 (see the discussion below), we generalize problem (3.15) by allowing a more complex inner problem which is bilinear in p and z, i.e., the recourse set for a given $z \in \mathscr{Z}$ is given by

$$MP(z) \doteq \{ p \in \mathbb{R}^m : C\, z \circ p^T + Bp + Az \le d, \ \ p \in \mathscr{P} \}, \tag{3.23}$$

and the bilevel problem to be solved is therefore

$$F^* \doteq \min F(z) \tag{3.24a}$$

$$\text{s.t. } \max\{g^T p : p \in MP(z)\} \le G_0, \tag{3.24b}$$

$$z \in \mathscr{Z}. \tag{3.24c}$$

Here $\mathscr{P} \subseteq \mathbb{R}^m$ is a given set, and $z \circ p^T \in \mathbb{R}^m$ is the vector with entries $z_j p_j$, $1 \le j \le m$. Notationally, we assume, for some $q \ge 1$, that $\mathscr{Z} \subseteq \mathbb{R}^m$; A, B, and C are $q \times m$ matrices; and $d \in \mathbb{R}^q$. Note that this requires the z and p vectors to be of the same dimension, but this is easily attained by adding "dummy" variables, if necessary.

This setup is more general than might first appear to be the case as it handles all cases of bilinear relationships as well as purely linear terms in both p and z. To see this, consider a setting where given $\mathscr{Z} \subseteq \mathbb{R}^n$ and $\mathscr{P} \subseteq \mathbb{R}^m$, for each $z \in \mathscr{Z}$ the set of recourse actions is described by the condition $p \in \mathscr{P}$, together with inequalities of the form

$$w_i^T z + \sum_{j=1}^{m} (v_{ij}^T z + b_{ij}) p_j \le d_i, \quad 1 \le i \le q, \tag{3.25}$$

where for each i, $w_i \in \mathbb{R}^n$, and for each i and j, $v_{ij} \in \mathbb{R}^m$ and $b_{ij} \in R$. This (completely) general setting can be reduced to a special case of (3.24). We outline how to carry out the reduction. First, for each pair $1 \le i \le q$ and $1 \le j \le m$ we extend the vector z to include an entry z_{ij} which is defined as $z_{ij} = v_{ij}^T z$ This "lifts" $\mathscr{Z}$ to a set $\mathscr{Z}' \subseteq \mathbb{R}^{n'}$ where

$n' \doteq n + mq$. The reduction is completed by likewise "padding" any vector in $\mathscr{P}$ to include entries of the form p_{ij} for each pair $1 \le i \le q$ and $1 \le j \le m$ together with the constraint $p_j = p_{ij}$. The C matrix will be 0/1-valued. Details are left to the reader.

It is also worth pointing out that (as above) when the set $\mathscr{P}$ describes a polyhedron, an application of linear programming duality transforms problem (3.24) into an equivalent, standard (i.e., single-level rather than bilevel) optimization problem. When $\mathscr{P}$ is a polyhedron, the constraints defining it can be directly incorporated as additional rows of the system $Cz \circ p + Bp + Az \le d$. We then have, as a generalization of (3.16),

$$\tilde{F} \doteq \min F(z) \tag{3.26a}$$

$$\text{s.t. } d^T \pi - \pi^T Az \le G_0, \tag{3.26b}$$

$$\sum_{i=1}^{m} (a_{ij} z_j + b_{ij}) \pi_i = g_j, \quad 1 \le j \le m, \tag{3.26c}$$

$$z \in \mathscr{Z}, \quad \pi \in \mathbb{R}^q_+. \tag{3.26d}$$

A qualification similar to that in Lemma 3.1.7 applies here, as well.

To further motivate the more general formulation (3.24) from the perspective of the N-K problem, consider, again, the standard DC approximation to the power flow equations on line (k, m),

$$p_{km} = y_{km}(\theta_k - \theta_m), \tag{3.27}$$

where $y_{km} = x_{km}^{-1}$ is the susceptance of (k, m). If we consider an adversary that can remove lines, then, using a binary variable z_{km} that indicates whether line (k, m) has been removed, the subformulation of the controller recourse formulation that handles line (k, m) could take the form

$$-u_{km}(1 - z_{km}) \le \quad p_{km} \quad \le u_{km}(1 - z_{km}), \tag{3.28a}$$

$$-M_{km} z_{km} \le p_{km} - y_{km}(\theta_k - \theta_m) \le M_{km} z_{km}. \tag{3.28b}$$

Here, u_{km} is the limit for line (k, m) and $M_{km} > 0$ is an appropriately large constant. A technical difficulty arises in choosing M_{km}. From an engineering standpoint, when line (k, m) is working then one would expect (in fact require) that $|\theta_k - \theta_m|$ would not be very large (say $\pi/4$). However, the constant M_{km} becomes significant in (3.28b) when $z_{km} = 1$, that is to say line (k, m) *is not* working. Hence M_{km}/y_{km} has to be chosen at least as large as the maximum phase angle difference between buses k and m in a grid where k and m are no longer directly connected by a line—in fact k and m may be in different components. It follows that M_{km} may have to be chosen quite large. This issue is taken up in [42], and we will return to it later. Hence, potentially, (3.28) could amount to a (much dreaded) "big M" formulation, which, as is well known in mixed-integer programming folklore and confirmed by experimentation, can lead to poor solvability. If, in fact, an explicit upper bound $\theta_{km}^{\max}$ on $|\theta_k - \theta_m|$ is known,[16] then (3.28a) can be tightened as follows:

$$-\tilde{u}_{km}(1 - z_{km}) \le p_{km} \le \tilde{u}_{km}(1 - z_{km}), \quad \text{where} \quad \tilde{u}_{km} \doteq \min\{u_{km}, y_{km}\theta_{km}^{\max}\}. \tag{3.29}$$

[16] Again this bound must be valid even if k and m are in separate components.

In contrast to this situation, one may choose a bilinear formulation:

$$|p_{km}| \leq u_{km}(1 - z_{km}), \tag{3.30a}$$

$$p_{km} = y_{km}(1 - z_{km})(\theta_k - \theta_m), \tag{3.30b}$$

which removes the explicit big M feature from (3.28b) at the not-insignificant cost of introducing bilinearities.

At this point we must consider what we have achieved by developing systems (3.28) and (3.30)—in fact all we have obtained is a linear, or bilinear, representation of constraints for the inner problem (3.23). This does have some value; the linear case, in particular, will be useful in algorithms based on the idea of branching which are detailed below. However, from the perspective of Benders decomposition, systems (3.28) and (3.30) are not quite what we needed. In fact what we should have considered is how to obtain equivalent linear representations of the bilinear constraints in (3.26). In the context of the N-K problem, and with slight abuse of notation, the constraints in (3.26) are versions of the *dual* constraints for the system of power flow equations (3.27) under interdiction, rather than versions for the primal system (3.27) under interdiction, which is what the exercise in the previous paragraph attained. This is a common phenomenon arising in interdiction models, namely the need to handle the interaction of interdiction variables and dual (rather than primal) variables. What the exercise above did achieve was to focus our attention on potential big M issues, which are pervasive in interdiction models (and the source of some errors in publications).

The following sections will present general algorithmic ideas that can be applied to the N-K problem and which are based on the above concepts. Below we present a branch-and-bound algorithm for addressing problems of the form (3.24). Then we present a method based on linearizing the bilinear terms in (3.24b). Finally, we describe a method based on Benders decomposition that can be applied when the set $\mathscr{X}$ describes binary (or, more generally, discrete) vectors, as would be the case in the N-K problem.

We note that the N-K problem can further be seen as a special case of more general interdiction problems arising in multiple contexts beyond power grid applications. See Section 3.4.1. To some extent, the technical difficulties alluded to above and examined below are echoed by similar challenges in the other applications. However, it appears that the N-K problems are especially challenging, for several reasons: (a) the multicommodity-like nature of power flow demands; (b) the thorny numerical characteristics of power grid data, which lead to big M formulations and generally to numerically poor behavior of convex relaxations; and, especially, (c) the complex nature of power flow physics, even in the DC approximation, which makes combinatorial reasoning difficult.

Branching

We will first describe an algorithm that is an adaptation of the familiar *branch-and-bound* methodology. Each *node* w in our branch-and-bound strategy will be equipped with a linear formulation $K_w z \leq \delta_w$. Intuitively, in the context of the N-K problem, when processing each ancestor of node w the branch-and-bound algorithm has discovered a controller recourse that must be defeated; the formulation $K_w z \leq \delta_w$ encodes this requirement. Furthermore, a *tolerance* $\epsilon > 0$ is specified. The scheme presented here is related to *constraint branching* [102], to the use of *no-good* cuts [11], and to *spatial branching* [219].

The manner in which nodes are processed is described by following Template 3.1.9. After processing a node, it is discarded. Steps 2(a), 3(a), and 4(a) of the template correspond to the traditional *fathoming* step (corresponding to relaxation-infeasibility, feasibility, and suboptimality, respectively).

Template 3.1.9. Branching method for problem (3.15): Processing a node

Input: A node w in the current branch-and-bound tree, with corresponding formulation $K_w z \leq \delta_w$, and the current upper bound U on the value of problem (3.15).

Step 1. Solve the problem

$$f^*(w) \doteq \min\{F(z) : K_w z \leq \delta_w, \, z \in \mathscr{Z}\}.$$

Step 2. (a) If the problem is infeasible, we label node w **infeasible** and **STOP**.
(b) Otherwise, let $\check{z}^w$ be an optimal solution.

Step 3. (a) If $MP(\check{z}^w)$ has value at most G_0, then we reset $U \leftarrow \min\{U, f^*(w)\}$ and **STOP**.
(b) Otherwise, let $\check{p}^w$ be an optimal solution to $MP(\check{z}^w)$.

Step 4. (a) If $U \leq f^*(w)$ we **STOP**.
(b) Otherwise, for $1 \leq j \leq q$, we create child node c_w^j from node w by appending to the system $K_w z \leq \delta_w$ the constraint

$$\sum_{i=1}^{m} c_{ji} \check{p}_i^w z_i + Az \leq d_j - B_j \check{p}_w - \epsilon e_j,$$

where B_j and $e_j \in \mathbb{R}^q$ are, respectively, the jth row of B and the jth unit vector.

A numerical example will be provided below. We would like first to focus attention on Step 4(b), where we create q new "children" nodes c_w^j for $1 \leq j \leq q$ (we will term this process *branching on constraint j*). Note that the added constraint can be equivalently restated as $[C(\check{p}^w \circ z) + Az]_j \leq d_j - B_j \check{p}_w - \epsilon e_j$. The principal attribute of each child c_w^j is that, because $\epsilon > 0$, when processing node c_w^j, vector z^w will be infeasible for the problem solved in Step 1. More concretely, $K_{c_w^j} z^w \not\leq \delta_{c_w^j}$. Essentially, we are branching by enumerating all possible ways in which we can prevent the recourse action represented by $\check{p}^w$: that is to say, since in Step 4 we have that $MP(\check{z}^w)$ has value greater than G_0, then we want to eliminate vector $\check{z}^w$ as a feasible candidate for a solution to the bilevel program; our branching strategy simply enforces this goal. Beyond the bilevel programming setting, other researchers have tackled the issue of branching in a continuous-variable setting. One of the better known approaches in this direction is that of spatial branching mentioned above.

As is usual in the context of branch-and-bound algorithms, the template would be incorporated in a tree-growing procedure (branch-and-bound proper). This procedure is initialized by instantiating one node r (the root) by setting $K_r = \emptyset$, $\delta_r = 0$, and $U = +\infty$. The procedure operates by picking, at any iteration, some unprocessed node w and processing it using Template 3.1.9, which will typically add a new set of unprocessed nodes (created in Step 4(b) of the template) to the tree; the procedure stops as soon as no unprocessed nodes remain. At termination of the procedure, the quantity U will be the value of the best solution z^w certified as feasible in Step 3(a). To provide an early termination criterion, the procedure will also maintain a quantity L (initially $-\infty$) which at any point would constitute a valid lower bound on the value of problem (3.15); this is attained by setting $L = \min\{f^*(w) : w \text{ unprocessed}\}$. This quantity can be used to terminate branch-and-bound (early) if $U - L$ is small enough.

At this point a number of questions regarding the overall branch-and-bound algorithm come up. Is the algorithm finite? If finite, could it be fast? In fact, is the algorithm even valid? The last question has a simple answer: no, the algorithm is not valid from a strict standpoint. This is due to our use of the tolerance $\epsilon > 0$; as a result of this feature, when processing a node w the family of children nodes created in Step 4(b) of Template 3.1.9 may fail to partition the set of feasible solutions represented by w, and some solutions where the attacker defeats the controller may be excluded. However, should any such solution $\bar{z}$ be excluded, it and the vector $\breve{p}^w$ discovered in Step 3(b) embody an ambiguous relationship when ϵ is small. See Figure 3.11. The manner in which children nodes are created in Step 4(b) resolves the ambiguity, and in that sense the algorithm *is* valid. We will return to a discussion of these points after the next result, which addresses finiteness of the branch-and-bound method and may already be folklore in the context of *spatial branching* methods.

Lemma 3.1.10. *Suppose $\mathscr{Z}$ and $\mathscr{P}$ are bounded. Then the branch-and-bound algorithm based on Template* 3.1.9 *terminates after enumerating* $O(\epsilon^{-1})$ *nodes, where the* $O()$ *notation "hides" instance-specific parameters.*

Proof. We will prove that the depth of any node in the branch-and-bound tree is bounded. Since $\mathscr{Z}$ and $\mathscr{P}$ are bounded, there are constants M and M' such that $||z|| \le M$ and $||p|| \le M'$ for each $z \in \mathscr{Z}$ and $p \in \mathscr{P}$. Define

$$\Delta \doteq \frac{\epsilon}{3MM' \sum_{ij} |c_{ij}|},$$

and, for $p \in \mathbb{R}^m$, define

$$\eta(p) \doteq \operatorname{argmin}\{\, ||v - p|| : v_j = r_j \Delta,\ r_j \text{ integral},\ 1 \le j \le m \,\},$$

i.e., the $\eta(p)$ is the vector nearest to p all of whose coordinates are integral multiples of Δ.

Let u and v be nodes produced in the branch-and-bound tree such that v is an ancestor of u. Further, let $1 \le j \le q$ be such that u is in the branch obtained at node v by branching on constraint j. Then, by definition of Δ and $\eta(\breve{p}^v)$ we have that

$$[C(\eta(\breve{p}^v) \circ \breve{z}^v) + A\breve{z}^v]_j \ge [C(\breve{p}^v \circ \breve{z}^v) + A\breve{z}^v]_j - \frac{\epsilon}{3}. \tag{3.31}$$

Thus, since

$$[C(\breve{p}^v \circ \breve{z}^v) + A\breve{z}^v]_j \ge [C(\breve{p}^v \circ \breve{z}^u) + A\breve{z}^u]_j + \epsilon \tag{3.32}$$

(because of the constraint imposed in Step 4(b) when processing v, and the fact that u is a descendant of c_v^j), it follows that

$$[C(\breve{p}^v \circ \breve{z}^u) + A\breve{z}^u]_j \le [C(\eta(\breve{p}^v) \circ \breve{z}^v) + A\breve{z}^v]_j \breve{z}_i^v - \frac{\epsilon}{3}. \tag{3.33}$$

As a consequence, for any enumerated node w and any vector $y \in \mathbb{R}^n$ which is an integral multiple of Δ, the number of ancestors $\hat{v}$ of w with the property that $\eta(\hat{v}) = y$ must

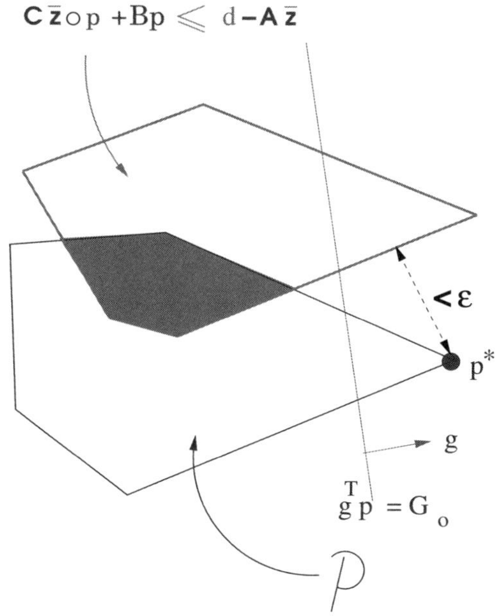

Figure 3.11. *Template* 3.1.9 *excludes a feasible solution.*

be finite (since $\mathscr{X}$ is bounded). But since $\mathscr{P}$ itself is bounded, the set $\{y \in \mathbb{R}^n : \eta(\hat{v}) = y \quad \text{for some } \hat{v} \in \mathscr{P}\}$ is finite. This concludes the proof. ■

The boundedness assumptions in Lemma 3.1.10 are routine in most applications of bilevel programming, in the sense that the sets $\mathscr{X}$ and $\mathscr{P}$ purportedly describe the actions available to a realistic "attacker" and "controller." Now we return to the issue of validity of the branch-and-bound scheme. As discussed above, our branching rule may cut off legitimate (attacker) solutions. In fact the branch-and-bound algorithm could exhibit two unsatisfactory behaviors: (a) it may terminate without producing a feasible solution to the bilevel program (3.24) even though that problem is feasible, and (b) it may report different values for the bilevel program if different optimization algorithms are used in Step 1 of Template 3.1.9. The second point becomes relevant when a problem solved in Step 1 has multiple optima; one algorithm may produce an optimal solution $\bar{z}$ that is accepted in Step 3(a), whereas the other algorithm may produce a different solution which generates branches that cut off $\bar{z}$. Should this event take place at the root of the branch-and-bound tree, different optimal values will likely be reported.

Figure 3.11 shows an example where our node-processing procedure cuts off a valid solution. The figure shows, for a given $\bar{z} \in \mathscr{X}$, the set $\mathscr{P}$ (the lower polyhedron) as well as the set $\{p \in \mathbb{R}^m : C\,\bar{z} \circ p + Bp \le d - A\bar{z}\}$ (the upper polyhedron). Thus, the shaded area indicates the feasible region available for the controller under attack $\bar{z}$. As shown in the figure, $\max\{g^T p : C\bar{z} \circ p + Bp \le d - A\bar{z},\ p \in \mathscr{P}\} < G_0$, and thus $\bar{z}$ is feasible for the bilevel program. However, $p^* \in \mathscr{P}$ satisfies $g^T p^* > G_0$, and it violates one of the constraints $C\bar{z} \circ p + Bp \le d - A\bar{z}$ by less than ϵ. Suppose that when processing the root r of the branch-and-bound tree we obtain some vector $\breve{z}^r \neq \bar{z}$ such that $p^* = \breve{p}^r$ in Step 3(b) of the template; then if we branch on the constraint highlighted in Figure 3.11 we will cut off $\bar{z}$. If p^* violates all constraints in $C\bar{z} \circ p + Bp \le d - A\bar{z}$ by less than ϵ, then $\bar{z}$ will be cut off in all branches, even though it is feasible.

Returning to the issue of efficiency, the construction in Lemma 3.1.10 highlights the risk that our branch-and-bound procedure could be extremely slow. To address this point, we will revisit Step 4(b) of Template 3.1.9 and create an alternative node-processing procedure that should prove more effective. To motivate this alternative approach let us first consider a numerical example showing Template 3.1.9 in action.

Example 3.1.11 (Example 3.1.6, continued). *Recall that in this example the attacker may disable up to two lines; the controller chooses which generators to operate and is successful if a yield of at least* 50% *can be feasibly attained. The objective function* $F(z)$ *for the attacker is arbitrary (we simply want to obtain a successful attack), but to fix ideas we consider* $F(z) = \sum_{ij} z_{ij}$.

Formulation (3.12) *constitutes the specific implementation of problem* $MP(z)$ *(equation* (3.23)*) given above. At the root* r *to branch-and-bound, we have (without loss of generality) that* $z = 0$*, i.e., no lines are disabled and* $f^*(r) = 0$*. As we saw in Example* 3.1.5*, the controller defeats this attack; in fact an optimal solution to formulation* (3.12) *for the case* $z = 0$ *attains a yield of* 100% *by operating generators at buses* 1, 4*, and* 5 *with the following outputs:* 6, 2*, and* 2*, respectively (other optimal solutions exist).*

Considering Template 3.1.9*, we can see that Steps* 2(a) *and* 3(a) *do not apply. Thus, Step* 3(b) *does apply, and the vector* $\breve{p}^r$ *is as just outlined; in particular the flow on line* $(1,6)$ *equals* 6 *units (with* $\theta_1 = 100$, $\theta_6 = 94$*). Since* $U = +\infty$ *at the start of branch-and-bound, Step* 4(b) *does apply, and we will now branch on linear constraints. Considering formulation* (3.12)*, we can see that there only the eight equations* (3.12i)–(3.12k) *incorporate the* z *variables—branching on any other constraint will produce a node that will be proved infeasible. So without loss of generality we will only branch on* (3.12i)–(3.12k)*. Now in Template* 3.1.9 *we implicitly assumed that all linear constraints were inequalities, and thus, in total, we have 16 possible branches. Consider, for example,* (3.12k) *applied to line* $(1,6)$*, which can be rewritten as two inequalities:*

$$p_{16} \leq (1-z_{16})(\theta_1-\theta_6), \quad -p_{16} \leq -(1-z_{16})(\theta_1-\theta_6). \tag{3.34}$$

Suppose $\epsilon = 0.01$*. Branching on the first inequality in* (3.34) *we will impose the inequality*

$$5.99 \geq (1-z_{16})(100-94) \;=\; 6(1-z_{16}), \tag{3.35}$$

which clearly enforces $z_{16} = 1$*. Let us denote the resulting node* $w(16)$*. If, on the other hand, we branch on the second inequality in* (3.34)*, we obtain*

$$-6 \leq -(1-z_{16})6-0.01, \quad or \quad -5.99 \leq -6(1-z_{16}). \tag{3.36}$$

This constraint, again, simply enforces $z_{16} = 1$ *and thus the corresponding node can be discarded.*

Using the example of line $(1,6)$ *as a guide, we see that branching on all inequalities arising from* (3.12i)–(3.12k)*, we will obtain, for each line* (i,j)*, a node* $w(ij)$ *that imposes the constraint* $z_{ij} = 1$.

Consider node $w(16)$*. Then, as argued in the previous paragraph, this node enforces* $z_{16} = 1$ *and so Step* 1 *will produce* $\breve{f}(w_{16}) = 1$*, with* $\breve{z}_{16}^{w(16)} = 1$ *and* $\breve{z}_{ij}^{w(16)} = 0$ *for all other* (i,j)*. We obtain the corresponding formulation* $MP(\breve{z}^{w(16)})$ *from formulation* (3.12) *by setting* $p_{16} = 0$ *and removing the version of constraint* (3.12k) *corresponding to line* $(1,2)$*. The resulting optimization problem has value* 4.00*—as argued in Example* 3.1.5*, the controller cannot succeed in meeting* 50% *of the demand, and the computed solution operates generators* 4 *and* 5 *only. Thus, Step* 3(a) *of Template* 3.1.9 *applies, and we now reset* U *to* 1 *and STOP.*

Continuing with branch-and-bound, the algorithm will next consider some other node $w(ij)$. Again, this node enforces the attack consisting of disabling line (i,j) alone, and as a result (as can easily be argued as in Example 3.1.5) *is unsuccessful. In other words, Template* 3.1.9 *will use Step* 4(b) *and branch. Thus, after enumerating all nodes $w(ij)$, branch-and-bound will have spawned a new generation of nodes, each of which enforces the condition that two distinct lines are disabled.*

Consider any such node. Let us denote it by $w(ij,hk)$, where (i,j) and (h,k) are the two disabled lines. When processing this node using Template 3.1.9, *we* will not *apply Step* 3(a): *this was argued in Example* 3.1.5 *(recall Figure* 3.10)*). But, by construction, $f^*(w(ij,hk)) =$ 2, and so Step* 4(a) *applies and the node will be fathomed. After enumerating every node $w(ij,hk)$, branch-and-bound will therefore stop.*

As the example highlights, our proposed branch-and-bound scheme might run very slowly. In the case that $\mathscr{Z}$ is discrete, it can amount to an organized form of enumeration. This is a somewhat unfair statement, in the sense that branch-and-bound can always be viewed as an "organized form of enumeration." However in Step 1 we are solving a linearly constrained discrete optimization problem; modern mixed-integer program solvers will apply branch-and-cut, which further (and automatically) generates stronger linear inequalities, which are implied by the explicit constraints $K_w z \le \delta_w$ together with the discreteness structure.

This provides an opportunity to strengthen our branch-and-bound approach. Rather than simply branching on the constraints that are directly present in the formulation (3.24b), it is appealing to branch on inequalities that are implied by those. As an example, in Example 3.1.11, one could branch on

$$p_{36} + p_{26} \le 100(1 - z_{26})(\theta_2 - \theta_6) + 100(1 - z_{36})(\theta_3 - \theta_6); \tag{3.37}$$

doing so will produce an inequality involving $z_{26} + z_{36}$, which may be more effectively leveraged by the underlying branch-and-cut solver used in Step 1 of Template 3.1.9. Below we address the issue of how to integrate branching on derived constraints such as (3.37) into Template 3.1.9. To obtain (3.37) we simply added (3.12i) and (3.12j). In fact, any nonnegative linear combination of valid inequalities for the recourse formulation in $MP(z)$ (see (3.23)) would yield an inequality that can be branched on. This is a technique familiar to mixed-integer programming researchers. In the current state of the art for mixed-integer programming, the choice of *which* inequalities to combine, and what coefficients to use, is arguably still more an "art" than a "science."

However, we need to investigate how to integrate branching on derived constraints into Template 3.1.9. Recall that in Step 4(b) of this template, given a (successful) controller recourse vector $\check{p}^w$ we branched on each of the inequalities to make such a recourse action infeasible. This was a valid step (modulo the tolerance ϵ) because any successful attack must make $\check{p}^w$ infeasible. However, given a derived inequality such as (3.37), we cannot guarantee that in an optimal attack, vector $\check{p}^w$ will violate the inequality—it may or it may not. However, we can still "branch" by creating two nodes that enforce, respectively,

$$f_{36}^w + f_{26}^w \ge 100(1 - z_{26})(\theta_2^w - \theta_6^w) + 100(1 - z_{36})(\theta_3^w - \theta_6^w) + \epsilon, \quad \text{and} \tag{3.38a}$$
$$f_{36}^w + f_{26}^w \le 100(1 - z_{26})(\theta_2^w - \theta_6^w) + 100(1 - z_{36})(\theta_3^w - \theta_6^w). \tag{3.38b}$$

Clearly, the second node *does not* enforce that $\check{p}^w$ be cut off, although it does apply a structurally significant constraint. (Alternatively, we could further remove ambiguity by adding a $-\epsilon$ term to (3.38b), and in that case $\check{p}^w$ will be cut off by all attacks feasible for that node.) We can therefore branch on (3.37) and then run branch-and-bound based on Template 3.1.9 on each of the two branches. From this perspective, branching on an implied inequality such as (3.37) can be viewed as an astute way of "prepping," or guiding, branch-and-bound.

Pushing this idea even further, in the particular example of (3.37), lines $(2,6)$ and $(3,6)$ form a graph-theoretic *cut* in the underlying network, and we would term (3.37) a *cut inequality*. Cuts are clearly important in optimization of many types of flow problems, and this particular cut separates a large generator from all the loads, and so it is a natural structural feature to branch on. In general many such graph-theoretic cuts may be available, and this is an example where the above-mentioned "art" would arise in the choice of a (graph-theoretic) cut inequality to branch on. However, one could develop a system for "prepping" branch-and-bound, where multiple nodes are created at the top of the tree by branching on different cut inequalities. This would create a small tree, each of whose leaves imposes a collection of inequalities of type (3.38a) or (3.38b) arising from different graph-theoretic cuts. We will return to these issues in the next subsection, which describes an enhanced version of Template 3.1.9.

Finally, in practice (in particular in the context of the N-K problem), subject matter experience would provide useful intuition about the likely structure of effective attacks. Such intuition can be used to guide the order in which nodes are evaluated, so as to favor the enumeration of branches that reflect such attacks. The strategy followed in Example 3.1.11 to enumerate nodes constitutes an example of such prior knowledge at work—we generated node $w(16)$ first, knowing that it would provide a feasible solution for the attacker. In general, insight into problem structure can be leveraged to obtain good solutions more quickly, a common strategy in the practice of optimization. In this context Template 3.1.9 may become an effective means to search for a successful attack, rather than a rigorous tool for solving the minimum effort attack problem.

Furthermore, in Example 3.1.11 we enumerate a total of 37 branch-and-bound nodes (including the root). Attaining a small node count is of course of paramount importance in all uses of branch-and-bound; it is especially important in the case of bilevel programming because the inner problem may be very difficult and hard to warm start. Considering the sequence of node creation in Example 3.1.11, we can see some obvious waste as a result of symmetry: for example, for any pair of lines $(i,j) \neq (h,k)$ (and both different from $(1,6)$), we generate both nodes $w(ij,hk)$ and $w(hk,ij)$, which are identical. If only one copy is enumerated, the node count will be reduced to 51 (in Example 3.1.11 we already applied a similar principle to prune redundant nodes). Again one wonders if a more effective symmetry-reducing methodology could be applied here.

Vector branching

The *vector branching* technique was introduced in [236] and also used in [38], both in the context of classical 0/1-integer programming (and in both cases under a different name). This technique pursues a more effective way of branching to exploit problem structure. To motivate this approach, consider Step 4(b) of Template 3.1.9. Here, each branch seeks to cut off $\check{z}^w$, and we simply enumerate all ways of strengthening one constraint to attain the exclusion.

When applying vector branching at a given node, we will generate the same number of branches; however we will do this in an ordered way that, in general, greatly reduces the search space. Template 3.1.12 details how vector branching would modify Step 4(b) of Template 3.1.9; additionally, there is a second tolerance δ.

Template 3.1.12. Vector branching modification of step 4(b) in Template 3.1.9

Additional input: Tolerance $0 \leq \delta \leq \epsilon$.

Step 4(b). For $j = 1,2,\ldots,q$, we create child node c_w^j from node w by appending to the system $K_w z \leq \delta_w$ the following constraints:

$$[C(\check{p}^w \circ z)]_i + Az \geq d_i - B_i \check{p}_w - \delta e_i \quad \forall\, 1 \leq i < j,$$

$$[C(\check{p}^w \circ z)]_j + Az \leq d_j - B_j \check{p}_w - \epsilon e_j.$$

In this template each branch includes information pertaining to all previous branching. In the strictest version, we set $\delta = \epsilon$. When, as in Example 3.1.11, branching under Template 3.1.9 is equivalent to setting a binary variable to 1, we can further strengthen the formulation. Further, when the template constraint i $(< j)$ is an *equation*, rather than an inequality, the first constraint imposed by the template would be replaced by two inequalities. The following example details these ideas.

Example 3.1.13 (Example 3.1.11, continued). *At the root, the controller vector $\check{p}^r$ specifies $\theta_1^r = 100$, $\theta_6^r = 94$, $\theta_2^r = \theta_3^r = 93.97$, $\theta_4^r = \theta_5^r = 92.47$, and $f_{16}^r = 6$, $f_{26}^r = f_{36}^r = -3$, $f_{24}^{ur} = f_{24}^{ud} = f_{35}^{ur} = f_{35}^{ud} = 1.5$, $f_{45}^r = 0$, with the generators at buses 1, 4, and 5 in operation. Suppose we enumerate lines in exactly the same order as in Example 3.1.11. As in Example 3.1.11, equation (3.12k) applied to line (1,6) yields two inequalities to branch on, but they will both produce two copies of the same node, denoted here by $v(16)$, which simply enforces $z_{16} = 1$.*

The second line to be enumerated is (2,6). Proceeding as in Template 3.1.12, the complementary inequality to (3.35) and (3.36) is $6(1-z_{16}) \geq 5.99$ in both cases. Thus we obtain a node, denoted $v(26)$, which incorporates the constraints

$$\begin{aligned}
&6(1-z_{16}) \geq 5.99,\\
&-3-\epsilon = f_{26}^r - \epsilon \geq 100(1-z_{26})(\theta_2^r - \theta_6^r) = 100(1-z_{26})(-0.03) = -3(1-z_{26}),\\
&\quad \textit{or} \quad 3(1-z_{26}) \geq 2.99.
\end{aligned}$$

In other words, node $v(26)$ simply enforces the conditions $z_{16} = 0$, $z_{26} = 1$. Continuing with the remaining lines, we will create nodes that enforce, respectively, the following sets of conditions:

$$\begin{aligned}
v(36)&: z_{16} = z_{26} = 0,\ z_{36} = 1,\\
v(24^u)&: z_{16} = z_{26} = z_{36} = 0,\ z_{24}^u = 1,\\
&\ldots\\
v(45)&: z_{16} = z_{26} = z_{36} = z_{24}^u = z_{24}^d = z_{35}^u = z_{35}^d = 0,\ z_{45} = 1.
\end{aligned}$$

Continuing with branch-and-bound, evaluation of node $v(16)$ will provide a feasible solution of value 1 (as with node $w(16)$ in Example 3.1.11). Consider the evaluation of the remaining nodes. Since each node specifies that a single z variable takes value 1, as in Example 3.1.11 we will reach Step 4. The exemption of node $v(45)$ will produce a new generation of nonisomorphic nodes, each of which enforces the condition that two z variables take value 1. For example, consider the evaluation of node $v(36)$.

We will have that $\check{z}_{36}^{v(36)} = 1$, and $\check{z}_{ij}^{v(36)} = 0$ for all other (i,j). As discussed before the controller can defeat this attack. Regardless of the particulars of the controller recourse action

$\breve{p}^{v(36)}$, the branching template will now branch on constraints where changing z away from $\breve{z}^{v(36)}$ renders the controller recourse infeasible. Since at node $v(36)$ we have fixed z_{16}, z_{26}, and z_{36}, it follows that a total of five new nodes will be created:

$$v(36,24^u): z_{16} = z_{26} = 0,\ z_{36} = 1,\ z^u_{24} = 1,$$
$$v(36,24^d): z_{16} = z_{26} = 0,\ z_{36} = 1,\ z^u_{24} = 0,\ z^d_{24} = 1,$$
$$\dots$$
$$v(36,45): z_{16} = z_{26} = 0,\ z_{36} = 1,\ z^u_{24} = z^d_{24} = z^u_{35} = z^d_{35} = 0,\ z_{45} = 1.$$

Evaluation of these second-generation nodes will complete the run of branch-and-bound. The node count equals 1 (for the root), plus 8 (the first generation nodes), plus $21 = 6+5+\dots+1$, to account for the children of nodes $v(26)$, $v(36), \dots, v(24^u)$, for a total of 30 nodes.

As was the case for the simple branching method in Template 3.1.9, Example 3.1.13 brings up a somewhat unsatisfactory behavior of the proposed branching scheme—it appears to amount to just an organized search method. More concretely, it does not seem to sufficiently employ problem structure in devising the search pattern. The remarks concerning inequality (3.37) are pertinent in the context of vector branching, as well. To consider these issues from a different perspective, we now address a minimum effort attack problem where the attacker actions involve continuous decisions.

First we provide some practical motivation. In a power grid, the temperature of a line can act as a constraint on the amount of power flow on the line. If the temperature is high, the line limit will be lowered. Line temperature is itself dependent on the power flow, and on ambient temperature, in a relationship approximately modeled using the standard heat equation. See Sections 4.3.2 and 5.3.2.

As a result, when ambient temperatures are high, power flows are (or should be) more constrained, and failure to adequately take ambient temperatures into account has contributed to recent cascading failures of power grids (see Section 4.2.1). In a geographically dispersed, large transmission system the ambient temperatures that different lines are exposed to might exhibit some broad correlation, but local variations could be significant (especially in the context of how ambient temperatures impact line temperatures).

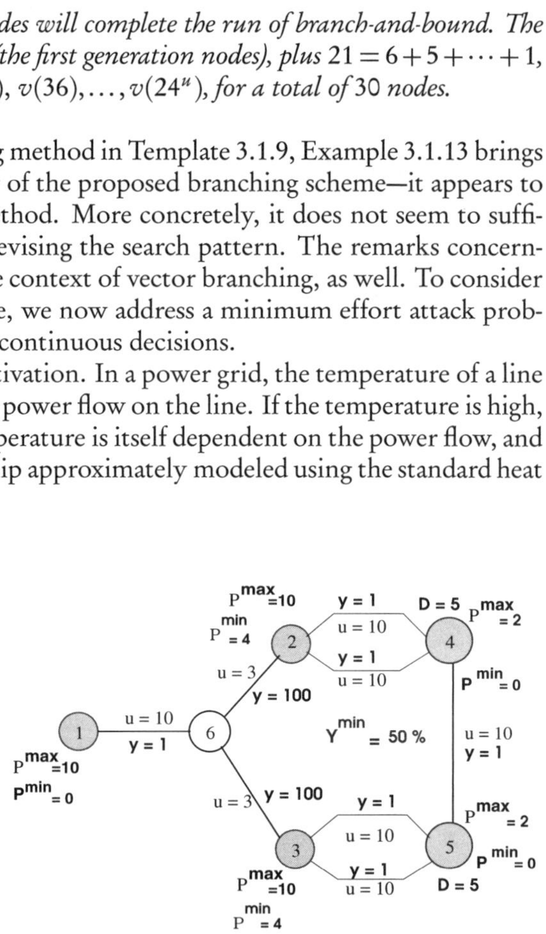

Figure 3.12. *Grid for Example* 3.1.14.

Thus, an interesting risk assessment of a grid would analyze how adversarially induced line limit reductions can affect the ability of the grid to deliver power to loads. Since it is unlikely that all (or even many) lines would be highly affected by ambient temperatures at the same time, we model the adversary as facing a nonlinear "cost" in attaining a line limit reduction. For related problems see Section 3.3.

Example 3.1.14 (modification of Example 3.1.5). *Here we again rely on the grid shown in Figure 3.9 modified so that the lower output limit for generators 2 and 3 has been reduced to 4 units (see Figure 3.12). We consider an attacker that reduces line limits. For each line (i, j),* *the attacker will choose a quantity* $0 \leq \gamma_{ij} \leq 1$*; this reduces the limit of* (i,j) *from its initial value* u_{ij} *to* $u_{ij}(1-\gamma_{ij})$*. The effort expended by the attacker in taking this action equals* $100\gamma_{16}^2 + 20\gamma_{26}^2 + 15\gamma_{36}^2$*, and we further assume that only lines* $(1,6)$*,* $(2,6)$*, and* $(3,6)$ *may be thus interdicted. An attack is successful if the controller cannot feasibly deliver more than* 50% *of the sum of demands. We consider the minimum effort attack problem. In contrast with Example* 3.1.5*, due to the reduced lower limits on generators* 2 *and* 3*, the controller can (ignoring for the moment the interdiction) operate just one of generators* 2 *and* 3 *(and not operate generator* 1*) and still deliver the desired amount of flow to the loads.*

We apply our branch-and-bound method to this problem, using Template 3.1.9 *modified to include vector branching as per Template* 3.1.12*. Note that even though the objective function faced by the attacker is convex, the feasible region may not be so. At the root of branch-and-bound, the attacker will choose* $\gamma_{ij} = 0$ *for every* (i,j)*. The controller recourse vector* $\breve{p}^r$ *will be as in Examples* 3.1.11 *and* 3.1.13*, with* $\theta_1^r = 100$*,* $\theta_6^r = 94$*,* $\theta_2^r = \theta_3^r = 93.97$*, and* $f_{16}^r = 6$*,* $f_{26}^r = f_{36}^r = -3$*. We apply strong branching by considering the constraints*

$$|p_{26}| + |p_{36}| \leq u_{26}(1-\gamma_{26}) + u_{36}(1-\gamma_{23}), \tag{3.39}$$

$$p_{16} \leq u_{16}(1-\gamma_{16}), \tag{3.40}$$

the first of which is implied. With $\epsilon = 0.01$*,* $\delta = 0$*, strong branching produces two nodes, which we denote by* v_1 *and* v_2*, corresponding to the systems*

$$v_1: \quad 3(1-\gamma_{26}) + 3(1-\gamma_{36}) \leq 5.99, \quad \text{i.e.,} \qquad 3(\gamma_{26}+\gamma_{36}) \geq 0.01, \tag{3.41}$$

and

$$v_2: 0 \leq 3(\gamma_{26}+\gamma_{36}) \leq 0.01, \tag{3.42}$$

$$v_2: 10(1-\gamma_{16}) \leq 5.99, \quad \text{i.e.,} \quad 10\gamma_{16} \geq 4.01. \tag{3.43}$$

The attacker objective function is $(100\gamma_{16}^2 + 20\gamma_{26}^2 + 15\gamma_{36}^2)$*. Thus the value of node* v_1 *is* 9.542×10^{-5}*, whereas the value of* v_2 *is* 1.6081×10^1*.*

Continuing with branch-and-bound, when we process node v_1 *we obtain the attack where* $\gamma_{16} = 0$*,* $\gamma_{26} = \gamma_{36} = \frac{0.01}{6}$*, i.e., lines* $(2,6)$ *and* $(3,6)$ *have their limits reduced to* $3(1 - 0.01/6.0) = 2.995$*. This attack is defeated by the controller by using generator* 1 *to supply* 2.995 *units of demand to loads* 4 *and* 5 *(each), while using the generators at* 4 *and* 5 *to supply* 2 *additional units (each) for a total of* $10 - 0.02$ *units (attained by setting* $\theta_1 = 100$*,* $\theta_6 = 94.01$*,* $\theta_2 = \theta_3 = 91.015$*, and similarly with the other angles; here* $p_{16} = 5.99$ *and* $p_{26} = p_{36} = -2.995$*). Thus, strong branching on* (3.39) *and* (3.40) *produces a node* v_3 *that enforces*

$$v_3: \quad 3(\gamma_{26}+\gamma_{36}) \geq 0.02,$$

as well as a node v_4 *that enforces*

$$v_4: 0.01 \leq 3(\gamma_{26}+\gamma_{36}) \leq 0.02,$$

$$v_4: 10\gamma_{16} \geq 4.01.$$

Notice the structural similarity between v_4 *and* v_2*. Processing* v_2*, however, will create a node that enforces* $3(\gamma_{26}+\gamma_{36}) \geq 0.02$ *in addition to* (3.42) *and is therefore infeasible, as well as a node* v_5 *that enforces*

$$v_5: 0 \leq 3(\gamma_{26}+\gamma_{36}) \leq 0.01, \tag{3.44}$$

$$v_5: 10\gamma_{16} \geq 0.02. \tag{3.45}$$

This node bears, again, a structural similarity to v_2. The branch-and-bound tree at this point consists of nodes v_3, v_4, and v_5. Consider the processing of node v_3. The solution of the formulation at this node attains $\gamma_{16} = 0$, $\gamma_{26} = \gamma_{36} = \frac{0.02}{6}$, and so it simply enforces that lines $(2,6)$ and $(3,6)$ have their limits reduced to 2.99. The controller again defeats this attack, and branch-and-bound will produce a node enforcing $3(\gamma_{26} + \gamma_{36}) \geq 0.03$ and another node enforcing $0.02 \leq 3(\gamma_{26} + \gamma_{36}) \leq 0.03$ as well as $10\gamma_{16} \geq 0.01$. We once again notice a similarity between this node and v_2.

In summary, branch-and-bound will produce two kinds of nodes. The first, denoted by $w(J,M)$, is parameterized by nonnegative integers J and M and enforces constraints

$$\begin{aligned} w(J,H): 0.01J &\leq 3(\gamma_{26}+\gamma_{36}) \leq 0.01(J+1), \\ w(J,H): 10\gamma_{16} &\geq 0.01H. \end{aligned}$$

The second, denoted by $u(M)$, is parameterized by an integer $M \geq 0$ and enforces

$$u(M): 3(\gamma_{26}+\gamma_{36}) \geq 0.01M.$$

These observations parallel the result in Lemma 3.1.10; i.e., branch-and-bound will terminate after examining a finite number of nodes.

Consider the processing of a node of the form $u(M)$. Clearly, the optimal solution γ^* to the optimization problem at this node will satisfy $3(\gamma^*_{26} + \gamma^*_{36}) = 0.01M$; consequently the cut made up of lines $(2,6)$, $(3,6)$ will have total capacity $6 - 0.01M$. If $M \leq 500$ the controller can defeat the attack by operating generators 4 and 5 (producing 2 units at each) and generator 1 (producing $6 - 0.01 * 500$ units), for a total of $6 - 0.01 * 500 + 4 = 5$ units. With $M = 501$, the optimal solution γ^* at node $u(M)$ has objective value 2.39×10^1 and is given by $\gamma^*_{16} = 0$, $\gamma^*_{26} = 0.716$, $\gamma^*_{36} = 0.954$; i.e., the limit of lines $(2,6)$ and $(3,6)$ is reduced to 0.852 and 0.138, respectively; in this case a computation (similar to the solution of problem (3.12)) shows that the maximum total demand that can be feasibly satisfied by the controller is 4.00; i.e., the attack defeats the controller. In summary, node $u(M)$ with $M \geq 501$ will be fathomed as feasible.

Now consider a node $w(J,H)$. As in the previous paragraph, when $J = 500$ the solution to this node attains $\gamma^*_1 = H/1000$, $3(\gamma^*_2 + \gamma^*_3) = 5$; i.e., the total capacity of cut $\{(2,6),(3,6)\}$ is reduced to 1 unit. Unless $\gamma^*_1 \geq 0.9$, i.e., $H \geq 900$, the controller will defeat this attack (again, by operating generators 1, 4, and 5), but in that case the value of node $w(J,H)$ exceeds 1.04×10^2, which is higher than the value of node $u(501)$. Moreover, the solution to a node $w(501,H)$ will defeat the controller regardless of the value of H. It follows that the minimum effort attack problem is solved by node $u(501)$.

As Example 3.1.14 makes clear, convergence of the basic branch-and-bound scheme can be expected to be slow, even if, in accordance with Lemma 3.1.10, when $\epsilon = 0.01$ termination is attained after enumerating "only" hundreds of nodes. On the positive side, we stress that the example handled a case involving a continuous (attack) problem in contrast to the traditional branch-and-bound setting. Additionally, the development of the example highlights the significant complexity of the N-K problem and bilevel programs in general.

Linearizing bilinear expressions

We now turn to a different methodology that can be employed to tackle the bilevel problem (3.24) by explicitly approximating each of the bilinear expressions in (3.23). We will describe two approaches in this direction. Each of these approaches will substitute the bilinear terms in (3.24b) with linear terms (but possibly adding binary variables). Note that because of the stipulation $p \in \mathscr{P}$ in the definition of sets $MP(z)$ in (3.23) this simplification does not completely yield a practicable formulation for problem (3.24) when $p \in \mathscr{P}$ stipulates that, e.g., some variables are integral. This feature certainly arose in the numerical examples discussed above. However, what the simplification will accomplish is to render (3.24) into a standard (linear) bilevel program (albeit one where the inner problem may be a mixed-integer program). In fact, current literature appears to indicate that this approach is in fact practicable.

Thus, consider a bilinear expression

$$x_1 x_2, \text{ where } \quad d_i \leq x_i \leq u_i, \ i = 1,2. \tag{3.46}$$

We will next provide a procedure to approximate (3.46) using linear expressions and auxiliary binary variables.

Powers of 1/2 approximation. Here we describe a procedure to obtain a provably accurate (mixed-integer) *linear* approximation to a bilinear expression, based on a method originally developed by Glover [105] and recently studied in a number of settings. We first simplify each bilinear system (3.46) by reducing it to an equivalent system over nonnegative variables. Defining $x_i' = x_i - d_i$ for $i = 1,2$, we have

$$x_1 x_2 = x_1' x_2' \ + \ d_1 x_2 + d_2 x_1 + d_1 d_2.$$

It follows that under the goal of approximating a bilinear expression using a linear system, we can assume, without loss of generality, that $d_1 = d_2 = 0$.

Under this assumption, we will now construct a binary mixed-integer system to approximate (3.46). Choose an integer $k \geq 1$, and consider the mixed-integer system given by the following inequalities:

$$0 \leq x_1 - u_1 \sum_{j=1}^{k} 2^{-j} z_j \ \leq \ u_1 2^{-k}, \tag{3.47a}$$

$$z_j \in \{0,1\}, \quad 1 \leq j \leq k, \tag{3.47b}$$

$$x_2 - u_2(1 - z_j) \leq x_{2j} \leq \min\{x_2, u_2 z_j\}, \quad 1 \leq j \leq k, \tag{3.47c}$$

$$0 \leq x_i \leq u_i, \quad i = 1,2, \tag{3.47d}$$

$$0 \leq x_{2j}, \quad j = 1, \ldots, k. \tag{3.47e}$$

Lemma 3.1.15. *Let $0 \leq \hat{x}_i \leq u_i$ for $i = 1,2$. Then there exists a feasible solution to* (3.47) *with $x_i = \hat{x}_i$, $i = 1,2$.*

Proof. We will first show that there exist binary $\tilde{z}_j$, $1 \leq j \leq k$, such that $(\hat{x}_1, \tilde{z}_1, \ldots, \tilde{z}_k)$ satisfy (3.47a). To do so, note that $\sum_{j=1}^{k} 2^{-j} = 1 - 2^{-k}$, and so the claim follows if $1 - 2^{-k} \leq \frac{\hat{x}_1}{u_1}$ since $\hat{x}_1 \leq u_1$. Otherwise we have $2^{k+1}\frac{\hat{x}_1}{u_1} < 2^{k+1} - 2$. Defining $p \doteq \lfloor 2^{k+1}\frac{\hat{x}_1}{u_1} \rfloor$, since $\sum_{j=1}^{k} 2^j = 2^{k+1} - 2$ it follows that there exist binary values $\tilde{z}_j$ $(1 \leq j \leq k)$ and $0 < r < 1$ such that

$$p = \sum_{j=1}^{k} 2^j \, \tilde{z}_{k+1-j} \quad \text{and}$$

$$\frac{\hat{x}_1}{u_1} = 2^{-(k+1)}(p + r).$$

From these two facts we obtain

$$\frac{\hat{x}_1}{u_1} = \sum_{j=1}^{k} 2^{-j} \tilde{z}_j + 2^{-(k+1)} r,$$

which proves the desired claim since $0 \leq r < 1$. Now define, for $1 \leq j \leq k$, $\tilde{x}_{2j} = \tilde{z}_j \hat{x}_2$. It follows that $(\hat{x}_2, \tilde{z}_1, \ldots, \tilde{z}_k, \tilde{x}_{21}, \ldots, \tilde{x}_{2k})$ satisfy (3.47c). ■

Lemma 3.1.16. *Suppose* $(\tilde{x}_1, \tilde{x}_2, \tilde{z}_1, \ldots, \tilde{z}_k, \tilde{x}_{21}, \ldots, \tilde{x}_{2k})$ *are feasible for* (3.47). *Then*

$$0 \leq \tilde{x}_1 \tilde{x}_2 - u_1 \sum_{j=1}^{k} 2^{-j} \tilde{x}_{2j} \leq u_1 \tilde{x}_2 2^{-k}.$$

Proof. Note that by (3.47c) and (3.47e), for $1 \leq j \leq k$ we have $\tilde{x}_{2j} = 0$ if $\tilde{z}_j = 0$ and $\tilde{x}_{2j} = \tilde{x}_2$ otherwise. The claim now follows by multiplying (3.47a) by $\tilde{x}_2$. ■

The primary consequence of Lemmas 3.1.15 and 3.1.16 is that if we have an optimization problem that includes the bilinear expression $x_1 x_2$ in constraints or the objective, we can replace each instance of $x_1 x_2$ with the (linear) expression $u_1 \sum_{j=1}^{k} 2^{-j} x_{2j}$ while also imposing the system (3.47). This strategy amounts to a *relaxation* (by Lemma 3.1.15) which is provably accurate by Lemma 3.1.16. More precisely, if, after the reformulation, we have $x_2 = 0$, then the approximation is *exact* (by Lemma 3.1.16, or (3.47c)). And if $x_2 > 0$ then

$$\frac{|x_1 x_2 - u_1 \sum_{j=1}^{k} 2^{-j} x_{2j}|}{x_2} \leq u_1 2^{-k};$$

in other words the error is small whether x_2 is large or small; by *small* we mean that we can choose k to make the error less than any desired threshold $\epsilon > 0$.

In summary, this approach approximates a product $x_1 x_2$ of box-bounded general variables by first approximately representing one of the variables as a (weighted) sum of binary variables and then using system (3.47b)–(3.47c) to describe the product of a binary variable and a continuous variable. As far as we can see, the first instance of this approach can be read in the work of Glover [105] for the case where at least one of x_1, x_2 is constrained to be integer valued. In the pure continuous setting it was used in [27]. Also see [78]. A recent, in-depth study of this methodology (in the pure integer case) can be found in [114].

The McCormick relaxation

When one of x_1, x_2 is binary to begin with, we can bypass the powers of $1/2$ step, leading to a much smaller formulation. The approach then constitutes a special case of the McCormick relaxation (see below following Example 3.1.17). Assuming without loss of generality that x_2 is binary, and as before that $d_1 \leq x_1 \leq d_2$, we can use the system given next, where π_{12} represents the product $x_1 x_2$:

$$x_1 + u_1 x_2 - u_1 \leq \pi_{12} \leq x_1 + d_i x_2 - d_1, \tag{3.48a}$$

$$d_1 x_2 \leq \pi_{12} \leq u_1 x_2. \tag{3.48b}$$

If $x_2 = 1$, then (3.48a) imposes $\pi_{12} = x_1$ and (3.48b) is inactive; whereas if $x_2 = 0$, (3.48b) forces $\pi_{12} = 0$ but (3.48a) is inactive. Hence in each case $\pi_{12} = x_1 x_2$. Note that system (3.48) is similar to (3.47c).

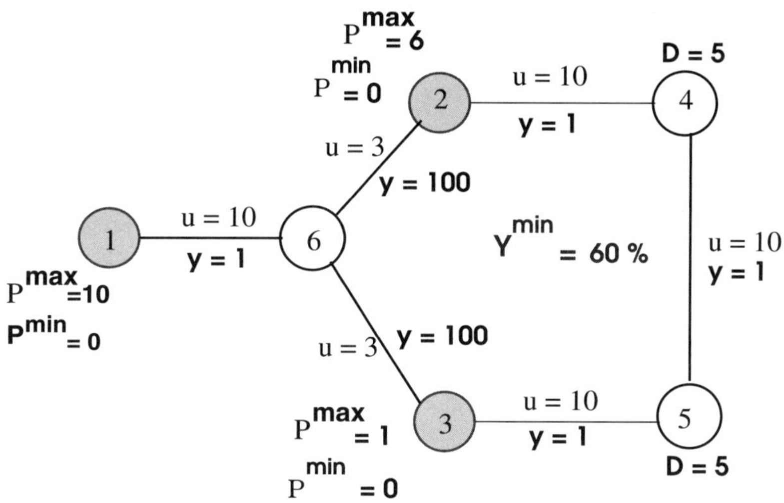

Figure 3.13. *Example* 3.1.17.

Example 3.1.17. *Consider the grid in Figure* 3.13. *The problem is given as follows:*

- *The attacker may remove any subset of lines chosen from among* $(1,6),(2,6),(3,6)$, *and* $(4,5)$.
- *Buses* 1, 2, *and* 3 *hold generators with the attributes shown.*
- *An attack defeats the controller if at most* 60% *of the demand can be feasibly delivered. The objective of the attacker is to remove as few lines as possible.*

Using the binary vector z to indicate a given attack, the optimization problem solved by the controller is as follows, where we precede each constraint by the corresponding dual variable (in bold):

$$
\begin{aligned}
\textbf{(C(z))}: \quad & \max p_{24} + p_{35} \qquad (3.49)\\
\text{s.t.}\quad & \\
& (\boldsymbol{b_6}) p_{16} + p_{26} + p_{36} = 0,\\
& (\boldsymbol{o_{16}}) p_{16} = (1 - z_{16})(\theta_1 - \theta_6) \quad (\boldsymbol{m_{16}}),\quad 0 \le p_{16} \le 10,\\
& (\boldsymbol{o_{26}}) p_{26} = 100(1 - z_{26})(\theta_2 - \theta_6) \quad (\boldsymbol{m_{26}}, \boldsymbol{d_{26}}),\quad |p_{26}| \le 3,\\
& (\boldsymbol{o_{36}})\ p_{36} = 100(1 - z_{36})(\theta_3 - \theta_6) \quad (\boldsymbol{m_{36}}, \boldsymbol{d_{36}}),\quad |p_{36}| \le 3,\\
& (\boldsymbol{o_{24}}) p_{24} = (\theta_2 - \theta_4) \quad (\boldsymbol{m_{24}}, \boldsymbol{d_{24}}),\quad |p_{24}| \le 10,\\
& (\boldsymbol{o_{35}}) p_{35} = (\theta_3 - \theta_5) \quad (\boldsymbol{m_{35}}, \boldsymbol{d_{35}}),\quad |p_{35}| \le 10,\\
& (\boldsymbol{o_{45}}) p_{45} = (1 - z_{45})(\theta_4 - \theta_5) \quad (\boldsymbol{m_{45}}, \boldsymbol{d_{45}}),\quad |p_{45}| \le 10,\\
& (\boldsymbol{m_2}, \boldsymbol{d_2}) 0 \le p_{26} + p_{24} \le 6 \quad (\boldsymbol{m_3}, \boldsymbol{d_3}),\quad 0 \le p_{36} + p_{35} \le 1,\\
& (\boldsymbol{m_4}, \boldsymbol{d_4}) 0 \le p_{24} - p_{45} \le 5 \quad (\boldsymbol{m_5}, \boldsymbol{d_5}),\quad 0 \le p_{35} + p_{45} \le 5.
\end{aligned}
$$

The last four constraints in this formulation express the generator bounds at buses 2 *and* 3, *and the demand requirements at buses* 4 *and* 5, *respectively. The dual of this linear program is as follows:*

$$\begin{aligned}\mathbf{V(z)} \doteq \min 10m_{16}+3m_{26}+3d_{26}+3m_{36}+3d_{36}+10m_{24}+10d_{24}+10m_{35}\\ +10d_{35}+10m_{45}+10d_{45}+6m_2+m_3+5m_4+5m_5\end{aligned} \tag{3.50}$$

$$\begin{aligned}
&\text{s.t.}\\
&(p_{24})m_{24}-d_{24}+o_{24}+m_2-d_2+m_4-d_4 = 1,\\
&(p_{35})m_{35}-d_{35}+o_{35}+m_3-d_3+m_5-d_5 = 1,\\
&(p_{16})b_6+m_{16}+o_{16} = 0,\\
&(p_{26})b_6+m_{26}-d_{26}+o_{26}+m_2-d_2 = 0,\\
&(p_{36})b_6+m_{36}-d_{36}+o_{36}+m_3-d_3 = 0,\\
&(p_{45})m_{45}-d_{45}+o_{45}-m_4+d_4+m_5-d_5 = 0,\\
&(\theta_1)\text{—}(1-z_{16})o_{16} = 0,\\
&(\theta_6)(1-z_{16})o_{16}+100(1-z_{26})o_{26}+100(1-z_{36})o_{36} = 0,\\
&(\theta_2)\text{—}100(1-z_{26})o_{26}-o_{24} = 0,\\
&(\theta_3)\text{—}100(1-z_{36})o_{36}-o_{35} = 0,\\
&(\theta_4)o_{24}-(1-z_{45})o_{45} = 0,\\
&(\theta_5)o_{35}+(1-z_{45})o_{45} = 0,\\
&\quad m_{16},m_{26},d_{26},m_{36},d_{36},m_{24},d_{24},u_{35},d_{35},u_{45},d_{45},m_2,d_2,m_3,d_3,m_4,\\
&\quad\quad d_4,m_5,d_5 \geq 0.
\end{aligned}$$

Suppose we compute $V^* \doteq \min\{V(z) : z \text{ binary}\}$. *Then there is an attack that defeats the controller if and only if* V^* *is at most* 60% *of the demand, i.e., at most* 6.

To address the computation of V^* *we must first linearize the bilinear expressions in* (3.50), *and there are three such expressions:* $z_{16}o_{16}, z_{26}o_{26}, z_{36}o_{36}$, *and* $z_{45}o_{45}$. *To apply the construction* (3.48) *we need to state valid upper bounds on* $|o_{16}|, |o_{26}|, |o_{36}|$, *and* $|o_{45}|$. *Toward this goal, the following observation proves useful. Suppose that* $z_{ij} = 1$ *for* $(i,j) = (1,6), (2,6), (3,6)$, *or* $(4,5)$. *This means that the attacker has disabled line* (i,j). *Thus, we obtain a formulation equivalent to* **(C(z))** *by* (i) *removing the constraint* $p_{ij} = y_{ij}(1-z_{ij})(\theta_i - \theta_j)$ *and* (ii) *setting* $p_{ij} = 0$. *From the perspective of the dual problem* (3.50) *we may therefore equivalently set* $o_{ij} = u_{ij} = d_{ij} = 0$. *We will use this observation below to bound the dual variables* o_{ij}.

Further, we note that the optimal objective value is never more than 10. *Hence* (p_{24}) *implies*

$$d_{24}+d_2+d_4 \leq u_{24}+m_2+m_4-1 \leq 1. \tag{3.51}$$

Also, using (p_{35}) *and* (θ_5) *we obtain (making use of the above observation) that*

$$o_{45} \leq u_{35}+m_3+m_5-1 \leq 5,$$

and using (p_{24}) *and* (θ_4) *we likewise obtain* $o_{45} \geq -5$ *and, thus,* $|o_{24}| \leq 5, |o_{35}| \leq 5$. *Next we bound* b_6 *by enumerating several cases. First, suppose that* $z_{16} = 0$. *Then using* (θ_1) *and* (p_{16}) *we obtain* $-1 \leq b_6 \leq 0$. *Likewise, if* $z_{26} = 0$, *using* (θ_3) *and* (p_{26}) *we obtain* $b_6 \leq d_{26}+d_2+o_{24} \leq 10/3+1+5 < 10$ *and* $b_6 \geq -u_{26}-m_2+o_{24} \geq -7$. *Finally, if* $z_{36} = 0$ *then a similar derivation again shows* $-7 \leq b_6 \leq 10$. *Thus the only case remaining is that where* $z_{16} = z_{26} = z_{36} = 1$. *In that case the attacker has disabled lines* $(1,6)$, $(2,6)$, *and* $(3,6)$. *From the controller's perspective buses* 1 *and* 6 *are isolated and an equivalent problem is obtained by restricting the grid to buses* 2, 4, 5, *and* 6. *Thus without loss of generality* $b_6 = 0$.

Hence in any case $-7 \leq b_6 \leq 10$*. We can now bound* $|o_{16}|$*,* $|o_{26}|$*,* $|o_{36}|$ *(we have already done so for* $|o_{45}|$*). Using* (p_{16}) *we obtain* $-11 \leq o_{16} \leq 7$*; using* (p_{26})*,* $-12 \leq o_{26} \leq 10/3 + 10 < 14$*; and using* (p_{36})*,* $-12 \leq o_{36} \leq 14$*.*

The resulting formulation has 34 *variables of which* 4 *are binary; using Cplex* 12.6 [125] *it is solved at the root by applying* 6 *cuts; the optimum solution removes lines* (3,6) *and* (4,5)*.*

We next review the general McCormick relaxation approach [160] for linearizing a general bilinear system (3.46), which relies on properties of the set $X \doteq \{(x_i, x_j, y_{ij})^T : y_{ij} = x_i x_j,\ d_k \leq x_k \leq u_k,\ k = i, j\}$. In fact, is known that the set conv(X) is described by the following inequalities:

$$y_{ij} \geq u_j x_i + u_i x_j - u_i u_j, \quad y_{ij} \geq d_j x_i + d_i x_j - d_i d_j, \tag{3.52a}$$

$$y_{ij} \leq u_j x_i + d_i x_j - d_i u_j, \quad y_{ij} \leq d_j x_i + u_i x_j - u_i d_j, \tag{3.52b}$$

together with $x_k \in [d_k, u_k]$ for $k = i, j$. A special case arises when (say) $d_i = 0$ and $u_i = 1$, in which case we obtain that conv(X) is given by

$$\max\{u_j x_i + x_j - u_j,\ d_j x_i\} \leq y_{ij} \leq \min\{u_j x_i,\ d_j x_i + x_j - d_j\}. \tag{3.53}$$

If in addition x_i is *binary*, then (3.53) together with $x_j \in [d_j, u_j]$ enforces clearly valid logical conditions: when $x_i = 1$ we have $y_{ij} = x_j$, whereas if $x_i = 0$ then $y_{ij} = 0$ as well. In fact, we see that system (3.47c) above is a special case of (3.52).

The system in (3.52) was proposed by McCormick [160], whereas the convex hull result was proved by Al-Khayyal and Falk [5]. Because of the prevalence of bilinear forms in optimization, this result has been the starting point of a rich line of research. Note, however, that the convex hull result applies in the case of a single bilinear form. In our sets $MP(z)$ described above we will have multiple bilinear forms, but involving the same set of underlying variables. Effectively the set of bilinear expressions defines an undirected graph, with a vertex corresponding to each underlying variable x_i and an edge $\{x_i, x_j\}$ if $x_i x_j$ appears in the optimization problem. In that case applying system (3.52) to every bilinear form will yield a linear formulation that is unlikely to be exact.

This fact is not surprising given that optimization problems with bilinear forms are (strongly) NP-hard. These issues are taken up in greater detail in [154]. To better understand the results in [154] we need some definitions. Suppose $f : \mathbb{R}^n \to \mathbb{R}$ is an arbitrary (finite-valued) function, and let $F \subseteq \mathbb{R}^n$ be a polytope. We are interested in deriving bounds on $f(x)$ for $x \in F$ that can be expressed by means of linear functions. Call a pair $(v, v_0) \in \mathbb{R}^n \times \mathbb{R}$ *valid* if $v^T x + v_0 \leq f(x)$ for all $x \in F$ (in other words $v^T x + v_0$ is an affine underestimator for $f(x)$ in F) and let

$$\text{vex}_F[f](x) \doteq \sup\{v^T x + v_0 : (v, v_0)\,\text{valid}\}.$$

At any $x \in F$ we have $v^T x + v_0 \leq f(x)$ for every valid (v, v_0), which implies that $\text{vex}_F[f](x) \leq f(x)$. Also one may show that $\text{vex}_F[f](x)$ is convex; in other words $\text{vex}_F[f](x)$ is the largest *convex underestimator* for $f(x)$. For related material see [47, Chapter 3]. Likewise, one can define the smallest *concave overestimator*, $\text{cav}_F[f](x)$, for $f(x)$ in F. Computation of the vex and cav functions is, in general, a nontrivial task. The case where $f(x)$ is a multilinear function is studied in [196].

To apply these concepts, let us consider a particular family of functions $f(x)$, namely

$$f(x) = \sum_{\{i,j\} \in S} b_{ij} x_i x_j, \tag{3.54}$$

where S is a set of pairs of indices and each b_{ij} is real. This case is of special interest since it arises in many settings. For example, it arises in our bilevel programming formulations for N-K problems (it also arises in the power flow and OPF problems). In this case the definitions of $\text{vex}_F[f]$ and $\text{cav}_F[f]$ can be sharpened. In the case of (3.54) the vex and cav functions have a simpler representation, which however requires exponential amount of work (in n) to compute. In [154] the authors propose an approximation. Given (3.54), suppose we apply the McCormick formulation (3.52) to every pair $\{i,j\}$ in the set S in (3.54). Denote by $\mathscr{P}$ the resulting polyhedron (on variables x and y), and write, for $x \in F$,

$$\text{mcl}_F[f](x) \doteq \min \left\{ \sum_{\{i,j\}\in S} b_{ij} y_{ij} \, : \, (x,y) \in \mathscr{P} \right\} \text{ and} \tag{3.55}$$

$$\text{mcu}_F[f](x) \doteq \max \left\{ \sum_{\{i,j\}\in S} b_{ij} y_{ij} \, : \, (x,y) \in \mathscr{P} \right\}. \tag{3.56}$$

Thus, $\text{mcl}_F[f](x)$ and $\text{mcu}_F[f](x)$ are the best lower and upper bounds, respectively, that we can prove for $f(x)$ using the McCormick relaxation. More precisely, if we use the McCormick relaxation we can expect to obtain an underestimate for $f(x)$ which is no lower than $\text{mcl}_F[f](x)$, and similarly with $\text{mcu}_F[f](x)$. A question of interest is therefore how large $\text{mcu}_F[f](x) - \text{mcl}_F[f](x)$ can be. Clearly, if $x \in F$, then the vector $y \in \mathbb{R}^S$ defined by $y_{ij} = x_{ij}$ for all $\{i,j\} \in S$ satisfies that $(x,y) \in P$. It follows that for each $x \in F$,

$$\text{mcl}_F[f](x) \le \text{vex}_F[f](x) \le f(x) \le \text{cav}_F[f](x) \le \text{mcu}_F[f](x).$$

Therefore we obtain $\text{mcu}_F[f](x) - \text{mcl}_F[f](x) \ge \text{cav}_F[f](x) - \text{vex}_F[f](x)$. The quantity on the right-hand side of this relationship is a natural lower bound for the weakness (or lack of tightness) of any (linear) relaxation used to approximate any function $f(x)$ (since as argued above it describes the best linear relaxation). A very interesting result proved in [154] states that the gap between the two quantities is never larger than two, provided that the b_{ij} quantities are positive. More precisely, we have the following result.

Theorem 3.1.18 (see [154]). *Let G denote the graph whose edge-set is the set S in* (3.54), *and let χ denote the chromatic number of G. Suppose $b_{ij} > 0$ for all $\{i,j\} \in S$. Then for any $x \in F$,*

$$mcu_F[f](x) - mcl_F[f](x) \le \left(2 - \frac{2}{\chi'}\right)(cav_F[f](x) - vex_F[f](x)),$$

where $\chi' = \chi + 1$ if χ is odd and otherwise $\chi' = \chi$.

As a remark, if $\chi = 2$ (G is bipartite), the theorem shows that the McCormick relaxation provides the convex envelope of $f(x)$, a previously known result. And in the general case the result shows that the McCormick relaxation is provably tight, in a precise sense. Additional results are provided in [154]. Despite the positive result in the above theorem, it is to be stressed that applying the McCormick relaxation to handle bilinear expressions will only result in an approximate formulation; additional techniques, such as branching, will be needed to actually solve optimization problems.

Now suppose we apply the McCormick reformulation to the bilinear equation (3.30b) arising in power grid interdiction models where there is an explicit upper bound $|\theta_k - \theta_m| \le \theta_{km}^{\max}$ which applies even if buses k and m are in different components, and of course z_{km} is binary. Using (3.53) we obtain

$$-y_{km}\theta_{km}^{\max}(1-z_{km}) \leq p_{km} \leq y_{km}\theta_{km}^{\max}(1-z_{km}), \tag{3.57a}$$

$$-y_{km}\theta_{km}^{\max}z_{km} \leq p_{km}-y_{km}(\theta_k-\theta_m) \leq y_{km}\theta_{km}^{\max}z_{km}, \tag{3.57b}$$

which is exactly of the form (3.28) we have seen before (also see (3.29)). We conclude as before that we would obtain a big M formulation unless a reasonably small bound on $\theta_{km}^{\max}$ is known. We see, again, that systems (3.28) and (3.30) are thus equivalent for binary z_{km}.

An effective use of Benders decomposition

Here we will outline the approach used in [42] to address a particular version of the N-K problem. Even though the approach is specific to that version of the problem, it is actually fairly generic, and, most important, it relies on a correct implementation of Benders decomposition.

The problem considered in [42] is a general version of that presented in Example 3.1.5. In this setting, the adversary seeks to disable a minimum number of lines of the grid to reduce yield to a value *strictly smaller* than a given percentage $Y^{\min}$. On the other hand, the controller must choose which generators to operate (a given generator g, if operated, must produce output between two values $P_g^{\min}$ and $P_g^{\max}$) to obtain a feasible power flow (under the DC approximation) that attains the desirable minimum yield value.

To explain the approach in [42], let us use the following notation. First, $\mathscr{G}$ and $\mathscr{D}$ are the sets of generator and load buses, respectively, which are assumed to be disjoint (a simple assumption under the DC model). For $i \in \mathscr{D}$, we denote by D_i^{nom} the demand at i preattack. Finally, for a bus i we denote by $\delta^+(i)$ and $\delta^-(i)$ the set of lines of the form (i,j) (resp., (j,i)). Thus, let $\mathscr{A}$ be an *attack*, i.e., a set of lines removed by the attacker, and define the binary vector $z^{\mathscr{A}}$ indexed by the lines as $z_{ij}^{\mathscr{A}} = 1$ if and only if $(i,j) \in \mathscr{A}$. Likewise, let $\mathscr{C} \subseteq \mathscr{G}$ be a set of generator nodes, and let the binary vector $y^{\mathscr{C}} \in \{0,1\}^{\mathscr{C}}$ be defined by $y_i^{\mathscr{C}} = 1$ if and only if $i \in \mathscr{C}$. Consider the following formulation [42]:

$$K_{\mathscr{C}}^*(\mathscr{A}): \quad t_{\mathscr{C}}^*(z^{\mathscr{A}}) \doteq \min t \tag{3.58a}$$

s.t.

$$(\alpha_i^{\mathscr{C}}) \quad \sum_{(i,j)\in\delta^+(i)} p_{ij} - \sum_{(j,i)\in\delta^-(i)} p_{ji} = \begin{cases} P_i, & i \in \mathscr{G}, \\ -D_i, & i \in \mathscr{D}, \\ 0 & \text{otherwise}, \end{cases} \tag{3.58b}$$

$$(\beta_{ij}^{\mathscr{C}}) \quad \theta_i - \theta_j - x_{ij}p_{ij} = 0 \quad \forall (i,j) \notin \mathscr{A}, \tag{3.58c}$$

$$(p_{ij}^{\mathscr{C}}, q_{ij}^{\mathscr{C}}) \quad u_{ij}\, t - |p_{ij}| \geq 0 \quad \forall (i,j) \notin \mathscr{A}, \tag{3.58d}$$

$$(\omega_{ij}^{\mathscr{C}+}, \omega_{ij}^{\mathscr{C}-}) \quad t - |p_{ij}| \geq 1 \quad \forall (i,j) \in \mathscr{A}, \tag{3.58e}$$

$$(\gamma_i^{\mathscr{C}+}, \gamma_i^{\mathscr{C}+}) \quad P_i^{\min}y_i^{\mathscr{C}} \leq P_i \leq P_i^{\max}y_i^{\mathscr{C}} \quad \forall i \in \mathscr{G}, \tag{3.58f}$$

$$(\mu^{\mathscr{C}}) \quad \sum_{j\in\mathscr{D}} D_j \geq Y^{\min}\left(\sum_{j\in\mathscr{D}} D_j^{\text{nom}}\right), \tag{3.58g}$$

$$(\Delta_j^{\mathscr{C}}) \quad D_j \leq D_j^{\text{nom}} \quad \forall j \in \mathscr{D}, \tag{3.58h}$$

$$P \geq 0, \quad D \geq 0. \tag{3.58i}$$

In this formulation, we have indicated on the left our naming convention for dual variables for the constraints. Note that constraint (3.58c) only enforces Ohm's law on lines not removed by the attacker, as should be the case. However, constraint (3.58e) follows a different approach—given a line (i,j) removed by the attacker, (3.58e) permits nonzero flow on that line; however, this will cause the auxiliary variable t to take value strictly greater than 1. This is an important element of the formulation; in particular it guarantees that problem (3.58) is *always* feasible, independent of $\mathscr{A}$, so long as $\sum_{i\in\mathscr{C}} P_i^{\max} \geq Y^{\min}\sum_{j\in\mathscr{D}} D_j^{\text{nom}}$. An additional and important reason for relying on (3.58e) will be given later.

We have the following result.

Lemma 3.1.19 (see [42]). *Attack $\mathscr{A}$ defeats the controller if and only if $t^*_{\mathscr{C}}(z^{\mathscr{A}}) > 1$ for all $\mathscr{C} \subseteq \mathscr{G}$.*

Proof sketch. $t^*_{\mathscr{C}}(z^{\mathscr{A}}) \leq 1$ if and only if there is a feasible solution to (3.58) with zero flow on lines $(i,j) \in \mathscr{A}$ and with $|p_{ij}| \leq u_{ij}$ for all (i,j), i.e., if the controller defeats attack $\mathscr{A}$ while only operating the generators in $\mathscr{C}$. ■

Thus, the minimum cardinality attack problem is that of finding an attack $\mathscr{A}$ with $|\mathscr{A}|$ minimum and such that $t^*_{\mathscr{C}}(z^{\mathscr{A}}) > 1$ for every subset $\mathscr{C}$ of the generators or, to put it differently, so that for every subset $\mathscr{C}$ of the generators the *dual* of $K^*_{\mathscr{C}}(\mathscr{A})$ has value greater than 1. In [42] the following strategy is employed to address the minimum cardinality problem:

(1) Given an attack $\mathscr{A}$, test whether the controller defeats $\mathscr{A}$ by searching for a subset $\mathscr{C} \subseteq \mathscr{G}$ such that $t^*_{\mathscr{C}}(z^{\mathscr{A}}) \leq 1$. This task can be performed by solving a binary mixed-integer program: this is the problem obtained from $K^*_{\mathscr{C}}(\mathscr{A})$ by replacing each quantity $y_i^{\mathscr{C}}$ with a binary variable y_i and setting $\mathscr{C}$ to the support of the resulting (optimal) vector y.

(2) Assuming $t^*_{\mathscr{C}}(z^{\mathscr{A}}) \leq 1$, use the dual of $K^*_{\mathscr{C}}(\mathscr{A})$ to find an inequality that cuts off $z^{\mathscr{A}}$.

At this point it is not clear how to perform step (2); however, putting this point aside for the moment, the combination of (1) and (2) yields an algorithm for the minimum cardinality attack problem, which, through its use of duality, can be seen as an example of Benders decomposition. Returning to (2), note that the attack $\mathscr{A}$ is reflected in formulation (3.58) through constraints (3.58c)–(3.58e). Consider, first, (3.58e), which gives rise to two nonnegative dual variables for each $(i,j) \in \mathscr{A}$. We can extend these to two dual variables for every (i,j), with the proviso that those dual variables arising from $(i,j) \notin \mathscr{A}$ are zero. In other words, we can write

$$\omega_{ij}^{\mathscr{C}+} + \omega_{ij}^{\mathscr{C}-} \leq z_{ij}^{\mathscr{A}}, \quad 0 \leq \omega_{ij}^{\mathscr{C}+}, 0 \leq \omega_{ij}^{\mathscr{C}-}. \tag{3.59}$$

The first inequality in (3.59) relies on the key observation that variable t only appears in constraints (3.58d) and (3.58e)—in both cases with positive coefficients, and with a coefficient of 1 in the objective (3.58a); thus $\omega_{ij}^{\mathscr{C}+} + \omega_{ij}^{\mathscr{C}-} \leq 1$. Likewise, (3.58d) gives rise to two dual variables for each line not in $\mathscr{A}$, but we can extend this set by writing

$$p_{ij}^{\mathscr{C}} + q_{ij}^{\mathscr{C}} \leq \frac{1}{u_{ij}}(1 - z_{ij}^{\mathscr{A}}), \quad 0 \leq p_{ij}^{\mathscr{C}}, 0 \leq q_{ij}^{\mathscr{C}},$$

and from (3.58c) we obtain

$$\beta_{ij}^{\mathscr{C}} \leq M_{ij}(1-z_{ij}^{\mathscr{A}}), \tag{3.60}$$

where $M_{ij} > 0$ is to be appropriately chosen (we will see below that this is *not* a big M approach). It follows that the dual to $K_{\mathscr{C}}^{*}(\mathscr{A})$ can be stated as

$$\max\left\{ w_{\mathscr{C}}^{T}\psi^{\mathscr{C}} : B\psi^{\mathscr{C}} \leq b + E\left(1-z^{\mathscr{A}}\right)\right\}, \tag{3.61}$$

where we use $\psi^{\mathscr{C}}$ to denote the vector of all dual variables, and w_C, B, b, and E are appropriately chosen (this includes modeling of nonnegativity requirements). Moreover, we note that in (3.61) the attack $\mathscr{A}$ is only represented through the vector $z^{\mathscr{A}}$—thus, in fact, for an arbitrary binary vector z if we write

$$D_{\mathscr{C}}(z) \doteq \max\left\{ w_{\mathscr{C}}^{T}\psi^{\mathscr{C}} : B\psi^{\mathscr{C}} \leq b + E(1-z)\right\},$$

we have that the minimum cardinality attack problem can be written as

$$\min \sum_{(i,j)} z_{ij} \tag{3.62a}$$

$$\text{s.t.} D_{\mathscr{C}}(z) > 1 \quad \forall \mathscr{C} \subseteq \mathscr{G}, \tag{3.62b}$$

$$z_{ij} = 0 \text{ or } 1 \ \forall\, (i,j). \tag{3.62c}$$

Note the strict inequality in (3.62b). Of course this is due to our definition of a successful attack; we could equally well have defined the problem setting so that the inequality in (3.62b) is not strict. These points are also related to our discussion of ambiguity at the start of Section 3.1.

Before proceeding with a solution methodology for problem (3.62), we return to the issue of why constraint (3.58e) was used in formulation $K_{\mathscr{C}}^{*}(\mathscr{A})$—surely (?) one could have used the seemingly simpler constraint

$$p_{ij} = 0 \quad \forall (i,j) \in \mathscr{A}. \tag{3.63}$$

Above we already provided one reason for relying on (3.58e) instead of (3.63): using (3.58e) guarantees feasibility of $K_{\mathscr{C}}^{*}(\mathscr{A})$. Another reason is that if we relied on (3.63), then the dual formulation would include, in the objective, a bilinear term of the form

$$\omega_{ij}(1-z_{ij}),$$

where ω_{ij} is the dual variable for (3.63). To linearize this expression by relying, e.g., on the McCormick relaxation we would need an upper bound on ω_{ij}. Computing such an upper bound appears difficult without relying on big M constructions, in particular when removing the set of arcs $\mathscr{A}$ leaves i and j in different components. In contrast, as discussed above, both $\omega_{ij}^{\mathscr{C}+}$ and $\omega_{ij}^{\mathscr{C}-}$ are upper bounded by 1.

Now we return to the minimum cardinality attack problem. We can use the dual representation of (3.61) to more conveniently characterize feasibility for (3.62b) of a binary vector z; namely a given binary z is feasible if and only if

$$\pi^{T}(b+E(1-z)) > 1 \tag{3.64}$$

for all $\pi \geq 0$ with $B^{T}\pi \geq w_{\mathscr{C}}$. This representation of constraint (3.62b) can now be used to develop a Benders decomposition mechanism (Algorithm 3.1.20) for solving the minimum cardinality attack problem, with the proviso that we replace the strict inequality in (3.64) with a relaxed version with right-hand side $\geq 1+\epsilon$ where $\epsilon > 0$ is a tolerance [42].

Algorithm 3.1.20. Algorithm for minimum cardinality attack problem

Initialization. The "master MIP" is initialized as $\min\{\sum_{ij} z_{ij} : z_{ij} = 0 \text{ or } 1\, \forall (i,j)\}$.

Iterate:

Step 1. Attacker: Solve master MIP, obtaining attack $\mathscr{A}$.

Step 2. Controller: Solve the controller's problem (3.58) to search for a set $\mathscr{C} \subseteq \mathscr{G}$ with $t^*_{\mathscr{C}}(z^{\mathscr{A}}) \leq 1$.

(2.a) If no such set $\mathscr{C}$ exists, **EXIT**: $\mathscr{A}$ is a minimum cardinality successful attack.

(2.b) Otherwise, suppose such a set $\mathscr{C}$ is found. Then do the following:
(2.b.1) Let $\pi_{\mathscr{A},\mathscr{C}}$ be the optimal dual solution to $D_{\mathscr{C}}(z^{\mathscr{A}})$.
(2.b.2) Add to the master MIP the corresponding cut (3.64), i.e.,
$\pi^T_{\mathscr{A},\mathscr{C}}(b + E(1-z)) \geq 1 + \epsilon$.
(2.b.3) Go to Step 1.

One should interpret Algorithm 3.1.20 as a generic procedure that can be modified and sharpened in multiple ways. In Step (2.b.2), instead of adding the cut as stated, we could add an entire dual subsystem

$$w^T_{\mathscr{C}} \psi^{\mathscr{C}} \geq 1 + \epsilon, \tag{3.65}$$

$$B\psi^{\mathscr{C}} \leq b + C(1-z), \tag{3.66}$$

where the $w_{\mathscr{C}}$ are new variables added to the formulation. This approach, which certainly cuts off the vector $z^{\mathscr{A}}$, subsumes all possible cuts (3.64) that arise from the set $\mathscr{C}$. The approach does come at a cost, of course, in that applying it too frequently will result in a very large formulation.

Another important point concerns the choice of ϵ. Because of the nonmonotonic behavior of the underlying model, the cut in (2.b.2) could have a significant impact on the value of the attack problem. Nevertheless, considering the nature of the cut (3.64) in (2.b.2) we can stipulate a (data-dependent!) choice for ϵ that is truly small, i.e., small compared to the magnitude of the dual variables $\pi_{\mathscr{A},\mathscr{C}}$, the vector b, and the matrix E. With ϵ thus chosen, the only vectors z excluded by the cut in (2.b.2) are indeed ambiguous, as stated before.

Folklore regarding Benders-like algorithms suggests that a verbatim implementation as given above may produce a large number of initial iterations where very little progress is made, in the sense that the master mixed-integer program remains a very weak relaxation. The (folklore) way to circumvent this (frequent!) difficulty is to precede the formal application of Step 1 with an application of many heuristics that attempt to compute "good" attack vectors $z^{\mathscr{A}}$. Each such heuristic attack is then subjected to Step 2. The outcome of this procedure is to initialize the formal algorithm with a family of strong cuts (3.64)—the cuts are good in the sense that they are triggered by intelligent heuristics which should produce near-optimal and near-feasible attacks (hence, the cuts are "deep"). From a practical algorithm engineering perspective, this is a very good idea since good, fast heuristics should always be part of a solution approach in any case.

Table 3.2. *Algorithm* 3.1.20 *on* 118-*bus test case.*

118 buses, 186 lines, 17 generators Entries show (iteration count), CPU seconds, attack status (**F** = no successful attack exists, **S** = attack success)			
	Attack cardinality (k)		
$Y^{\min}$	**2**	**3**	**4**
0.92	(4), 18, **S**		
0.90	(5), 180, **F**	(6), 193, **S**	
0.88	(4), 318, **F**	(6), 595, **S**	
0.84	(2), 23, **F**	(6), 528, **F**	(148), 6562, **S**
0.80	(2), 18, **F**	(5), 394, **F**	(7), 7755, **F**
0.75	(2), 14, **F**	(4), 267, **F**	(7), 6516, **F**

This approach was implemented in [42], where additionally it is shown that the constant M_{ij} in (3.60) can be chosen as

$$M_{ij} = x_{ij}^{-1/2} \max_{(k,m)} \left\{ \frac{1}{\sqrt{x_{km}}\, u_{km}} \right\}. \tag{3.67}$$

In particular, M_{ij} effectively *does not grow* as a function of network size, which justifies our characterization of the method as not a big M procedure.

We can provide here an outline of some of the experimental results in [42].[17] Table 3.2 displays results using the 118-bus test case mentioned before; the focus of these experiments is to evince algorithm performance as a function of the minimum yield $Y^{\min}$. Each entry in the table shows the iteration count for Algorithm 3.1.20 as well as the CPU time. Additionally, for each chosen value of $Y^{\min}$, we show when the algorithm *proved* a new lower bound on the minimum cardinality of a successful attack. This particular goal is a significant optimization challenge which only a strict algorithm can be expected to yield (our use of tolerances notwithstanding).

Note the rapid increase in computational time when K (the attack size) is increased from 3 to 4. However, it is worth noting that a pure enumerative approach will face significant challenges. In particular, consider the $Y^{\min} = 0.75$, $K = 4$ cell of the table, and note that in the case of this particular cell, the algorithm proved that no successful attack exists. In order for an enumerative algorithm to attain the same result, it would have to examine all combinations of 4 items out of 186, which number 48277230. And, for each such combination, all possible choices of which generators to operate would have to be enumerated. Of course this could be done by formulating an appropriate mixed-integer program, but continuing on an enumerative vein, we would examine $2^{17} - 1 = 65535$ combinations. In total, therefore, in order to examine this case, an enumerative algorithm would need to consider $3163848268050 \approx 3 \times 10^{12}$ separate cases, each of which requires the solution of a linear program. Experiments show that each such linear program, roughly, requires on the order of 10^{-3} seconds. Hence, conservatively, the result in Table 3.2 is one hundred thousand times faster than what could be achieved by a pure enumerative approach!

Table 3.3 (also from [42]) reports on the same type of experiment as Table 3.2, but on a network with more lines, and furthermore using different sets of generator buses.

In the case of Table 3.3, the $Y^{\min} = 0.75$, $K = 4$ case, using similar estimations as above we conclude that Algorithm 3.1.20 is roughly thirty thousand times faster than

[17]Performed using Cplex v. 10.01 on a 3.4 GHz Xeon workstation.

Table 3.3. *Algorithm* 3.1.20 *on larger network.*

98 buses, 204 lines Entries show (iteration count), CPU seconds, attack status (**F** = cardinality too small, **S** = attack success)			
10 generators			
	Attack cardinality		
$Y^{\min}$	**2**	**3**	**4**
0.89	(2) 177, **F**	(30) 555, **S**	
0.86	(2), 195, **F**	(12), 5150, **F**	(14), 5184, **S**
0.84	(2), 152, **F**	(11), 7204, **F**	(35), 223224, **F**
0.82	(2), 214, **F**	(9), 11458, **F**	(16), 225335, **F**
0.75	(2), 255, **F**	(9), 5921, **F**	(17), 151658, **F**
0.60		(1), 4226, **F**	N/R
12 generators			
	Attack cardinality		
$Y^{\min}$	**2**	**3**	**4**
0.92	(2), 318, **F**	(11), 7470, **F**	(14), 11819, **S**
0.90	(2), 161, **F**	(11), 14220, **F**	(18), 16926, **S**
0.88	(2), 165, **F**	(10), 11178, **F**	(15), 284318, **S**
0.84	(2), 150, **F**	(9), 4564, **F**	(16), 162645, **F**
0.75	(2), 130, **F**	(9), 7095, **F**	(15), 93049, **F**
15 generators			
	Attack cardinality		
$Y^{\min}$	**2**	**3**	**4**
0.94	(2), 223, **F**	(11), 654, **S**	
0.92	(2), 201, **F**	(11), 10895, **F**	(18), 11223, **S**
0.90	(2), 193, **F**	(11), 6598, **F**	(16), 206350, **S**
0.88	(2), 256, **F**	(9), 15445, **F**	(18), 984743, **F**
0.84	(2), 133, **F**	(9), 5565, **F**	(15), 232525, **F**
0.75	(2), 213, **F**	(9), 7550, **F**	(11), 100583, **F**

a completely enumerative method. Another interesting feature revealed by Tables 3.2 and 3.3 (and replicated in all other experiments in [42]) concerns the running time of the algorithm, for each fixed value of K, and as a function of $Y^{\min}$. As $Y^{\min}$ decreases, running time initially increases. However, running time will then decrease as $Y^{\min}$ drops below the critical value for which no successful attack exists. This is intuitively justified (informally, the problem becomes easier once $Y^{\min}$ crosses the critical threshold), but the point here is that Algorithm 3.1.20 is able to capitalize on this feature by discovering appropriate cuts. Moreover, for a fixed $Y^{\min}$, as K increases the computational time increases until the algorithm reaches the smallest K for which a successful attack is found. This behavior can be explained by the fact that as K increases the algorithm must work harder to prove that no successful attack exists, until of course K does become large enough that an "S" outcome is attained. Experimental evidence (not provided in the tables) shows that for K very large the problem becomes much easier (because many successful attacks exist, and all we have to do is find one to terminate the algorithm).

Table 3.4. *Minimum cardinality algorithm on one-configuration problem.*

$Y^{\min}$	Min. attack size	Time (sec.)
0.95	2	2
0.90	3	20
0.85	4	246
0.80	5	463
0.75	6	2158
0.70	6	1757
0.65	7	3736
0.60	7	1345
0.55	8	2343
0.50	8	1328

On the negative side, it is clear that the analysis of a transmission system with tens of thousands of lines, buses, and generators may be beyond the reach of Algorithm 3.1.20, or at least beyond the (quite careful) implementation in [42]. In Sections 3.2 and 3.3 we will consider two well-founded methodologies that can shed light on the N-K problem.

3.1.2 ▪ The one-configuration case

Consider the special case of the minimum cardinality attack problem studied in the previous section where the controller is constrained by the use of a single configuration of generators, i.e., $\mathscr{C} = \mathscr{G}$. This is a relevant case to study in the sense that a grid operator may be reluctant or unable to trip (turn off) or start up a generator on very short notice. Using (3.62) (and notation from that formulation) we can state the problem as

$$\min \sum_{(i,j)} z_{ij}$$
$$\text{s.t. } B\psi \leq b + E(1-z), \quad w^T \psi \geq 1+\epsilon,$$
$$z_{ij} = 0 \text{ or } 1 \ \forall (i,j),$$

where $\epsilon > 0$ is a tolerance parameter. Since this is a standard binary mixed-integer program, we can forgo Algorithm 3.1.20 and instead rely on a standard mixed-integer program solver. This is the approach used in [42], from which Table 3.4 is drawn, using the same network as that in Table 3.3. Note that, broadly, CPU time first increases as a function of $Y^{\min}$ but then seems to flatten out (and, in fact, decrease). This can be explained by the fact that, in order to make $Y^{\min}$ "small," a larger attack cardinality is needed, but in that case many successful attacks exist. And when $Y^{\min}$ is "large," then a small attack cardinality is needed, and, effectively, an enumerative approach is likely to find some successful attack. The intermediate cases are thus the most difficult.

Aside from such issues, we highlight the precise metering for attack impact (i.e., "risk") that Table 3.4 provides. This is a capability that only a rigorous optimization framework can provide—at a computational cost, certainly. As a final remark here, note that formulation (3.68) computes a minimum cardinality set of lines whose removal will either disconnect the network or cause a line flow to exceed the line's limit. As an alternative, we can formulate a problem where the number of lines that can be removed is upper bounded by a given quantity K and we seek such a set of lines of maximum impact. Using the above notation, this can be formulated as

$$
\begin{aligned}
t_K^* \doteq{} & \max t \\
\text{s.t.} \quad & B\psi \le b + E(1-z), && (3.69a)\\
& w^T \psi - t \ge 0, && (3.69b)\\
& \sum_{(i,j)} z_{ij} \le K, && (3.69c)\\
& z_{ij} = 0 \text{ or } 1 \ \forall\, (i,j). && (3.69d)
\end{aligned}
$$

Denoting the network by $\mathcal{N}$, in (3.69) we have $t_K^* > 1$ if and only if there is a set with at most K lines whose removal from $\mathcal{N}$ either makes the DC system infeasible or causes some line flow to exceed its limit. *However*, suppose $t_K^* > 1$ and let L^* be the set of lines (i,j) with $z_{ij} = 1$, i.e., the optimal set of lines to remove. If $\mathcal{N} - L^*$ is connected, *it does not follow* that the maximum line load in $\mathcal{N} - L^*$ equals t_K^*. This is due to constraints (3.58e) in the primal version of formulation (3.69), which allow nonzero flow on the lines in L^*. In other words, t_K^* will be a (typically strict) lower bound on the maximum line load in $\mathcal{N} - L^*$. We will return to this point in Section 3.3.

3.2 ▪ A cut problem related to N-K analysis

In light of the inherent difficulty associated with the N-K problem, an intelligent strategy would be to abstract combinatorial features associated with an optimal attack and formulate an optimization problem that seeks to evince those features. This is the approach followed by Pinar, Meza, Donde, and Lesieutre in [187]. Also see [90].

We begin with the *network inhibition problem*, studied in [186], and presented in somewhat generalized form here in order to bring it closer to power flow problems. Consider a directed graph G with vertex-set V and arc-set A. Write $|V| = n$, $|A| = m$. Additionally, we have the following:

- For each arc $(i,j) \in A$ we have a *capacity* $u_{ij} \ge 0$ and a *cost* $c_{ij} \ge 0$.
- For each vertex i we have a *supply* b_i; if $b_i > 0$ we say i is a source and if $b_i < 0$ we say i is a destination. Further, $\sum_i b_i = 0$. Finally, we write $S \doteq \{i : b_i > 0\}$ to denote the set of *supply* nodes, and $D \doteq \{i : b_i < 0\}$ to denote the *demand* nodes.

Let N denote the $(n \times m)$ node-arc incidence matrix of G. Then [3] a *feasible flow in* G is a vector $f \in \mathbb{R}^A$ such that there exists $\hat{b} \in \mathbb{R}^V$ with the properties that

$$
\begin{aligned}
& Nf = \hat{b}, && (3.70a)\\
& |f_{ij}| \le u_{ij} \quad \forall (i,j) \in A, && (3.70b)\\
& 0 \le \hat{b}_i \le b_i \quad \forall i \in S, && (3.70c)\\
& b_i \le \hat{b}_i \le 0 \quad \forall i \in D, && (3.70d)\\
& \hat{b}_i = 0 \quad \text{otherwise}. && (3.70e)
\end{aligned}
$$

Informally, a feasible flow in G routes a portion of each supply to the demand nodes. We denote the *throughput* of f by

$$\Theta_G(f) \doteq \sum_{i \in S} \hat{b}_i.$$

As is well known the sum of all rows of N is the zero vector. Hence we equivalently have $\Theta_G(f) = -\sum_{i\in D} \hat{b}_i$. Since we assume $\sum_i b_i = 0$ we have that if f is feasible then

$$\Theta_G(f) \leq \sum_{i\in S} b_i \leq -\sum_{i\in D} b_i.$$

We now have the following definition.

Definition 3.2.1. *Let* $0 \leq T^{\min} \leq \sum_{i\in S} b_i$ *be given. The* **network inhibition problem** *concerns the task of computing a subset of arcs* A^*, *of minimum cost, such that*

$$\Theta_{G-A^*}(f) \leq T^{\min} \quad \textit{for any } f \textit{ feasible in } G - A^*.$$

This problem is certainly of interest in the context of network survivability, broadly interpreted. It was introduced by Phillips in [186] (also see [229]), where it is proved that the problem is NP-hard but special cases can be solved in polynomial time; furthermore, in the case that G is planar there is a fully polynomial-time approximation scheme (that is to say given $0 < \epsilon < 1$ we can compute a solution within relative error ϵ of the optimum, in time polynomial in n, m, and ϵ^{-1}).

The classical max-flow problem provides a useful perspective on the network inhibition problem, which we will justify in Section 3.2.2. Let vector $f^* = \text{argmax}\{\Theta_{G-A^*}(f) \leq T^{\min}$ for any f feasible in $G - A^*\}$. Then there is a partition of the nodes into two classes S and T, with the property that each arc $(i,j) \notin A^*$, where one of i,j is in S and the other is in T, satisfies $f^*_{ij} = u_{ij}$.

In the following sections we will discuss motivation given in [187] for studying the network inhibition problem, from the perspective of the minimum effort attack problem introduced in Section 3.1—in fact the model for power flows will be the lossless AC model (see Section 2.3.1). Then we will describe mixed-integer programming formulations of the network inhibition problem.

3.2.1 ▪ Motivation for studying the network inhibition problem

Consider the minimum cardinality attack problem under the lossless AC power flow model restricted to active power flows. The motivation for using this model is that, while it does not completely account for the physics of AC power flows, it is clearly more complete than the DC model, in particular by partially accounting for phase angles. This increased accuracy would lend more credence to an N-K assessment.

We assume that the attacker removes lines, whereas the controller may shed load and reduce generator output; furthermore, an attack succeeds if the sum of demands that can be feasibly satisfied is *at most* a given quantity $T^{\min}$. We assume for simplicity that any bus holds a generator (the set of generator buses is indicated by $\mathcal{G}$) or is otherwise a load bus (the set of load buses is indicated by $\mathcal{L}$), possibly with zero load. For $i \in \mathcal{G}$ we indicate by $P_i^0 > 0$ its output, preattack, and if $i \in \mathcal{D}$ we indicate by $D_i^0 > 0$ its load, again preattack. We assume $\sum_{i\in\mathcal{G}} P_i^0 = \sum_{i\in\mathcal{D}} D_i^0$ (flow balance, preattack) and $T^{\min} < \sum_{i\in\mathcal{D}} D_i^0$.

Let us indicate an attack, i.e., a subset of lines that is disabled, by using a binary vector z so that for each line $\{k,m\}$ we set $z_{\{k,m\}} = 1$ if and only if line $\{k,m\}$ is disabled; we further denote the set of disabled lines by $\mathcal{X} = \mathcal{X}(z)$. Thus, the controller problem corresponding to z can be stated as

$$T(z) \doteq \max_{D,P,\theta} \sum_k D_k \tag{3.71a}$$

$$\text{s.t.} \quad \forall \text{ bus } k, \quad \sum_{\{k,m\} \notin \mathscr{Z}} b_{\{k,m\}} \sin(\theta_k - \theta_m) = \begin{cases} P_k, & k \in \mathscr{G}, \\ -D_k, & k \in \mathscr{D}, \end{cases} \tag{3.71b}$$

$$|\theta_k - \theta_m| \leq \theta^{\max}_{\{k,m\}} \quad \forall \text{ line } \{k,m\} \notin \mathscr{Z}, \tag{3.71c}$$

$$0 \leq P_k \leq P^0_k \quad \forall k, \tag{3.71d}$$

$$0 \leq D_k \leq D^0_k \quad \forall k. \tag{3.71e}$$

In this formulation D_k is the load satisfied at load bus k and P_k is the output of generator k (both are variables). The sum in (3.71b) is over all lines incident with bus k which are not disabled by the attack z; thus (3.71b) is a flow balance constraint. Constraint (3.71c) is a maximum phase angle constraint; $\theta^{\max}_{\{k,m\}} \leq \pi/2$ is an input parameter which in [187] is assumed to equal $\pi/2$. Note that in the above formulation we make use of the undirected-edge notation in order to represent the grid; below, with some abuse of notation, we will bypass this convention. Also note that the sum of left-hand sides of (3.71b) equals zero, and thus $\sum_{k\in\mathscr{G}} P_k = \sum_{k\in\mathscr{D}} D_k$, i.e., the system is "lossless."

Using this notation, the minimum cardinality attack problem can be succinctly stated as

$$\min \sum_{\{k,m\}} z_{\{k,m\}} \tag{3.72a}$$

$$\text{s.t.} \quad T(z) \leq T^{\min}, \tag{3.72b}$$

We note, in passing, that due to the trigonometric constraints in problem (3.71), instances of problem (3.72) where the optimal attack z^* is unique and is such that $T(z^*) \approx T^{\min}$ will give rise to the ambiguity issues we have discussed above. In fact, from a more technical standpoint, it is not even clear that verifying *feasibility* of an instance of a binary vector z for constraint (3.72b) is even in the class NP.

Problem (3.72) is clearly a bilevel optimization problem; the primary contribution in [187] is an analysis of the structure of optimal solutions to the controller problem (3.71) with the goal of justifying the use of the network inhibition problem as a suitable proxy for problem (3.72). This task is carried out in two steps.

First, clearly problem (3.71) is a nonconvex, nonlinear optimization problem. However, it is shown in [187] that by suitably adjusting the network data, one can (arbitrarily closely) approximate any instance of problem (3.72) with another which has the useful property that, for any attack z, problem (3.71) satisfies a *constraint qualification property* [47], and, as a result, first-order (KKT) optimality conditions will always hold at the global optimum for (3.71). Moreover, it should be the case that the instance of problem (3.72) using the adjusted network will have the same value as that on the original network.

Putting these issues aside, let us consider what the KKT conditions for (3.71) imply at optimality. Let the vector λ indicate the duals for (3.71b), γ^L_k, γ^U_k the duals for the lower and upper bounds (respectively) in (3.71d), and δ^L_k, δ^U_k the duals for the lower and upper bounds (respectively) in (3.71e). Here λ is unrestricted in sign but $\mu^L, \mu^U, \delta^L, \delta^U$ are all nonnegative. The first-order conditions applied to a load bus k yield

$$1 - \lambda_k - \delta^U_k + \delta^L_k = 0, \quad k \in \mathscr{D}. \tag{3.73}$$

Suppose $\lambda_k < 1$. It follows that $\delta_k^U = 1-\lambda_k > 0$ and $\delta_k^L = 0$, that is to say, no load is shed at bus k. If on the other hand $\lambda_k > 1$ then the lower bound in (3.71e) holds with equality, meaning all load at bus k is shed by the controller.

We can likewise analyze the behavior at a generator bus k; the corresponding condition is

$$\lambda_k - \mu_k^U + \mu_k^L = 0. \tag{3.74}$$

If $\lambda_k > 0$ then $\mu_k^L = 0$ and $\mu_k^U > 0$, indicating that $P_k = P_k^0$ (generation is not reduced at bus k, postattack), whereas if $\lambda_k < 0$ then $\mu_k^U = 0$ and $\mu_k^L > 0$, indicating that generator k has reduced its output to zero, postattack.

In summary, therefore, we can partition the buses into two classes:

(a) Buses k with $\lambda_k < 1$. Demand buses in this class do not shed any load. We indicate this set by $\bar{S}$.

(b) Buses k with $\lambda_k \geq 1$. Generators in this class do not reduce their output. We indicate this set by $\bar{T}$.

Note that since $\sum_{k\in\mathscr{G}} P_k = \sum_{k\in\mathscr{D}} D_k$, we therefore have

$$\sum_{k\in\mathscr{G}\cap\bar{S}} P_k + \sum_{k\in\mathscr{G}\cap\bar{T}} P_k^0 = \sum_{k\in\mathscr{D}\cap\bar{S}} D_k^0 + \sum_{k\in\mathscr{D}\cap\bar{T}} D_k \tag{3.75a}$$

$$= \sum_{k\in\mathscr{D}} D_k^0 \; - \text{ amount of load shed.} \tag{3.75b}$$

Using this equation we can make an important observation about the partition $(\bar{S},\bar{T})$. Since we are assuming that $T^{\min} < \sum_{i\in\mathscr{D}} D_i^0$, some load *must* be shed; by (a) the shedding is done at buses in $\bar{T}$. Since $\sum_{i\in\mathscr{G}} P_i^0 = \sum_{i\in\mathscr{D}} D_i^0$, an equal amount of generation must be reduced; by (b) the reduction is done at buses in $\bar{S}$. This observation is valid modulo the assumption that the KKT conditions hold (given the optimal attack); as discussed above this assumption requires a small perturbation to the grid data which does not affect the value of the minimum cardinality attack problem.

At this point [187] introduces an additional assumption, which may not necessarily hold. Under this additional assumption, every line $\{k,m\} \notin \mathscr{X}$ with $k \in \bar{S}$ and $m \in \bar{T}$ satisfies (again, at the primal-dual optimum) $|\theta_k - \theta_m| = \theta_{\{k,m\}}^{\max}$.[18] Thus, every line $\{k,m\}$ across the cut defined by the partition $(\bar{S},\bar{T})$ is "saturated" in that the absolute value of the flow on that line, $|b_{\{k,m\}} \sin(\theta_k - \theta_m)|$, reaches its maximum possible value, $b_{\{k,m\}} \arcsin(\theta_{\{k,m\}}^{\max})$.

Thus, recalling our remarks toward the end of Section 3.2, we see that the set $\mathscr{X}$ is an optimal solution to the instance of the network inhibition problem with arc capacities $b_{\{k,m\}} \arcsin(\theta_{\{k,m\}}^{\max})$, supplies P_k^0, and demands D_k^0. Thus, indeed, there is some justification for studying the network inhibition problem, though we stress that the additional assumption that fuels this justification may not necessarily hold.

3.2.2 ▪ Solving the network inhibition problem

We now return to the network inhibition problem in the general setting given above. Consider an instance of the problem on a directed graph G. Let us create an auxiliary directed graph, $\bar{G}$, with vertex-set $\bar{V}$ and arc-set $\bar{A}$, defined as follows:

[18]When the values $\theta_{\{k,m\}}^{\max}$ all equal $\pi/2$, it can be shown that at least one line has this property [187].

(a.1) For each arc $(i,j) \in A$, $\bar{A}$ will contain arcs (i,j) and (j,i). Further we will set $u_{ji} = u_{ij}$.

(a.2) $\bar{V} = V \cup \{\tilde{s}, \tilde{d}\}$, where $\tilde{s}$ and $\tilde{d}$ are new vertices. For each $i \in S$, $\bar{A}$ includes the arc $(\tilde{s}, i)$ with capacity $u_{\tilde{s}i} = b_i$, and for each $i \in D$, $\bar{A}$ includes the arc $(i, \tilde{d})$ with capacity $u_{i\tilde{d}} = -b_i$.

Now we have the following result.

Lemma 3.2.2. $\max\{\Theta_G(f) : f$ *feasible in* $G\}$ *is equal to the maximum* $s-t$ *flow in* $\bar{G}$.

Proof sketch. The proof is routine; see [3].

It follows that the optimal f saturates every arc in every minimum capacity cut in $\bar{G}$—this justifies our remarks toward the end of Section 3.2. Based on Lemma 3.2.2, we can also now give a mixed-integer programming formulation for the network inhibition problem:

$$\min_{y,v,w} \sum_{(i,j)\in A} c_{ij} y_{ij} \tag{3.76a}$$

$$\text{s.t.} \sum_{(i,j)\in A} u_{ij} v_{ij} + \sum_{i \in S} b_i (1 - w_i) - \sum_{i \in D} w_i b_i \leq T^{\min}, \tag{3.76b}$$

$$\max\{0, w_i - w_j\} + \max\{0, w_j - w_i\} \leq v_{ij} + y_{ij} \quad \forall (i,j) \in A, \tag{3.76c}$$

$$y_{ij} = 0 \text{ or } 1 \quad \forall (i,j) \in A, \tag{3.76d}$$

$$v_{ij} \geq 0 \quad \forall (i,j) \in A, \tag{3.76e}$$

$$w_i = 0 \text{ or } 1 \quad \forall i \in V. \tag{3.76f}$$

Lemma 3.2.3. *Suppose* (y^*, v^*, w^*) *is an optimal solution to problem* (3.76). *Then* $A^* \doteq \{(i,j) \in A : y^*_{ij} = 1\}$ *is an optimal solution to the network inhibition problem.*

Proof. Define $A^1 = \{(i,j) \in A : |w^*_i - w^*_j| = 1\}$. Since $u_{ij} \geq 0$ for all $(i,j) \in A$, (3.76b) and (3.76c) imply that, without loss of generality,

$$\text{for each } (i,j) \in A: \ v^*_{ij} = \begin{cases} 1 & \text{if } (i,j) \in A^1 - A^*, \\ 0 & \text{otherwise.} \end{cases} \tag{3.77}$$

With this proviso, consider the cut in the graph $\tilde{G} - A^*$ defined by the partition of $\tilde{V}$ into the two sets

$$X^1 \doteq \tilde{s} \cup \{i \in V : w^*_i = 1\},$$
$$X^0 \doteq \{i \in V : w^*_i = 0\} \cup \tilde{d}.$$

By construction of $\tilde{G}$, the capacity of this cut equals

$$\sum_{i \in S \cap X^0} b_i - \sum_{i \in D \cap X^1} b_i + \sum_{(i,j) \in A - A^*} \{u_{ij} : (i \in X^1 \text{ and } j \in X^0) \ \ \textbf{or} \ \ (j \in X^0 \text{ and } i \in X^1)\}$$

$$= \sum_{i \in S} b_i (1 - w^*_i) - \sum_{i \in D} b_i w^*_i + \sum_{(i,j)\in A} u_{ij} v^*_{ij}, \tag{3.78}$$

where the expression for the last term follows from (3.77). By (3.76b), the quantity in (3.78) is at most $T^{\min}$. Now an application of Lemma 3.2.2 to the graph $\tilde{G} - A^*$ yields that A^* is a feasible solution to the network inhibition problem, of equal cost. One similarly shows that a feasible solution to the network inhibition problem yields a feasible solution to problem (3.76), again of equal cost. ■

Formulation (3.76) renders the network inhibition problem as a binary mixed-integer program with $|A| + |V|$ binary variables. However, the combinatorial structure of the formulation places it close to the minimum cut problem and as such we expect that some computational success may be expected (recalling, however, the NP-hardness of the problem, some caution is called for). In [187] it is stated that realistic instances on networks with tens of thousands of nodes and arcs were solved in "tens of seconds"; however, further details are not provided.

Remark 3.2.4. *The network inhibition problem is a bilevel program. As such, formulation* (3.76), *which relies on the (linear programming) duality arising between the max-flow and the min-cut problem, brings to mind formulation* (3.16), *which we developed for general bilevel programs. That formulation was nonlinear. In the case of* (3.76), *the nonlinearities are resolved by resorting to binary variables (the w_i).*

3.2.3 ▪ The network inhibition problem for DC power flows

The formulation given above for the network inhibition problem applies to generic flows. Here we discuss how to modify the above approach to handle DC power flows. We begin by modifying system (3.70) as follows:

$$Nf = \hat{b}, \tag{3.79a}$$

$$\text{for each } (i,j) \in A: \left\{ \begin{array}{l} |p_{ij}| \leq u_{ij}, \\ p_{ij} = y_{ij}(\theta_i - \theta_j), \end{array} \right. \tag{3.79b}$$

$$0 \leq \hat{b}_i \leq b_i \quad \forall\, i \in \mathcal{G}, \quad b_i \leq \hat{b}_i \leq 0 \quad \forall\, i \in \mathcal{D}, \quad \hat{b}_i = 0 \quad \text{otherwise}. \tag{3.79c}$$

Here, $\mathcal{G}$ and $\mathcal{D}$ are, respectively, the set of generator and load buses; $\hat{b}_i$ is the net output of bus i; y_{ij} is the susceptance of line (i,j); and $p_y ij$ and θ_i are, respectively, the flow on line (i,j) and the phase angle at bus i. We stress that in this system, the vectors f, θ, and $\hat{b}$ indicate variables. One wonders if there is an extension of formulation (3.76) that applies in this case. However, correctness of (3.76) relies on the max-flow min-cut theorem as shown in Lemma 3.2.3. The max-flow min-cut theorem is in fact a case where linear programming duality takes a combinatorial flavor, and we are not aware of any related result that applies when there are side constraints on the flows, such as Ohm's law.

However, there *is* a compact formulation of the network inhibition problem in this case, at least in an approximate sense. In fact, this is none other than the one-configuration problem discussed in Section 3.1.2, with a simple modification. The reason for our qualification is that in the problem discussed in that section and in Section 3.1.1, an attack was considered successful if the fraction of demand that could be feasibly satisfied (i.e., the throughput), postattack, was strictly less than $T^{\min}$ (appropriately defined). In the context of the network inhibition problem (and in [187]), instead we require that the throughput be at most $T^{\min}$. Modulo this difference (which touches on our discussion of ambiguous cases above), we obtain a formulation for the network inhibition problem, for DC power flows, by using formulation (3.68). We further theorize that there is a formulation that applies without any qualification; this would follow an approach similar to the one described following Lemma 3.1.19 and culminating in formulation (3.62), i.e., an approach that "combinatorializes" the dual of a formulation describing the controller problem.

3.3 ▪ A continuous model related to the N-K problem

In the above sections we have primarily considered network interdiction models that are primarily of a discrete nature; that is to say the adversary operates by, e.g., removing lines from a network. The classical N-K problem is quintessentially of such a nature, and when $K = 1$ this focus is easily justified, because in a large grid the outage of a single component can be a fairly common event.

However, in the case of larger values of K, justification for studying the standard N-K problem may be more difficult and has received some challenges due to its potential conservatism. And as we have seen above, the computational obstacles are significant, and overcoming these, in the context of large grids, would require additional fundamental developments. Finally, and possibly most important, N-K security for $K \geq 3$ addresses very rare events. The events are so rare that perhaps "exotic" may be a better term, and a valid likelihood estimate is difficult to come by. From this perspective one should therefore also focus on other equally exotic events that are not easily classified under the N-K metric. As a simple example, some contingencies with $K = 10$ outages may be just as likely as some contingencies with $K = 3$, because "likely" becomes a slippery term when dealing with extremely unlikely events.

Other, more complex, examples of the unsuitability of the N-K metric come to mind. Consider the causes leading to the 2003 large-scale cascading power grid failure in the Northeast U.S. and Canada, discussed in Section 4.2.1. Even though this event was initiated by the outage of a relatively small number of lines (two, three, or four, depending on arguments related to timing and causality), it was preceded by a longer period with (roughly defined) "stress" for the grid, caused by warm summer weather, which adds to the temperature stress of power lines, and some voltage decay, which makes power transmission more difficult. One could argue that the cascade was caused by the combination of both types of agents (i.e., stress and line outages). Very likely, the stress, rather than being generic, probably had focused impact on specific parts of the grid. In summary, here we have an example of a complex, compound event that is difficult to categorize using N-K criteria.

Below we describe an attack model where the adversary interdicts a grid in a continuous manner. This will be given in Section 3.3.3. Note that Definition 3.1.3 given above allows for such models; however the particular problems and models we have studied so far have been of a discrete nature. The switch to a continuous model has two benefits. First, we obtain a possibly better mechanism for addressing the exotic events discussed above. Second, we (potentially) obtain a family of optimization problems that are more easily handled than the discrete models. In particular, the reliance on continuous models makes it possible to leverage powerful, modern techniques derived from convex optimization and reviewed in the following two sections.

3.3.1 ▪ Interior point methods applied to general continuous nonlinear programs

Consider a general (equality-constrained) continuous optimization problem

$$\min_{x \in \mathbb{R}^n} \ f(x) \tag{3.80a}$$

$$\text{s.t. } c(x) = 0, \tag{3.80b}$$

$$x \geq 0, \tag{3.80c}$$

where $c(x)$ is vector valued, and no assumptions are made regarding convexity of f or c. A great deal of research has been aimed at producing algorithms that compute *local optima* for problem (3.80), with strong theoretical underpinnings and leveraging advances originally developed in the context of interior point methods (primarily for linear programs). The high-level summary we provide here concerns IPOPT [225]. Also see [223], [222].

An approximate solution for problem (3.80) can be obtained by solving a sequence of problems of the form

$$\min_{x \in \mathbb{R}^n} f(x) - \mu^{(k)} \sum_{i=1}^{n} \ln x_i \tag{3.81a}$$

$$\text{s.t.} \quad c(x) = 0 \tag{3.81b}$$

for $k = 1, 2, \ldots$. In this formulation, $\mu^{(k)} > 0$. Further, it is implicitly assumed that $x_i > 0$ for all i, i.e., x is an "interior point." This requirement is fulfilled by developing an (iterative) interior point for problem (3.81) which always maintains the interior point condition. As $\mu^{(k)} \to 0$, the objective in (3.81) converges, in a precise sense, to that in (3.80). The numerical attributes of this convergence are well known in the context of general logarithmic-barrier interior point methods. Roughly speaking, the overall algorithm (approximately) solves a sequence of problems (3.81) to attain an approximate solution to the first-order (KKT) optimality conditions for (3.80).

We note that problem (3.81) will in general be nonconvex, except in special cases, for example if $f(x)$ is convex and $c(x)$ is linear. In that setting, the algorithms described here will converge to a (global) optimum for problem (3.80). For a given $\mu^{(k)}$ the algorithm described in [225] converges to an approximate optimum for (3.81), more precisely, a solution that approximately satisfies primal feasibility and the KKT conditions. The algorithm iterates by performing Newton-like iterations aimed at attaining the KKT conditions and then performing a line search that updates the primal solution and the Lagrange multipliers (this line search is particularly needed in the nonconvex case). In [225], a line-search filter method is given for this purpose; this method seeks to balance improvement in terms of $f(x)$ and also violation of $c(x) = 0$ (while maintaining the interior point condition). A central ingredient of the filter method is a technique designed to prevent looping, that is to say, a behavior of the algorithm where the iterates return to the same vector x, an essential ingredient in finite convergence. See [225] for details.

We note that, of course, problem (3.80) is strongly NP-hard. From a rigorous standpoint, we cannot therefore demand fast convergence to even an approximate optimizer. Accordingly, [225] does not make any such claims and the class of algorithms we have outlined could pessimistically be viewed as educated heuristics. However, this is indeed a pessimistic outlook; even if the methodology cannot provide any hard guarantees, its experimental behavior on realistic problem settings in science and engineering has been quite good. See, e.g., Section 2.4.

3.3.2 ▪ Trust-region and derivative-free methods

Trust-region methods are a general class of algorithms for problem (3.80) which operate by solving a sequence of problems, each of which minimizes a quadratic approximation, or "model," for $f(x)$, over a Euclidean ball in $\mathbb{R}^n$ (the "trust region") in which the model is approximately valid. After each optimization, the ball, and the model, are updated. The basic underlying algorithmic step (termed the trust-region subproblem) has been outlined in Section 2.2.1; however, numerous technical details must be addressed beyond this step, in particular the construction of the model and of the trust region. Under appropriate conditions (e.g., convexity), the sequence can be shown to converge to a global optimum for problem (3.80), and convergence to a local optimum will take place under more general conditions. See [73]. The algorithms are intuitively appealing and benefit from strong underlying theoretical foundations. Software implementations have proved robust and have attained success in several domains in science and engineering. An open question at this point is whether trust-region methods are successful on AC OPF problems, in particular, difficult instances where nonconvexities are especially notorious.

The implementation of a trust-region algorithm requires derivative information (of $f(x)$ and $c(x)$) to construct the models. In many applications in science and engineering, however, such information may be unavailable. For example it may be the case that $f(x)$ can only be computed by running a simulation. One could still approximate derivatives (by using partial differences), but the resulting estimations may prove inaccurate, noisy, and in general computationally expensive.

Derivative-free methods [72] attempt to circumvent these difficulties by maintaining models that only require functional value estimation at selected sample points and then construct approximations to, e.g., $f(x)$ by means of interpolation, for example. A thorough description of derivative-free methods is beyond the scope of this work; however, it is worth pointing out that modern versions of this idea are geared toward achieving good practical performance while at the same time enjoying strong theoretical attributes. Under appropriate conditions, the methods converge to a stationary point. See [72] for complete background and details.

Many software implementations of the ideas in the above paragraphs, and related methods, are available, including commercial versions, and have attained wide use, in particular (see Section 2.4) MINOS [211], SNOPT [212], IPOPT [225], CONOPT [74], and KNITRO [174], and also LOQO [222]. It is useful to understand how these implementations would be used to address a problem of the form (3.81); one such use is described below in Section 3.3.3. For simplicity we will assume a linearly constrained case, i.e., the constraints are of the form $Ax \geq b$.

To put it in a simple way (and to simplify the text below), solving problem (3.81) using the methodology we have described amounts to a collaboration between two algorithms: an *agent* algorithm and a generic *engine* algorithm, with different degrees of computational responsibilities. LOQO, IPOPT, and the other implementations described above are examples of engines, whereas the agent algorithm must be developed with a given application in mind. Broadly speaking, the combined algorithm will sample points $\hat{x}$ (chosen by the engine algorithm) and at such points compute or estimate

(1) the functional value $f(\hat{x})$;

(2) the gradient $\nabla f(\hat{x})$; and, possibly,

(3) the Hessian $\nabla^2 f(\hat{x})$.

In one version of the algorithm the engine requests from the agent the values (1), (2), (3) at each sampled point. In a second version, the agent instead provides (at the start of the algorithm) a symbolic representation of the function $f(x)$ as well as the constraints, and the engine computes or estimates the partial derivatives in (2) and (3) (e.g., using finite differences or symbolic computations) at the sampled points. In this alternative version, the vast majority of the computational overhead lies with the engine algorithm.

Which of the two approaches is preferable? Certainly, the second approach will save development time and may even prove more efficient than the first. However, it is also possible that one could exploit implicit information regarding the function $f(x)$ to obtain faster and possibly more compact representations of items (1)–(3) above, in particular leading to faster computation than would be possible through symbolic or finite-difference estimations. We will describe such a case next.

3.3.3 ▪ A continuous impedance interdiction model

Here we discuss the model provided in [42], where the adversary interdicts a grid by increasing the impedance of each line, subject to constraints. More precisely, we have the following:

(a) We rely on the DC approximation model; for each generator bus g its output $P_g > 0$ is fixed and for each load i its demand $D_i = -P_i > 0$ is likewise fixed. We denote by $\mathcal{A}$ the set of lines and by u_{km} the limit for line (k,m).

(b) The adversary changes the reactance x_{km} of each line (k,m) subject to constraints

$$x^L_{km} \leq x_{km} \leq x^U_{km}. \tag{3.82}$$

Often, we will have x^L_{km} set to the nominal value of the reactance, and thus the range in (3.82) simply indicates a possible increase. In terms of line susceptances this is the same as stating

$$y^L_{km} \leq y_{km} \leq y^U_{km} \quad \text{for each line } (k,m), \tag{3.83}$$

with appropriately defined y^U_{km}, y^L_{km} parameters. We will impose additional constraints on the allowable y_{km} (specified later). We denote the set of allowable *susceptance* vectors by Γ.

(c) The goal of the adversary is to solve

$$\max_{y \in \Gamma} \max_{(k,m) \in \mathcal{A}} \left\{ \frac{|p^y_{km}|}{u_{km}} \right\}, \tag{3.84}$$

where the p^y_{ij} are the power flows corresponding to the given loads and generation under susceptances specified by y. In other words, the adversary seeks to impede lines to maximize the maximum line overload.

A relevant example of a constraint to impose on the susceptances is

$$\sum_{(k,m)} y^{-1}_{k,m} \leq B, \tag{3.85}$$

where $B > 0$ is a given "budget"; this constrains the sum of all reactances. Constraint (3.85) together with (3.83) would constitute an interesting set Γ over which to solve problem (3.84). A related (and more general) class of models would reply on nominal values $\bar{y}_{k,m}$ for the susceptances and impose a constraint of the form

$$\|y^{-1} - \bar{y}^{-1}\|_p \leq B, \tag{3.86}$$

where the inverse operator is interpreted componentwise and $p \geq 0$. Each of these formulations gives rise to a continuous interdiction model. Below we provided generic motivation for studying this form of attack that allows the adversary to distribute an attack over many lines. Of course, a model like (3.85) or (3.86) also allows the adversary to concentrate the attack on a focused subset of lines (small, if B is small in some sense). In this regard, [42] presents the following result.

Theorem 3.3.1. *Let L be a set of arcs whose removal does not disconnect the underlying network G, assumed to be connected. Let $0 < \epsilon$. Define $y_{ij}^{L,\epsilon} = 1/x_{ij}$ for each line $(i,j) \notin L$ and $y_{ij}^{L,\epsilon} = \epsilon$ for each line $(i,j) \in L$. Let $(p^{L,\epsilon}, \theta^{L,\epsilon})$ denote the solution to the resulting DC system on G, and $(\bar{p}, \bar{\theta})$ be the DC solution on $G - L$, in both cases assuming that the phase angle of a fixed bus is set to zero. Then*

(a) $\lim_{\epsilon \to 0} p_{ij}^{L,\epsilon} = 0$ *for all* $(i,j) \in L$;

(b) *for any* $(u,v) \notin L$, $\lim_{\epsilon \to 0} p_{uv}^{L,\epsilon} = \bar{p}_{uv}$;

(c) *for any bus* i, $\lim_{\epsilon \to 0} \theta_i^{L,\epsilon} = \bar{\theta}_i$.

We will discuss the implications of this result below, but first, as a (somewhat flawed) argument toward a proof, let

$$Y^{L,\epsilon} = \text{diag}\{y_{ij}^{L,\epsilon}\}, \quad \mathcal{B}_{L,\epsilon} = N Y^{L,\epsilon} N^T,$$

i.e., $\mathcal{B}_{L,\epsilon}$ is the bus susceptance matrix using reactances $y^{L,\epsilon}$. Likewise let $\mathcal{B}_{\bar{L}}$ be the bus susceptance for $G - L$. Then, clearly,

$$\lim_{\epsilon \to 0} \mathcal{B}_{L,\epsilon} = \mathcal{B}_{\bar{L}}. \tag{3.87}$$

However, this is not enough to prove the theorem. The reader will note that (3.87) does not require $G - L$ to be connected. If $G - L$ is disconnected, $\mathcal{B}_{L,\epsilon}$ has rank $n - 1$ for $\epsilon > 0$, whereas $\mathcal{B}_{\bar{L}}$ has smaller rank (here n is the number of buses). Thus one needs a further (and combinatorial) argument to show that, e.g., some pseudoinverse of $\mathcal{B}_{L,\epsilon}$ converges to a pseudoinverse of $\mathcal{B}_{\bar{L}}$ when L is not a cutset. (Also see Theorem 3.0.3.) In fact if $G - L$ is disconnected, (a)–(c) above will in general not hold, e.g., in general the DC system on $G - L$ will be infeasible (when at least one of the components of $G - L$ has a demand/supply imbalance), whereas for arbitrary $\epsilon > 0$, $(p^{L,\epsilon}, \theta^{L,\epsilon})$ is feasible on G using reactances $y^{L,\epsilon}$, and thus in particular (b) does not hold. A correct proof is provided in [42].

Theorem 3.3.1 provides a different perspective on the continuous attack models we study in this section. Namely, using (3.85) or (3.86) with B large, the adversary can approximate the removal of a set of lines that does not constitute a cutset; essentially this provides an alternative approach to addressing the N-K problem (again in the nondisconnection setting; see the discussion in Section 3.0.1). Thus, to some extent the continuous attack model we consider here should provide a more conservative security outlook.

In any case, problem (3.84) is nonconvex and continuous. To apply the techniques outlined in Sections 3.3.1 and 3.3.2, we must first remove another attribute of problem (3.84), namely its *nonsmooth* nature. Here [42] applies a classical (and seemingly folklore) optimization technique: problem (3.84) can be equivalently restated as

$$\max_{y,\pi,\nu} \quad \sum_{(k,m)\in\mathcal{A}} \frac{p_{km}^{y}}{u_{km}} (\pi_{km} - \nu_{km}) \tag{3.88a}$$

$$\text{s.t.} \quad \sum_{(k,m)\in\mathcal{A}} (\pi_{km} + \nu_{km}) = 1, \tag{3.88b}$$

$$y \in \Gamma, \quad \pi, \nu \geq 0. \tag{3.88c}$$

This problem is certainly smooth. Further, if (y^*, π^*, ν^*) is optimal for (3.88), then the optimal value for (3.88) is at most $\max_{(k,m)\in\mathscr{A}}\left\{\frac{|p^{y^*}_{km}|}{u_{km}}\right\}$ (because of constraint (3.88b)); however, this is the objective attained by y^* in problem (3.84). This argument shows that the value of problem (3.88) is a lower bound for that of problem (3.84), and the converse is proved similarly.

Hence we can focus on problem (3.88), which [42] addresses by relying on LOQO, using problem structure to speed up computations. We stress, however, that this methodology is only guaranteed to produce a local optimum for (3.88) and should thus be viewed as a heuristic, albeit one with good pedigree.

To address this question in the context of problem (3.88), consider the objective function in that problem. To apply either of the approaches we just outlined, either the agent or engine algorithms would need to compute or estimate the first- and second-order partial derivatives of the function p^y_{km} for each line (k,m). Recall (see (3.2)) that, given y,

$$p^y = Y_y \tilde{N}^T \tilde{\mathscr{B}}_y^{-1} \tilde{P}, \tag{3.89}$$

where N is the node-arc incidence matrix of the network, $Y_y = \mathrm{diag}(y_{km})$, $\mathscr{B}_y = N^T Y_y N$, and $\tilde{N}$ and $\tilde{\mathscr{B}}_y$ are obtained by striking out a given row from N and $\mathscr{B}_y$, respectively. The derivative-computing version of the agent algorithm would ideally rely on (3.89) to compute $\frac{\partial p_{uv}}{\partial y_{ij}}$ for any (u,v) and (i,j) and $\frac{\partial^2 p_{uv}}{\partial y_{ij}\partial y_{hk}}$ for any (u,v), (i,j), and (h,k). *However,* (3.89) includes the factor $\tilde{\mathscr{B}}_y^{-1}$, which would need to be computed. This appears difficult, because such a capability seems outside the intended scope of the agent-engine collaborations. In fact (3.89) is just an implicit representation of the vector p^y, one that is difficult to directly provide to an engine algorithm, except perhaps as a highly nonlinear constraint, if $\tilde{\mathscr{B}}_y^{-1}$ has been symbolically computed by the *agent*. Moreover, and most important, one can show that $\mathscr{B}_y^{-1}$ will be fully dense for any vector of positive y values assuming (as we do) that the underlying network is connected.[19]

What one *could* do, in order to use the agent-engine framework, is provide an implicit representation of p through the usual DC approximation system:

$$Np = P, \quad p = YN^T\theta, \tag{3.90}$$

and then run the agent-engine framework, with partials computed by the engine, in the higher-dimensional (p,θ) space to address a problem equivalent to (3.88) (or even (3.84)).

Instead, [42] relies on the first version of the agent-engine framework, using the fact that (3.2) is not simply a linear system, but a very special linear system involving Laplacians. We have already stressed this point repeatedly above; see in particular Section 3.0.1. In fact, [42] shows that this special structure can be leveraged so that (in a network with n nodes and m arcs) the computation of all first- and second-order partials requires a workload of $O(n^2 + nm)$ plus the solution of m sparse linear systems on $n+m$ variables and constraints. As an example of the constructions in [42], let $y > 0$ and consider the solution (p^y, θ^y) to system (3.90) (where we fix one degree of freedom to obtain unique angles θ^y). To compute the first-order partial derivatives $\frac{\partial p_{uv}}{\partial y_{ij}}$ we proceed as follows. Let $\underline{1}$ be

[19] For any given network, $\mathscr{B}_y^{-1}$ will have the same off-diagonal density pattern as the adjacency matrix for the *transitive closure* of the network; see [202] for background and useful algorithmic constructions.

the vector of n 1's, and define

$$\mathcal{B}_y = NYN^T, \quad J_y = \mathcal{B}_y + \frac{1}{n}\underline{1}\underline{1}^T.$$

It can be shown that $J_y \succ 0$. In fact the $n-1$, positive eigenvalues of $\mathcal{B}_y$ are also eigenvalues of J_y (with the same eigenspaces) and $\underline{1}$ is an eigenvector for both matrices, in the first case with eigenvalue 0, and in the latter case with eigenvalue 1 (see [42] for further background). Further, suppose that for each arc (i,j) we denote by n_{ij} the column of N corresponding to (i,j). Then, setting

$$\widetilde{\nabla}^y_{ij} \doteq -J_y^{-1} n_{ij} n_{ij}^T \theta^y \quad (\in \mathbb{R}^n), \tag{3.91}$$

one can prove that [42]

$$\frac{\partial f_{uv}}{\partial y_{ij}} = \begin{cases} y_{uv}\, n_{uv}^T \widetilde{\nabla}_{ij} & \text{if } (u,v) \neq (i,j) \\ \\ n_{ij}^T \widetilde{\nabla}_{ij} + y_{ij} n_{ij}^T & \text{if } (u,v) = (i,j). \end{cases} \tag{3.92}$$

Recall that the vectors n_{uv} have two nonzeros. Thus, both of the partial derivatives can be quickly evaluated using (3.92), provided that the vector $\widetilde{\nabla}^y_{ij}$ has been computed. However, we *do not* want to directly rely on (3.91), since as noted above J_y^{-1} will be fully dense. In fact, for each arc (i,j), consider the linear system on variables x, λ:

$$N^T x - Y^{-1}\lambda = 0, \tag{3.93a}$$

$$N\lambda = n_{ij}. \tag{3.93b}$$

(Note: This is simply a DC system with flows λ and phase angles x.) It is a simple exercise to show that any x solving this system is of the form $x = J_y^{-1} n_{ij} + \delta \underline{1}$ for some real δ. Thus, for any line (u,v) (including $(u,v)=(i,j)$) we will have

$$n_{uv}^T x = n_{uv}^T J_y^{-1} n_{ij}.$$

Using (3.91), we can see that the right-hand side in this expression appears as a factor in $n_{uv}^T \widetilde{\nabla}_{ij}$. In other words, given x we can efficiently compute all quantities $\frac{\partial f_{uv}}{\partial y_{ij}}$ using (3.92). A similar analysis applies to the second-order partials.

In summary, thus, we can reduce the computation of first- and second-order to a set of very sparse matrix-vector products, plus the solution of one system (3.93) per derivatives line (i,j). In [42] this was done by relying on Cplex (so as to solve a nominal linear program), thereby benefiting from warm starts, and thus matrix factorizations, as different lines (i,j) are enumerated. Each such computation can be performed quite quickly—on a network with several thousand lines, and using a modern, fast sparse linear algebra package, the computation of all partial derivatives should only take on the order of one CPU second. Furthermore, and quite clearly, the entire process could easily be parallelized with complete scalability. We also want to stress at this point that the closed-form nature of equations (3.92), together with the factorization in (3.91) and the closed-form nature of solutions to (3.93), is what makes it attractive to rely on an agent-engine framework where the agent computes the partial derivatives.

Table 3.5. *Solution histogram for continuous attack (*from [42]*).*

(49,90)	II, $\Delta B = 57$	(300,409)	I, $\Delta B = 27$	(600,990)	I, $\Delta B = 36$
Range	**Count**	**Range**	**Count**	**Range**	**Count**
[1,1]	8	[1,1]	1	[1,1]	14
(1,2]	72	(1,2]	405	(1,2]	970
(2,3]	4	(2,9]	0	(2,5]	3
(5,6]	1	(9,10]	3	(5,6]	0
(6,7]	1			(6,7]	1
(7,8]	4			(7,9]	0
(8,20]	0			(9,10]	2

Next we provide an example (from [42]) focusing on the qualitative nature of solutions to problem (3.84).[20] We consider a family of two classes of experiments (Cases **I** and **II** below), as follows:

- For each line (k,m), $x^L_{km} = 1$. Further, in Case **I**, $x^U_{km} = 10$ for each line (k,m), and in Case **II**, $x^U_{km} = 20$ (see (3.82)).
- The constraint on the adversary is of the form $\sum_{(i,k)} x_{km} \leq \sum_{(i,k)} x^L_{km} + \Delta B$, where ΔB is a parameter (see (3.85)).

In other words, the adversary can increase the reactance x_{km} of any line by as much as a factor of 10 in Case **I** and a factor of 20 in Case **II**. Hence, if, e.g., $\Delta B = 45 = 9 \times 5$ in Case **I** the experiment allows the adversary to increase the reactance of five lines to the maximum allowable amount. But the adversary is of course also able to increase the reactance of many more lines by much smaller amounts. Table 3.5 describes the outcomes of several experiments under this setup: the table indicates experiments with three artificial networks with 49 buses and 90 lines, 300 buses and 409 lines, and 600 buses and 900 lines, respectively (indicated by pairs of parameters in each of the three sections of Table 3.5). Each section is also designated by "**I**" or "**II**" according to the case studied, and the corresponding parameter ΔB. Thus, for example, in the third column we are considering a network with 990 lines under Case **I** with $\Delta B = 36$. Under this setup, should the adversary choose to simply increase the reactance of a subset of lines to their maximum allowed value (which is 10, under Case **I**), then only *four* lines can thus be interdicted—though, as discussed above, other actions are also allowed to the adversary. In the experiment in the first and second columns, only three lines can be maximally interdicted.

However, Table 3.5 shows a different form of attack. Each column in the table shows the histogram of values taken by reactances in the *local* optimum actually computed by the above algorithm, which relies on LOQO to carry out the underlying search. We can see that in each column the corresponding attack does not maximally interdict the maximum possible set of lines (for example, only two lines are near-maximally interdicted in the third experiment out of four possible), though indeed a few lines, in all three experiments, show a high degree of reactance increase. On thc other hand, in all three experiments many lines are partially interdicted, with a very pronounced pattern. Thus in all three experiments we effectively see attacks made up of two modes, one focusing sharply on a small set of lines, and the other distributing the "budget" across many lines. Notice that in all three experiments, some lines are not at all interdicted. One wonders about the significance of this two-pronged structure: does one of the two modes have negligible impact?

[20]For the experiments in this section we used a 2.66 GHz Xeon workstation and LOQO 4.05.

Table 3.6. *Discrete versus continuous attacks.*

σ	N-3		CONTINUOUS				
	3-OVL	Attack	C-OVL	Top 6 lines	3-C-OVL	3I-10%	C-10%
1.0	1.44088	29;32;45	2.14967	29(7.79), 27(7.20), 41(7.03), 67(7.02), 54(6.72), 79(5.71)	1.71758	1.33454	1.67145
1.2	1.43132	27;29;41	1.78687	29(8.28), 27(7.72), 41(7.32), 67(7.19), 54(6.92), 79(5.78)	1.43132	1.11211	1.38642
1.4	1.22685	27;29;41	1.55634	29(8.31), 27(7.74), 41(7.53), 67(7.48), 54(7.18), 79(6.15)	1.22685	0.95324	1.21329
1.6	1.07349	27;29;41	1.35995	29(8.18), 27(7.58), 41(7.53), 67(7.58), 54(7.22), 79(6.25)	1.07349	0.83409	1.05458
1.8	0.692489	18;57;60	1.20271	29(8.43), 27(7.90), 41(7.53), 67(7.48), 54(7.18), 79(6.12)	0.95421	0.74141	0.93595
2.0	0.68630	20;89;45	1.07733	29(7.87), 27(7.29), 41(7.04), 67(7.01), 54(6.70), 79(5.63)	0.85889	0.66727	0.83878

A different outlook on the structure of continuous attacks can be obtained by comparing to the N-K setup we considered in Section 3.1.2. Formulation (3.69) considered the problem where for a given K we wish to compute a set with at most K lines whose removal either renders the problem infeasible or causes some line flow to exceed its limit. As discussed in the remarks around Theorem 3.3.1 above, this goal is related to what an attacker operating under the continuous attack model *could* attain. Of course, the method used in Section 3.1.2 computed *optimal* solutions as opposed to the local optima we have discussed here.

Thus, we have two qualitative questions that arise in the context of the continuous attack model: (1) in view of the histograms above, is it the case that the lightly interdicted lines do not contribute much to the attack, that is to say, could we simply restrict the attack to the highly interdicted lines, and (2) how does the set of highly interdicted lines compare to optimal sets of lines computed in the context of the N-K problem for meaningful values of K? To address these issues, [42] performed the experiment detailed in Table 3.6, which compares a discrete $N-3$ attack with a continuous attack.

In this table, we have the following:

(1) The experiments involve the artificial network on 49 buses and 90 lines described above, with $x^L_{ij} = 1$ and $x^U_{ij} = 20$ for each line (i, j). This is the same basic setup as in the first column of Table 3.5. In each row of the table, the experiment was carried out by scaling all line limits by the given quantity σ. Larger values of σ yield a network that is more difficult to interdict. The network is such that no three lines constitute a cutset.

(2) The columns labeled "3-OVL" and "C-OVL" show the maximum line load under the computed N-3 and continuous attacks, respectively.

(3) Under "N-3" the column labeled "Attack" shows the *optimal* set of three lines. Under "CONTINUOUS," the table shows the six lines whose reactance was maximally increased by the continuous algorithm we discussed above, together with their final reactance value.

(4) The column headed "3-C-OVL" indicates the maximum line load obtained by removing the three lines with largest computed reactance while leaving all other line reactances unchanged.

(5) We also consider load shedding. Suppose the three lines with maximum computed resistance are removed. The column labeled "3I-10%" shows the maximum line load that ensues should the controller optimally shed 10% of the load.

(6) As a contrast to (4), suppose we do not remove any lines but instead change the reactance of all lines as computed by the continuous algorithm. Further suppose that the controller is allowed to shed 10% of the load with the goal of minimizing the maximum line load. The column labeled "C-10%" shows the value of the resulting min-max load.

Comparing the entries in the "Attack" and "Top 6 lines" columns we can see a clear correlation between the computed attacks when $\sigma \leq 1.6$, despite the widely different methodologies that were employed. When $1 < \sigma \leq 1.6$, furthermore, the top three lines interdicted in the continuous model are identical to those identified by the (exact) N-3 approach. Comparing the "C-OVL" and "3-C-OVL" columns, we see that the two modes in the continuous attack *do* play a role—simply removing the most interdicted lines has a markedly poorer impact on line loads than the increase of all reactances as computed by the continuous model. This takes place in all experiments.

Another interesting item is revealed by columns "3I-10%" and "C-10%." Here the controller is allowed some recourse. One should allow such recourse because potentially a high line overload could be attained by having a line with very small limit carry large enough load. If it is only this effect that is responsible for the high values on the above table, then a controller could perhaps shed a small amount of load and return to a secure system. However, the table shows that this is not the case. In fact, the attack proves robust even though it was computed without knowledge of the possible load shedding recourse. Finally, a comparison of the entries in columns "3I-10%" and "C-10%" echoes the remarks above regarding the two-mode distribution of the continuous attack: the small increases of reactances on many lines play a clear role.

A final item concerning Table 3.6 involves the cases $\sigma = 1, 1.8, 2.0$. Notice that in these cases the "3-OVL" value is smaller than the "3-C-OVL" value. In other words, the continuous model computed a more effective attack that the exact, mixed-integer program based $N-3$ model. This is due to the fact that the mixed-integer program model (3.69) does not necessarily identify the three lines whose removal has the maximum impact on line loads. It only finds one that causes a load greater than 1 (see the remarks following formulation (3.69)).

Extension to AC power flow models

An interesting extension of the methods in the above section is that to the context of AC power flow models. Such an extension would allow not only a more realistic modeling of grid operations, but also a more complete description of potential vulnerabilities, such as voltage decay.

These points have recently been taken up in [134]. Here the means by which a hypothetical adversary interdicts a transmission system is the same as in the continuous attack model in [42]: impedances are scaled, subject to budget constraints, both global and on a per-line basis. However, [134] does use an AC power flow model, and moreover the metrics by which the success of an attack is measured include

- voltage magnitude decay, in L_∞-, L_1-, and L_2-norm variants;
- maximum phase angle difference across lines, again in several L_k-norm versions; and

- maximum load loss, in a recourse model where the controller seeks to attain a physically valid solution under the modified impedances, while minimizing loss of load.

A technical challenge encountered in the pursuit of these goals is that, unlike the case for DC power flows as discussed above, there are no closed-form solutions to AC power flow systems, such as the characterization in, e.g., Lemma 1.2.3 available for DC approximations, or Theorem 1.2.5. Among several implications, we will not have a closed-form expression for the gradient and Hessian of the merit function for the attacker, such as (3.91). This implies that a straightforward application of the agent-engine framework may be problematic.

Instead, [134] resorts to the classical Frank–Wolfe method [50]. It is worth explaining this method using a concrete example. Let us denote by x the vector of reactances that will be chosen by the adversary, and let x^0 denote the vector of initial reactances in the transmission system. An example of a constrained adversary is given by the following system, where $\mathscr{A}$ is the set of all lines and $\lambda^{\max} < \Lambda$ are given constants:

$$x_{pq} = \lambda_{pq} x^0_{pq} \quad \forall (p,q) \in \mathscr{A}, \tag{3.94a}$$
$$1 \le \lambda_{pq} \le \lambda^{\max} \quad \forall (p,q) \in \mathscr{A}, \tag{3.94b}$$
$$\sum_{(p,q)\in\mathscr{A}} \lambda_{qp} \le \Lambda. \tag{3.94c}$$

Here, (3.94b) regulates the per-line increase in reactance, whereas (3.94c) constrains the total increase. For example, if $\lambda^{\max} = 3$ and $\Lambda = 21$, then the reactance of any line can be increased by at most a factor of 3, and the sum of increases is at most 21 (so up to seven lines may have their reactance increased by a factor of three, though of course the adversary may allocate the total budget of 21 among more than three lines). Let us denote by $\mathscr{B}$ the system (3.94).

Suppose further that the adversary seeks to maximize the sum of squares of phase angle differences among lines. Denoting by $\theta_p(x)$ the phase angle at bus p using reactances x, the problem we wish to solve can be written as

$$\max \sum_{(p,q)\in\mathscr{A}} (\theta_p(x) - \theta_q(x))^2 \tag{3.95a}$$
$$\text{s.t.} \quad x \in \mathscr{B}. \tag{3.95b}$$

Generally, we will want to solve a problem of the form

$$\max \mathscr{F}(x) \tag{3.96a}$$
$$\text{s.t.} \quad x \in \mathscr{B}, \tag{3.96b}$$

where $\mathscr{F}(x)$ is any desirable metric (e.g., a measure of voltage loss). The alert reader may already identify a problem with this approach. For example in (3.95) the notation $\theta_p(x)$ seems problematic, because it is not necessarily the case that, given reactances x, there is a *unique* AC power flow solution under those reactances. There may be multiple solutions—or there may be none. Let us, for the moment, put aside these concerns, and instead focus on how to solve, at least locally, an optimization problem of the form (3.96) using the Frank–Wolfe method, which is an iterative procedure. Iteration k is given in outline by Algorithm 3.3.2.

Algorithm 3.3.2. Iteration k of Frank–Wolfe applied to problem (3.96).

Step 1. Compute the gradient $g_{k-1} \doteq \nabla \mathscr{F}(x^{k-1})$.

Step 2. Solve the linear optimization problem $g^* \doteq \min\left\{g_{k-1}^T y : y \in \mathscr{B}\right\}$, with solution y^k.

(2.a) If $g^* \leq \epsilon$, then **STOP**.

(2.b) Otherwise set $x^k = x^{k-1} + \alpha y^k$, where $0 \leq \alpha \leq 1$ is chosen to maximize $\mathscr{F}(x^{k-1} + \alpha y^k)$.

In Step (2.a) of Algorithm 3.3.2, the quantity $\epsilon > 0$ is a given tolerance; the stop condition indicates that the algorithm has converged to a local optimum. Also note that the optimization problem in (2.a) is a linear program when (as is the case in the above example) $\mathscr{B}$ is a polyhedron. Finally, Step (2.b) is a line search.

In our grid attack setting, there are several challenges that this outline needs to overcome. One is that the gradient computation is expensive: here the derivatives need to be estimated (for example through finite differences), which will require multiple power flow computations, which may experience convergence difficulties. Another challenge was already brought up before, namely the nonuniqueness of power flow solutions. In terms of this last issue, [134] applies the standard criterion found in the power engineering literature: given a power flow system, either a power flow solver will find a "good solution," or no such solution exists.

Of course, nonconvergence of power flow computations will impact the estimation of gradients. One version of the methodology in [134] formulates an ACOPF-like problem (minimize L_2-norm of infeasibilities of power flow equations) that proves more robust in the sense that solvers (usually) converge to *some* solution. We refer the reader to [134] for additional details.

In [134] the basic approach we have outlined, with several additional enhancements, is able to analyze Polish grid examples from MATPOWER [234] in CPU time of the order of one hour to a few hours. The experimental observations in [134] match those in the previous section: the continuous attack model produces (locally optimal) attacks that are focused on a small number of lines, plus a "tail" which is itself impactful.

The following example is typical. Consider the 2383-bus ("winter peak") Polish grid example from Matpower. Suppose the attacker has a total interdiction budget consistent with doubling the reactance of at most three lines. The goal of the attacker is to maximize the L_2-norm of voltage magnitude drops, relative to the given steady-state solution. Figure 3.14, from [134], displays the result from the optimization.

In the figure, the solid black curve shows the original disposition of voltage magnitudes, whereas the dots are the voltages postattack. We see that the attack magnifies the voltage drop among buses whose original voltage was already low, an intelligent feature. Moreover, some buses attain voltages of value close to 0.7 (per unit), a dangerous condition. In fact we can count six buses of voltage approximately 0.8 or lower. An additional feature of this (locally optimal) attack is that it concentrates the reactance increase among just four lines.

Summary We view the line of research in [42] and [134] as holding great promise. It bypasses the overt enumeration required by the traditional N-K criterion while leveraging efficient tools of modern optimization, such as fast linear programming solvers. In implementations, the algorithms prove fast on large grids and obtain solutions that have

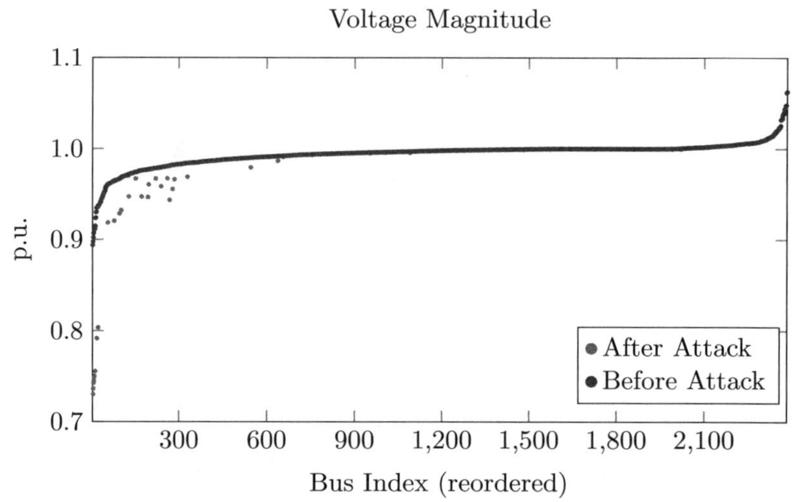

Figure 3.14. *Effect of optimal reactance attack on voltage magnitudes, Polish grid.*

interesting and meaningful structure. We expect future research to focus on additional tools of continuous optimization, such as filter and derivative-free methods.

3.4 ▪ Multistage models

The importance and complexity of the N-K problem has spurred a great deal of research in a number of communities. The problem, which arose historically as an extension of the N-1 criterion, can clearly be extended to encompass additional, realistic criteria. Whereas the models we considered above defined postcontingency "risk" as any situation where some equipment overload takes place, in power grid practice the actual picture is more nuanced. For example, some measure of recourse on the part of the controller could be incorporated, and it can be argued that in power engineering practice pure enumerative methods for verifying N-1 compliance do take some measure of recourse into account. We have already seen in the general attacker/controller framework introduced in Section 3.1 (see, e.g., Example 3.1.1 and Definition 3.1.3) a formal modeling structure for incorporating recourse into optimization models. Salmeron, Wood, and Baldick [198], [199] discuss several problems along these lines. An interesting ingredient in their work is the modeling of time required to recover from a contingency and how this would impact an adversary's decision. They develop heuristics and algorithms based on Benders decomposition and report experiments on realistic system with 5000 buses and 5000 lines.

There is an additional and far more complex perspective to the N-K problem which gives the problem a true multistage flavor. As we have stated above, a line overload is normally deemed a risky condition, with corrective action necessary. If a line is overloaded (in particular, overloaded in the sense that the flow it carries exceeds the line's "emergency rating"), it will, eventually, trip. If the overload is very large, this event will happen rapidly. After the line trip, power flows will be reestablished—at different values than before, and largely through automatic decisions (further background in Chapter 4). In the new realization of the power flows, some new overloads (of lines or other equipment such as transformers) may now take place, with, possibly, additional trips taking place at a later time. If this sequential process of line trips (or other equipment trippings) and power flow realizations is repeated, we have what is termed a "cascade." In the following chapters in this work we will study such phenomena in greater detail; however, at this point a cascade is best seen as a sequential process which is punctuated by equipment tripping events (such events literally cause equipment, such as lines, to disappear from the functioning grid). Such a process can cause significant islanding to take place and, with it, loss of demand, as generators become topologically separated from loads. In fact, under the AC power flow a system may be unable to deliver all demands even without any islanding, but solely due to the availability of a reduced set of lines.

A more comprehensive approach to "risk" would then consider the risk of a severe cascading outage as a consequence of a size-K contingency. Additionally, one could even attempt to model potential recourse actions to be taken at various points of the cascade. Yet another interesting problem concerns the evaluation of contingencies where the adversary's actions are sequential, and possibly taken as a response to the controller's actions. The so-called N-1-1 security criterion mentioned above attempts to model this sequential process. In any of these multistage settings the resulting risk evaluation can be quite complex; it requires modeling of numerous operational and engineering details, in particular modeling of island dynamics and modeling of the timing of equipment tripping. Necessarily, such an evaluation will be both data and CPU intensive. In the following chapters we will analyze these issues in greater detail. Nevertheless, a relevant and compelling research goal is the study of the N-K problem in such broader contexts.

Grid security analysis under cascading failures

An interesting study is provided by Eppstein and Hines in [92] (see Chapter 4, Section 4.3.2 of this monograph). A significant part of the modeling in [92] is devoted to representing cascades, and we will discuss their models in Section 4.3.2. To tackle the N-K problem with $2 \leq K \leq 5$ they additionally present a search heuristic motivated by an application to chemistry, which they term the random chemistry (RC) method. This heuristic produces a sequence of contingencies which, when evaluated under the cascade model, prove "risky" in a predefined manner, i.e., the cascade terminates with a large loss of load. (We will use the term "risky" assuming such a precise meaning, next.)

The RC method initially creates a *random* contingency of size $K' > K$ which is risky. If K' is chosen large enough, then (as can be experimentally verified) a risky contingency of size K' can be found relatively rapidly, or to put it more accurately, can be found after a relatively small number of enumerations. Let S be such a contingency. The RC method will next attempt to find a proper subset $S' \subset S$ which also constitutes a risky contingency. For example, this can be done by randomly enumerating subsets of S of cardinality $\alpha|S|$ for a given $\alpha < 1$, e.g., $\alpha = 0.5$. Note that if $|S|$ is relatively small (e.g., $|S| \leq 20$), this random search component could plausibly be extended to near-full enumeration, or at any rate the chance of success is not negligible. Assuming that a risky contingency is found, this size-reduction step is repeated until either a risky contingency of size K is found or the process stops with failure, i.e., it stops with a risky contingency S with $|S| > k$. In the latter case the RC method then sequentially attempts to prune one element at a time from S until a risky contingency is found that is either of size K or inclusion-minimal but of size greater than K.

This methodology is, of course, heuristic. In addition to the hazards related to random search of (potentially large) spaces, there is the issue of nonmonotonicity that was raised above (in particular see Example 3.1.5), i.e., one could have two risky contingencies X, Y with $X \subset Y$ such that each contingency Z with $X \subset Z \subset Y$ is *not* risky. Nevertheless, [92] reports some computational success in investigating one of the Polish grid examples in MATPOWER [234]. Here, $N = 2896$, and [92] reports on nearly 750,000 runs of the RC method, which result in the discovery of tens of thousands of distinct, risky contingencies of size $2 \leq K \leq 5$. They further report that for $K = 5$ their method requires, on average, some 1800 contingency evaluations (i.e., cascade simulations) and even fewer for smaller values of K.

Another perspective on multistage models is provided in [37]. A common practice in power grid security analysis is that of identifying, through a set of computational expedient procedures such as those described in this chapter and in the citations, a set of contingencies "of interest," i.e., contingencies that appear to pose some risk. This set of contingencies is then analyzed in further detail using a more comprehensive model of the grid. In the setting used in [37], an arbitrary set Σ of contingencies is thus given as an input. The analysis of each contingency then relies on a model of a cascading process as outlined above, including a model for the thermal process that results in tripping of overloaded lines, and another model for power dynamics in freshly created islands. For a given $R > 1$ and $\mu \geq 0$, the grid is said to survive a given contingency if it reaches the Rth step of the cascade while still attaining yield (fraction of served load) at least μ. In addition, the main problem considered in [37] studies an optimal investment plan to upgrade the grid; namely, for any line (k,m), by investing a given amount w_{km} one can upgrade line (k,m)—in particular by increasing its rating or by installing a parallel line. The problem of interest becomes how to invest a minimum amount so that the grid survives the cascade initiated by each contingency in the given list Σ. This problem is addressed as a mixed-integer program and handled by a combination of Benders decomposition and cutting-plane techniques.

3.4.1 ▪ Stochastic programming and sample average approximation methods

The models discussed above have all been of a deterministic nature. A classical technique in the optimization literature is that of *stochastic programming with recourse*. In the context of the N-K problem in the cascading failures setting, it might be possible to introduce stochastics to *approximate* multiple sequential trajectories of a cascading failure (discussed in the following chapters), with recourse actions taken by the controller. Stochastic programming has evolved over the last decades into a very mature discipline, supported by rich theoretical underpinnings as well as high-performance computing capabilities. There is a vast literature on the topic; see [43], [152], and the website of the Stochastic Programming Society, *stoprog.org*.

The techniques and overall thrust of stochastic programming methods would seem to make an ideal toolset for analyzing power grid security models, with the possible exception of a strict interpretation of the N-K problem, which would not allow for stochastic criteria. Such an interpretation, however, is not the most frequently found in the literature, with the primary focus instead being on discovering underlying weaknesses or errors in operations and standards. We are not aware of any rigorous work in this direction, though some may be in progress. A hazard that would be encountered concerns the stochastic model itself, for example a (reasonable) stochastic model that one could use to address "noise" in the line tripping process. We will return to this topic in Chapters 4 and 5 (Section 5.5). In contrast, stochastic programming has been extensively applied in the context of a different topic in power engineering: the unit commitment problem. For a recent reference, see [204] and the citations therein. In this setting, the use of stochastics is easily justified as a means to model the evolution of demand over some time horizon. Further, the generation of adequate scenarios (and respective probabilities) in this context is a task that can be successfully fulfilled, since normal demand patterns and their variation are well understood.

Stochastic programming techniques have also been applied to network interdiction models outside a power grid setting. Here we will focus on the model considered in [131], because it provides a good example of the application of stochastic optimization, and because of the intellectual proximity of the problem studied therein to the N-K problem.

In [131] the authors consider an adversary that interdicts a (traditional) capacitated network to minimize the maximum flow that can be sent from a given node r to another node t. The adversary attempts to remove arcs from the network; however, the attempt succeeds only stochastically. Furthermore, the adversary's attempts are also constrained; for example, the number of arcs that the adversary can attempt to remove is limited, and, more generally, there is a knapsack constraint limiting the possible removal attempts. The goal of the attacker is to find an allowed attack which minimizes the expected max flow from r to t. Clearly, this interesting problem is similar to an N-K problem, of course without the power engineering details.

The interdiction problem in [131] can be stated as a stochastic, mixed-integer optimization problem, as given next, where S is a given set of *scenarios*. This particular modeling detail is explained below, but for now we may think of a scenario as a "state" of the network which is beyond the control of the adversary. The probability of each scenario $s \in S$ is denoted by p_s. Thus the interdiction problem can be written as follows:

$$F^*(p,S) \doteq \min_x \sum_{s\in S} p_s \phi^s(x) \tag{3.97a}$$

$$\text{s.t.} \sum_{i,j} h_{ij} x_{ij} \leq K, \tag{3.97b}$$

$$x_{ij} \in \{0,1\} \ \forall (i,j). \tag{3.97c}$$

In this formulation, for each arc (i,j) the binary variable x_{ij} indicates the decision to interdict (i,j); (3.97b) is the constraint faced by the attacker (the h_{ij} and K are given parameters). Furthermore we indicate the max r-t flow under scenario s by $\phi^s(x)$. We now explain the notion of scenarios. In [131], these are used to quantify the probability that an interdiction attempt will succeed at removing a line. Specifically, a scenario s specifies a subset $V(s)$ of arcs, and under scenario s an interdicted arc (i,j) is removed if and only $(i,j) \in V(s)$ (and of course noninterdicted arcs are not removed). We might thus think of the set $V(s)$ as the "vulnerable" arcs under scenario s that the adversary aims to exploit. The adversary knows each set $V(s)$ but of course does not know which scenario will be realized. Note that this setting allows for a broad class of models, for example models where each arc (i,j) becomes vulnerable with a certain known probability p_{ij} independent of all other arcs, and models with correlations in the vulnerability process. In terms of the attacker-controller paradigm we have relied on above (see Definition 3.1.3), here we introduce a third party, which one could term "nature" and is beyond the control of both attacker and controller. Thus, in problem (3.97) the attacker chooses an interdiction while aware of the set of scenarios S and their probabilities; then nature acts to randomly choose a scenario, and finally the controller observes both the interdiction and the scenario and chooses a max flow. Closely related problems were considered in [229], [75].

In [131] problem (3.97) is addressed using the *sample average approximation* (SAA) method [138], which will be discussed in more detail below. The first step is to describe each quantity $\phi^s(x)$ as an appropriate optimization problem. This is done relying on the familiar observation that a max-flow problem can be restated as an optimal circulation problem using linear programming duality. Let $\mathscr{A}$ denote the set of arcs in the network, and let N denote the node-arc incidence matrix of the graph obtained by adding (t,r) to the set of arcs. For any arc $(i,j)\in\mathscr{A}$ let u_{ij} denote its capacity (assumed finite), and for any scenario s and arc (i,j) we write

$$\delta^s_{ij} = \begin{cases} 1 & \text{if } (i,j)\in V(s), \\ 0 & \text{otherwise.} \end{cases} \tag{3.98}$$

Then the interdiction problem can be written as

$$F^*(p,S) = \min \sum_{s\in S} p_s f^s_{tr} \tag{3.99a}$$

$$\text{s.t.} \quad \sum_{i,j} h_{ij} x_{ij} \le K, \tag{3.99b}$$

$$x_{ij}\in\{0,1\} \quad \forall (i,j)\in\mathscr{A}, \tag{3.99c}$$

and, for each $s\in S$,

$$Nf^s = 0, \tag{3.99d}$$

$$0 \le f^s_{ij} \le u_{ij}(1-\delta^s_{ij}x_{ij}) \quad \forall (i,j)\in\mathscr{A}, \tag{3.99e}$$

$$v^s_i - v^s_j + \sigma^s_{ij} \ge 0 \quad \forall (i,j)\in\mathscr{A}, \tag{3.99f}$$

$$v^s_t - v^s_r \ge 1, \tag{3.99g}$$

$$0 \le v^s_i \le 1 \; \forall i, \tag{3.99h}$$

$$0 \le \sigma^s_{ij} \quad \forall (i,j)\in\mathscr{A}, \tag{3.99i}$$

$$f^s_{rt} - \sum_{(i,j)\in A} u_{ij}(1-\delta^s_{ij}x_{ij})\sigma^s_{ij} = 0. \tag{3.99j}$$

Here, constraints (3.99d)–(3.99e) indicate primal feasibility under the interdiction implied by x of the scenario-dependent flow vector f^s, where as desired the capacity of arc (i,j) is zero if $(i,j)\in V(s)$ and $x_{ij}=1$; (3.99f)–(3.99i) are similar dual feasibility conditions [3]; and (3.99j) states primal-dual optimality. Not surprisingly, this condition includes bilinear terms (the $x_{ij}\sigma^s_{ij}$). In terms of the notation used in (3.97) at an optimal solution for (3.99) we will have $\phi^s(x) = f^s_{tr}$ for each $s\in S$ with $p_s>0$.

To render the above formulation into practicable form one must therefore replace (3.99j) with an equivalent linearization. Above, in the context of the N-K problem, we considered the case of a power line (k,m) subject to the DC flow equation $p_{km} = y_{km}(\theta_k - \theta_m)$ and subject to interdiction by an adversary, with a binary variable z_{km}. A direct representation of this interdiction was given in system (3.30), restated here for convenience:

$$|p_{km}| \le u_{km}(1-z_{km}), \quad p_{km} = y_{km}(1-z_{km})(\theta_k - \theta_m),$$

where u_{km} is the capacity of the line. This system also includes a bilinearity; system (3.28) provided an equivalent, linear representation, which however could potentially amount to a big M formulation. One wonders if a similar fate awaits a linearization of (3.99j).

However, formulation (3.99) enjoys a significant advantage. Using (3.99f) and (3.99h), we have that without loss of generality $\sigma^s_{ij}\le 1$ for all $s\in S$ and $(i,j)\in\mathscr{A}$. As a consequence, by applying the McCormick reformulation method to each bilinear form, constraint (3.99j) is equivalent to the system

$$f_{rt}^s - \sum_{(i,j)\in A} u_{ij}(\sigma_{ij}^s - \delta_{ij}^s \chi_{ij}^s) = 0, \tag{3.100a}$$

$$\sigma_{ij}^s - (1 - x_{ij}) \leq \chi_{ij}^s \leq \sigma_{ij}^s, \quad \chi_{ij}^s \leq x_{ij}, \tag{3.100b}$$

which is clearly not a big M formulation. This successful linearization relied on the properties of dual solutions to max-flow problems; in the context of an extension of the interdiction problem to a *multi*commodity flow setting no such bounds on dual variables are known. In fact, upper bounds on dual variables may grow as a function of network size, and the linearization could be more problematic.

It is also worth comparing this success to the approach used in [42] when dealing with the N-K problem; see Section 3.1.1. In that setting, formulation (3.58) was used to describe the optimal recourse action to a given interdiction (or attack, in the language of Section 3.1.1) $\mathscr{A}$. This was a minimization problem, which we termed $K^*_{\mathscr{C}}(\mathscr{A})$, stating that the controller seeks an action to minimize the impact of the attack $\mathscr{A}$. Thus, in principle, one could use the dual of $K^*_{\mathscr{C}}(\mathscr{A})$ to obtain a valid statement for the N-K problem, using binary variables to represent the interdiction $\mathscr{A}$ and stating that the attacker seeks to find an attack $\mathscr{A}$ which maximizes the value of the *dual* of $K^*_{\mathscr{C}}(\mathscr{A})$. In fact this was the approach used in [42] and described in Section 3.1.1—the key design point in the construction of formulation (3.58) was to avoid producing bilinearities. However, in formulation (3.58) several constraints in that formulation (in particular (3.58c), (3.58b), and (3.58e)) applied to lines either not interdicted or interdicted by the adversary. For example, consider (3.58c), which is repeated here:

$$\theta_i - \theta_j - x_{ij} p_{ij} = 0 \quad \forall\, (i,j) \notin \mathscr{A},$$

where x_{ij} is the reactance of line (i,j). When passing to the dual, this constraint will have a dual variable β^C_{ij}; however the constraint only applies if $(i,j) \notin \mathscr{A}$, and in order to obtain a formulation for the dual of $K^*_{\mathscr{C}}(\mathscr{A})$ which is valid for any $\mathscr{A}$ we need a constraint of the form

$$\beta^{\mathscr{C}}_{ij} \leq M_{ij}(1 - z_{ij})$$

(i.e., (3.60)). The quantity M_{ij} could potentially be a big M quantity. This is precisely the same situation as discussed in the previous paragraph; once again we can see how bounding dual variables for an interdiction-related linear program tends to become a pressing issue. In [42] this issue was resolved by relying on equation (3.67), which provides an upper bound on the $\beta^{\mathscr{C}}_{ij}$ which is independent of network size. A similar issue, already discussed above, concerns inequality (3.58e),

$$|p_{ij}| \leq t \quad \forall\, (i,j) \in \mathscr{A},$$

for which, by construction, the variable is upper bounded by 1 (see the discussion concerning the alternative inequality (3.63)). In both cases therefore we see "well-behaved" dual variables.

Returning to problem (3.97), what we have obtained so far is an exact linear mixed-integer representation given by formulation (3.99) with (3.100) substituting for (3.99j). Even though this formulation is certainly correct, it will prove far too large for direct solution if the set of scenarios is large enough. As a simple example, consider the case where the scenarios are obtained by drawing the δ^s_{ij} quantities in (3.98) from a Bernoulli distribution, that is to say for each (i,j), $\delta_{ij} = 1$ with some fixed positive probability independent of all other arcs. The number of scenarios is $2^{|\mathscr{A}|}$, far too large when $|\mathscr{A}| > 50$ (or even smaller).

For $|S|$ of intermediate size one can attempt a solution of the formulation by resorting to the Benders decomposition (with a mixed-integer master formulation restricted to the x_{ij} variables, and estimates for the value f^s_{tr} for each $s \in S$). For $|S|$ large enough these methods prove inadequate, and to handle this issue, [131] applies techniques of modern stochastic programming, specifically the SAA method.
To motivate the SAA method, for $x \in \{0,1\}^{\mathscr{A}}$ we write

$$X \doteq \{x \in \{0,1\}^{\mathscr{A}} : h^T x \leq K\},$$
$$\Phi(x,p,S) \doteq \sum_{s \in S} p_s \phi^s(x),$$

so that using the notation in (3.97),

$$F^*(p,S) = \min_{x \in X} \Phi(x,p,S).$$

Given an integer $N \geq 1$, suppose we *sample* (independently) N times from the distribution p obtaining scenarios $s_1, s_2, \ldots, s_N$. Then clearly for any $x \in \{0,1\}^{\mathscr{A}}$,

$$\mathbb{E}\left[\frac{\phi^{s_1}(x) + \cdots + \phi^{s_N}(x)}{N}\right] = \Phi(x,p,S), \tag{3.101}$$

since the samples are independent and identically distributed (i.i.d.), and so $\mathbb{E}\phi^{s_i}(x) = \Phi(x,p,S)$ for each i. Now, *if* we knew an optimal solution x^* to the problem (3.97), i.e., if $\Phi(x^*,p,S) = F^*(p,S)$, we could use this observation to estimate $F^*(p,S)$, perhaps relying on the central limit theorem.

The difficulty of course is that we do not know an optimal solution to problem (3.97). Nevertheless, given the observation, we theorize that by constructing a "large enough" sample of scenarios, each drawn from the distribution p, and solving the interdiction problem restricted to that sample, we would be able to approximately solve (3.97). Also see [159], [171], [91]. This line of thinking forms the basis of the SAA method. To make these ideas more precise, let

$$J_N = \{s_1, \ldots, s_N\} \tag{3.102}$$

be obtained as above, i.e., by sampling N times from the distribution p.[21] Consider the instance of the interdiction problem restricted to the subfamily J_N endowed with the *uniform* distribution, that is to say the distribution q^{J_N} where $q^{J_N}_s = N^{-1}$ for each $s \in J_N$; we have

$$F^*(q^{J_N}, J_N) = \min_{x \in X} \frac{1}{N} \sum_{s \in J_N} \phi^s(x). \tag{3.103}$$

Now recall that J_N was obtained by sampling from the original distribution p. Thus, for any $x \in X$, by taking expectations (under the random sampling distribution), we therefore

[21]Since we are sampling *without* replacement, J_N as defined in (3.102) may include repetitions; however, in what follows we may describe it as a "set" for convenience.

obtain

$$\begin{aligned}\mathbb{E}\frac{1}{N}\sum_{s\in J_N}\phi^s(x) &= \frac{1}{N}\sum_{s\in J_N}\mathbb{E}\phi^s(x) = \frac{1}{N}\sum_{s\in J_N}\sum_{s'\in S}p_{s'}\phi^{s'}(x)\\ &= \sum_{s'\in S}p_{s'}\phi^{s'}(x) = \Phi(x,p,S).\end{aligned} \tag{3.104}$$

Now suppose that $x^* \in X$ solves problem (3.97), i.e.,

$$\Phi(x^*,p,S) = F^*(p,S). \tag{3.105}$$

Then it follows that for any sample J_N of scenarios as constructed above,

$$\frac{1}{N}\sum_{s\in J_N}\phi^s(x^*) \geq \min_{x\in X}\frac{1}{N}\sum_{s\in J_N}\phi^s(x) = F^*(q^{J_N},J_N). \tag{3.106}$$

Since J_N is randomly chosen, both expressions in this inequality are random variables. By taking expectations, and using (3.104) and (3.105), we thus obtain the following result.

Lemma 3.4.1 (see [159]). *Let the subset J_N be constructed by sampling N times from the distribution p. Then*

$$\mathbb{E}F^*(q^{J_N},J_N) \leq F^*(p,S).$$

Notice that in the above arguments the quantity N is *arbitrary*—the lemma holds even if $N = 1$. However, since it can be argued (see [131], and for a more complete treatment, see [138]) that

$$\mathbb{E}F^*(q^{J_N},J_N) \to F^*(p,S) \quad \text{as} \quad N = |J_N| \to +\infty,$$

or more precisely [138]

$$F^*(q^{J_N},J_N) \to F^*(p,S) \quad \text{as} \quad N \to +\infty \quad \text{with probability 1},$$

we have that Lemma 3.4.1 establishes an important property of the sampling approach. Namely, writing

$$L^N \doteq \mathbb{E}F^*(q^{J_N},J_N) \tag{3.107}$$

(where the expectation is taken over random sets J_N with $|J_N| = N$), an accurate estimation of L^N will be tantamount to estimating a *lower bound* for $F^*(p,S)$, the quantity of interest, which will be tight assuming N is large enough. More precisely, for any $0 < \epsilon < 1$ we can construct a $(1-\epsilon)$-confidence interval

$$[L_\epsilon^{N,1}, L_\epsilon^{N,2}] \tag{3.108}$$

for L_N. For $M > 1$ appropriately chosen, this construction will require, for $1 \leq i \leq M$, the computation of $F^*(q^{J(i)},J(i))$, where J_i, with $|J(i)| = N$, is randomly chosen (from the distribution p), and then an application of the central limit theorem. Further, for each given set $J(i)$ we can compute $F^*(q^{J(i)},J(i))$ for example by using a Benders-like method. In [131], a version of the "L-shaped" method of [208] is applied. If N is not very large this deterministic computation will be practicable. In short, the computation of the interval in (3.108) may be within reach, especially if the quantities $F^*(q^{J_N},J_N)$ are obtained in parallel.

A more complete algorithmic structure would be attained if a similar confidence interval could be constructed for an *upper bound* for $F^*(p,S)$. We will then clearly obtain a high-probability range for $F^*(p,S)$. (It remains to be seen if this range is actually small.) Such an upper bound is in fact readily produced by a closely related sampling process, as follows. Given any family H of scenarios and

$$x^H \in \operatorname{argmin} F^*(q^H, H), \tag{3.109}$$

since x^H is feasible for problem (3.97) but not necessarily optimal, it follows that

$$F^*(p,S) \leq \sum_{s \in S} p_s \phi^s(x^H) = \Phi(x^H, p, S). \tag{3.110}$$

For arbitrary H the gap between the left- and right-hand sides in this inequality could be large. However, if H is randomly drawn from the distribution p, and if $|H|$ is large enough, one would expect that

$$F^*(q^H, H) \approx F^*(p,S)$$

(at least with high enough probability), and so given (3.109) the gap in (3.110) should not be very large. In fact, we can use the sets $J(i)$ above to obtain the interval in (3.108); thus

$$\min_{1 \leq i \leq m} \Phi(x^{J(i)}, p, S)$$

should provide a high-quality upper bound on $F^*(p,S)$. The weakness in this argument is that the quantities $\Phi(x^{J(i)}, p, S)$ could prove difficult to compute if $|S|$ is large. A more accessible approach is as follows. Given N, let U_N be the random variable defined by the following process:

(1) Let J_N be obtained by sampling N times from the distribution p, and let $x^{J_N} \in \operatorname{argmin} F^*(q^{J_N}, J_N)$.

(2) Write

$$U_N \doteq \Phi(x^{J_N}, p, S). \tag{3.111}$$

As just argued in (3.110), $F^*(p,S) \leq U_N$. We contrast this to the random variable $F^*(q^{J_N}, J_N)$ in (3.106), which as argued in Lemma 3.4.1 provides a lower bound on $F^*(p,S)$. We note here that the particular choice for x^{J_N} in (1) is actually *needed*—no matter how $x^{J_N} \in X$ is chosen we will always have $F^*(p,S) \leq \Phi(x^{J_N}, p, S)$. But as we argued above the choice does make sense.

Since $F^*(p,S) \leq \mathbb{E} U_N$, we can now obtain a confidence interval for $\mathbb{E} U_N$. Let $J(i)$, $x^{J(i)}$, $1 \leq i \leq M$, be as above (which satisfies criterion (1)). For a given i, let U^i be a family of scenarios obtained by randomly sampling N times from the distribution p. Then, by taking expectations over this random family, we have

$$\begin{aligned} \mathbb{E}\frac{1}{N}\sum_{s \in U^i} \phi^s(x^i) &= \frac{1}{N}\sum_{s \in U^i} \mathbb{E}\phi^s(x^i) = \frac{1}{N}\sum_{s \in U^i}\sum_{s' \in S} p_{s'}\phi^{s'}(x^i) \\ &= \sum_{s' \in S} p_{s'}\phi^{s'}(x^i) = \Phi(x^i, p, S). \end{aligned}$$

It follows that

$$\frac{1}{M}\sum_{i=1}^{M}\frac{1}{N}\sum_{s \in U^i}\phi^s(x^i)$$

has expectation equal to $\mathbb{E}\, U_N$, and another application of the central limit theorem will yield, given $0 < \epsilon < 1$, a $(1-\epsilon)$-confidence interval

$$[U_\epsilon^{N,1}, U_\epsilon^{N,2}] \tag{3.112}$$

for $\mathbb{E} U_N$, as desired. Together with (3.108) we obtain that

$$[L_\epsilon^{N,1}, U_\epsilon^{N,2}] \tag{3.113}$$

constitutes a $(1-2\epsilon)$-confidence interval for $F^*(p,S)$. For further details we refer to [131], which contains additional implementation details, such as variance reduction techniques.

To summarize the above procedure, we can identify four critical components for the estimation of the value of a stochastic optimization method using the SAA method:

(a) the construction of a confidence interval for a lower bound on the optimal value of the problem,

(b) the construction of a confidence interval for an upper bound on the optimal value of the problem,

(c) the capability to solve a sampled problem of the form $\min_{x\in X} \frac{1}{N} \sum_{s\in J} \phi^s(x)$ where N is large enough, and

(d) the availability of ample computing power to carry out multiple tasks (c) in parallel.

Additional techniques used to speed up convergence, in particular variance reduction methods, are described in [131]. The numerical experiments in [131] concerned networks with up to approximately 400 nodes and 800 arcs, of which approximately 25% were interdictable; in the experiments inequality (3.97b) was a cardinality constraint on the number of interdicted arcs (i.e., $h_{ij} = 1$ for all interdictable (i,j)) with value of K ranging from 10 to 20. In the larger examples the number of scenarios amounts to more than 10^{76}. The experiments used $50 \leq N \leq 5000$ and relied on a pool of approximately 40 heterogeneous CPUs. The CPU time required to compute 95% confidence intervals (3.108) and (3.112) ranged from approximately 100 seconds (on the smallest examples) to 23000 seconds. From a qualitative perspective the resulting confidence intervals were reasonably tight; for example, on a 400-node case, the confidence intervals (3.108) and (3.112) on the lower bound and upper bound for the problem were, respectively,

$$[180.66, 184.10] \quad \text{and} \quad [181.39, 187.95],$$

meaning that a likely pessimistic 90% confidence interval on the problem value is [180.66, 187.95]. The results in [131] can thus be considered successful; using a much larger number of current processors will clearly result in improvements. At the same time, some limitations emerge, not surprisingly. The application of the method to much larger networks (ten or twenty times larger, say) will possibly result in an exponential increase in runtime. A more significant issue is that of controlling the range in the joint confidence interval (3.113). Using a small value for ϵ will obviously help here (though resulting in more extensive computations); however, the underlying issue is actually more combinatorial and concerns the gap between the exact values for $\mathbb{E} L_N$ and $\mathbb{E} U_N$ (see (3.107) and (3.111), resp.).

Summary

The study in [131] and references therein make a compelling case for the use of stochastic optimization in the context of the N-K problem in power grid security analysis. Several technical hurdles must be overcome. One is the need for a very fast solver for a static (i.e., deterministic) network analysis, relying on linear programming duality to express an adversarial problem in convenient form. In fact, such a reliance is found in other work cited in this chapter, in a deterministic setting, i.e., the model in Section 3.1. Possibly one could use an approximate, second-order cone AC power flow model, and apply SOCP duality; though of course that would amount to a convex representation of a model where the nonconvexities are a key feature.

A second hurdle concerns the underlying stochastic model. As discussed above, the rationale for the N-K model is *deterministic*—the stated goal is to provide a worst-case assessment of the safety of a given grid and the set of policies used to govern it. The traditional focus of stochastic optimization problems has been that of estimating (or rather, optimizing) expectations of relevant quantities. Value-at-risk (VAR) [197] analyses might prove more pertinent and practicable under similar methodologies as outlined above. Chance-constrained optimization is now finding a role in other areas of power transmission analysis [35].

The classical stochastic optimization framework could furnish a key role in power grid security analysis. Rather than focusing on the use of stochastic scenarios as a means of quantifying system-wide uncertainty (and, thus, "risk"), stochastic optimization could be used to help make deterministic analysis more robust. In this context, randomness is used to help find deterministic solutions to optimization problems whose quality does not suffer from extreme sensitivity to small changes in numerical parameters. In other words, we would compute successful attacks (in the language of Definition 3.1.3) that prove successful even if the data are subject to (small) "noise." Similar considerations apply to controller actions. This approach would help address the issue of ambiguity that we have raised above. We will describe an application of these ideas below, in the context of cascading failures.

Chapter 4

Modeling transmission system cascades: A survey

We are interested in developing a rigorous mathematical theory, based on optimization, for mechanisms that arrest cascading failures, and for risk assessments of potential cascades. This is a complex undertaking partly because of the difficult nature of the underlying mathematics, and partly because it is difficult to arrive at a precise and intellectually clean statement of what a cascade is. It is also especially taxing to arrive at a satisfactory definition of risk. A model that comprehensively addresses many features of cascading failures will rapidly escalate in complexity and become completely unmanageable from the perspective of optimization. It is therefore important to invest some thought in understanding several modeling issues arising in the context of cascades.

As a result, in this chapter, and in Chapter 5, we will describe progressively more refined and complex models of cascades. Even though the discussion will take us away from an optimization focus, the objective of the exercise is to abstract a notion of cascades in such a way that algorithms for dealing with and optimizing over (abstract) cascades become possible.

This chapter is organized as follows. First we provide a brief introduction to the topic of cascading failures, followed by a survey of recent cascading failures (Section 4.2). Then we review some of the original cascading failure research (Section 4.3) to bring focus on two of the critical modeling features that need to be adequately addressed: line tripping and islanding. This review will suggest areas of modeling where more detail would be helpful, if appropriately implemented; this line of analysis is taken up in Chapter 5. In Chapter 6 we will describe some initial attempts at providing rigorous algorithms for dealing with cascades; these algorithms rely on common threads found in the research described in this chapter.

4.1 ▪ Introduction

Power grid cascades can be superficially described by a common narrative. A cascade originates with an initiating event, or "contingency," where one or several grid elements are disabled due to exogenous reasons. The affected elements are, typically, power lines, though transformer or generator failures can also take place.

After this event has taken place, the grid continues to operate—unless mitigating action is immediately taken, loads will continue drawing power at the same level as just before the event, and generators will continue providing power. However, due to the altered grid topology, power flows and generator behavior will change, and how they will

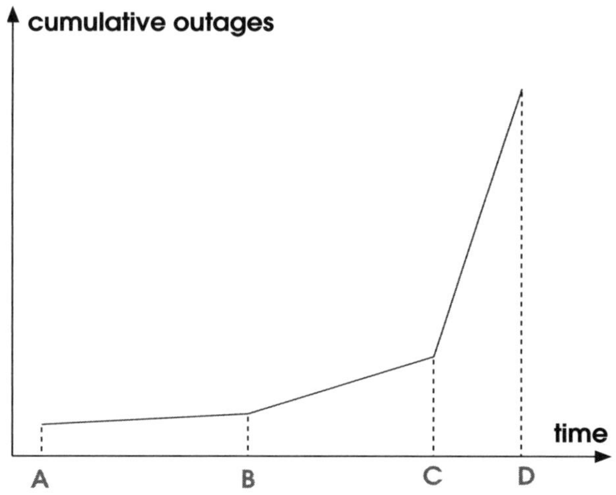

Figure 4.1. *Outline of a cascade.*

change is a function of the underlying physics. Potentially, the new power flows will exceed the capability (rating) of some other line or set of lines or other pieces of equipment, such as transformers. If the excess is significant enough, eventually, those overloaded lines will protectively trip (to avoid damage). Or, if the initiating event is severe enough (post-contingency), selected generators may experience frequency instability, which may result in their being taken out of service. In any case, grid structure and possibly topology will again be altered, triggering yet another change in power flows.

In an extremely unlikely but not impossible scenario, this process repeats, feeding on itself and possibly accelerating. Eventually, the acceleration may become explosive, with large numbers of outages that cripple the system. This is the formal "cascade" phase, which is essentially impossible to contain and as a result of which widespread power outages will occur. During this accelerating stage, and sometimes before, the grid will separate into several "islands" (sometimes referred to as "electrical islands") or, in graph-theoretic language, components. Some of these islands will suffer from a severe excess of demand over available power; in these islands the grid may no longer operate in steady-state condition, and huge oscillations of power flows, voltage, and frequency may be experienced. This will likely result in generators shutting down for protection, thereby creating island-wide blackout conditions. The entire process can be very fast or may take hours. In particular, the initial phases where relatively few lines are progressively tripping can be slow, since line tripping may primarily be linked to thermal processes which last on the order of several minutes (or longer).

When does a cascade start?

Figure 4.1 displays the progression of a prototypical cascade, showing the total number of element outages as a function of time. At point A the initial, exogenous, set of faults take place, with subsequent outages slowly accumulating with some perceptible acceleration at point B; at point D we have an effective collapse of the underlying system, though as stated above significant loss of load may take place before that point. A semantic point where disagreement is seen in the literature is a formal definition of the start of a cascade. One point of view is that the cascade starts at point C in the sense that up to that point some form of control should have been possible to avert a complete collapse, and that after point C the rate of element outages is so rapid that no control is possible. However, from the perspective of this work there is a causal relationship between all events in the process, and thus all outages, including those in the slow stage, are part of the same cascade. For example, in the August 2003 event (discussed in Section 4.2.1 below) the overall timeline (between points A and D) lasted approximately four hours, whereas the time span between C and D arguably was measured in just a few minutes.

Remark 4.1.1. *Throughout this monograph we will view point A as the start of the cascade.*

4.2 ▪ Analysis of recent cascades

The simple narrative provided above masks a huge amount of complexity in every detail: how, exactly, do the exogenous events conspire to attain extremely unlikely scenarios; what is the exact process by which lines or generators trip; and what types of control actions are available? These issues will be examined in greater detail later, and in Section 4.3 we will survey existing work in the literature.

When studying cascade failures of transmission systems, an important detail to keep in mind is that these are rare events and thus it is difficult to make statements that are valid from a strict statistical perspective. Nevertheless, a number of such failures have taken place in the last decade alone, across the world. Are cascades more likely as population grows and grids evolve, primarily in response to economic forces? What should be done to prevent large-scale failures of power grids, and at what cost? What is the likelihood that a grid could be disabled by a large exogenous event focused on a particular geographical location, such as an earthquake, a solar flare, or even a malevolent attack? An analysis of recent cascading events helps pinpoint a number of technical issues of significance.

4.2.1 ▪ The 2003 Northeast U.S.–Canada blackout

On August 14, 2003, a large cascading blackout affected the northeast of the United States and Canada, with some 50 million people losing power, some for several days. This event affected the transmission system known as the Eastern Interconnection, described in Section 1.4, which spans a very wide geographical area (covering the U.S. Midwest, the entire eastern United States including the southern states, and Ontario). The causes for this blackout were quite complex and highlight the need for sophisticated, agnostic analysis of weaknesses in transmission systems. Much valuable material can be found in the official report [101] from which we will quote. Also see [1].

The pure cascade part of the blackout was preceded by a mix of conditions, failures, and problems, none of which caused the blackout itself but on the other hand impeded the ability of grid operators to effectively respond to the cascade once it was initiated. The particular day of the blackout was warm and humid, though not exceptionally so. It is worth repeating what has been stated before: warm weather places stress on the grid, because as demand for power increases, so do power flows, causing power lines to carry larger amounts of power (and thereby increasing power losses). Additionally, high ambient temperatures increase thermal stress on heavily loaded power lines. As power loads increase, it is critical to use reactive power to support voltage levels at the loads; should such reactive power become scarce, the grid becomes more reliant on generators to supply active power, thereby stressing the system even more, as just indicated.

As [101] indicates, on the day of the blackout the grid was approaching a condition of scarce reactive power, thus running a risk of falling voltage levels (and in an extreme case, "voltage collapse"). During the period running (approximately) from noon to 3 PM, the grid experienced a number of unrelated conditions: several lines going out of service, a key generator tripping out, human errors, and software problems. While none of these problems were "electrically related" to the actual blackout, they did on the other hand limit the capability of the system to react to deteriorating conditions later on. The actual failure process started at approximately 15:05 EDT, with a sequence of 345-kV lines in northern Ohio failing. First, the 345-kV Harding-Chamberlin line tripped due to a tree contact (presumably arising because of inadequate maintenance). This is a major line, and its failure caused large flows to migrate to other, nearby lines. Significantly, the fact that the Harding-Chamberlin had failed was not known until *after* the blackout [101]. Some 30 minutes later (15:32 EDT), another 345-kV line (Hanna-Jupiter) tripped. It is notable that Hanna-Jupiter was the line onto which most of the Harding-Chamberlin flows had migrated. Increased current on this line, and resulting temperature increase, caused the line to sag and thereby contact tree branches and fail, while only carrying 88% of its normal power rating ([101, page 57]).

The next event in the cascade, according to [101], is interesting because more structure is known. At 15:41 EDT, the Star-South Canton 345-kV line tripped. This is an important line, which on previous occasions had experienced overload conditions, resulting in relief actions. On the day of the cascade, this line was running overloaded (but within emergency limits) as a result of the Hanna-Jupiter trip, and it failed while relief attempts were being arranged. The line failed after a sequence of tree contacts, which can be explained as follows. As a result of increasing current the line would overheat, and over a period of several minutes, sag, causing a tree contact, which would temporarily take the line out of operation and partially destroy the tree limbs. Upon return to operation (and, according to the report, carrying increased current), the line would resume sagging, thereby causing the next contact. The last contact permanently tripped the line, with power flow at 93% of the (emergency) line rating, at 15:41 EDT.

Following this key event, the system steadily deteriorated, and human operators became increasingly aware that a significant crisis in operations was at hand, one for which no precomputed contingency plan existed [101, page 66]; furthermore, operators lacked exact knowledge of the state of the overall system and were hampered by computer system difficulties (e.g., bugs). Between 15:39 EDT and 15:59 EDT, seven 138-kV lines in northern Ohio tripped (in many cases, due to sagging which caused contacts with other lines). At approximately 16:06 EDT, the 345-kV Sammis-Star line tripped for purely electrical reasons (not tree contacts, and not thermal failure, but related to current overload). This critical event caused the rapid cascade stage to start. In Figure 4.1 this would correspond to point C.[22]

As a result of the previous failures, flows surged into Michigan, causing line trips there, which, in turn, severed existing path flows and caused large flows in the *reverse* direction than normally expected, in order to serve loads. In typical cascade fashion, this resulted in a new large wave of line trips. Quickly (by 16:10 EDT) the entire Northeast U.S. and Ontario became an electrical island separated from the rest of the grid; this island lacked sufficient generating power to meet its demand. At this point the AC power system became unstable, forcing multiple generators to trip. By 16:12 EDT a large blackout condition had developed, with on the order of 63 gigawatts of lost demand; the Eastern Interconnection grid split into multiple islands, with several hundred lines out of operation and several hundred power plants shut down.

[22]As argued above in Section 4.1, we regard the *entire* process as the "cascade," as there is a causal chain of events starting from the initial outages.

4.2.2 ▪ The 2003 Italy blackout

On September 28, 2003, Italy experienced a complete blackout (this material is based on [1]). This event followed a classical cascading failure pattern. Italy is a net importer of power; on the day of the blackout the import flows amounted to some 6400 MW (originating in France and Switzerland).

The cascade began when a tree "flashover" (electrical charge jumping from a line to nearby vegetation) caused a 380-kV line to fail on the border between Italy and Switzerland. Due to the resulting overload, a parallel line soon failed; at this point the power shortage in Italy quickly caused a condition known as "synchronism loss," which then caused the remaining lines at Italy's border to quickly trip as well. This left Italy with a 6400-MW, deficit which caused a complete system collapse. The Italy blackout closely followed a less drastic cascading failure of the southern Sweden–Denmark grid on September 23, 2001 (with approximately 4 million people losing power). Other European events included a 2006 blackout in Germany, with 15 million people losing power.

4.2.3 ▪ The 2011 San Diego blackout

On September 8, 2011, a large, fast-moving blackout affected southern California and nearby parts of Mexico, eventually leaving nearly 7 million people without power in some cases for approximately 24 hours. See [71], [200], and the linked document "San Diego Blackout" (also available from this book's website[23]). Also see [227] and references therein.

It seems clear that the cascade started with the incorrect maintenance-related tripping of a major power line, the 500-kV Hassayampa-North Gila line tying Arizona to California. This line plays a critical role in feeding power to the region, in particular the San Diego area and the Imperial Irrigation District, or IID. According to [71], the WECC interconnect (which includes the area concerned in this event) was not being operated in (N-1)-secure condition. As a result, within one-third of a second, many other events took place, including that of power plants in nearby northern Mexico being forced to trip (meaning, shut down) and the nearby Imperial Valley-El Centro line experiencing problems, as well. Notably, a major line (Path 44) saw its load increase by 84% as a result of the Hassayampa-North Gila trip.

There followed an 11-minute process of steadily worsening conditions as the system attempted to adjust to the decrease of power available to the region by relying instead on flows from the north. This process, however, caused other lines and equipment to overload, and trip, for protective reasons. The SONGS nuclear power plant (a large supplier of power located north of San Diego), as well as other plants, separated from the grid, resulting in the blackout as the San Diego area effectively became cut off.

The report [71] cites inadequate "operational planning" and "situational awareness" as the leading causes for this cascading failure and indicates that similar observations had been made in the analysis of the 2003 Northeast cascade.

4.2.4 ▪ 2012 India blackout

On July 30 and 31, 2012, two separate cascading blackouts affected India; see [62]. The second event caused nearly 600 million people to lose power for several hours. According to the published reports, these events followed the standard outline for a cascading failure. The power grid in India comprises four separate regional grids, which import or export power from each other using special connecting lines. Just prior to the events, the north-

[23] www.columbia.edu/~dano/books/grid/materials/CPUC_briefing_on_San_Diego_blackout.pdf.

ern region was overdrawing power on such lines, in particular the Agra line. In addition several lines had been taken out of operation for repair or upgrade. The power overdraft caused a sequence of line trips (beginning with the Agra circuit) in cascading fashion, eventually causing system frequency to drop and power to be lost in several areas.

According to the report [62] the system should have tolerated the loss of a single line. Instead, the report blames improper operation ("grid indiscipline") for the event.

4.2.5 • Summary

In the above accounts we see as agents of cascades a combination of the relentless pressure placed by the physics of power flows; incomplete real-time information; and, possibly most important, inadequate control algorithms to handle unplanned-for contingencies. In [101], the first stated cause of the blackout was "inadequate system understanding." It would be incorrect, however, for a reader to view the operation of the grid as haphazard or primitive. It is not—in fact the grid is operated by deploying a great deal of sophistication and insight; the development of transmission systems stands as one of the most significant advances of twentieth-century engineering.

While some common threads can be found among the initiating causes for some of the events described above, for example, tree contacts or human error, it would be incorrect to state that a complete description for the causes of cascading blackouts exists. The "low-hanging fruit" that emerges from the above is the need for better real-time sensor information (or, as [101] puts it, better "situational awareness"). Certainly this would prove very useful; however it would not eliminate the need for intelligent control algorithms. What the accounts of previous incidents do highlight is that resource-constrained operations of large, complex systems give rise to risk. When such risk can have a large economic or social impact, it should be approached in a scientific manner, including, especially, an agnostic viewpoint in examining the costs and benefits of control actions and, most important, the analysis of potential contingencies.

A methodology found in the power industry (and the literature) is to perform grid vulnerability studies by considering a fixed set of cases, or contingencies, and performing a simulation of system response in the event that any such contingency is realized. Since each simulation likely entails a complex sequence of computations, and since the set of *all* potential contingencies is exponentially large, some (heuristic) screening is used to focus on a smaller set of cases. Enlarging this set through adversarial analysis should be helpful in this regard. A technical approach to this issue is the so-called N-K problem discussed in Chapter 3, which systematically investigates whether the simultaneous failure of K grid components (lines, for example) creates a potential hazard. Typically, K is a small integer—current operation requires that the grid remain (N-1)-safe, but considering larger values would help in identifying unknown problems or errors in operation under different grid conditions. As seen in Chapter 3, even under assumptions that greatly simplify the setting, the N-K problem is a notoriously difficult combinatorial problem.

A related need arises in the context of real-time operation. Since grid conditions change constantly, such adversarial analysis would also need to be constantly performed. To some extent practices based on this observation are in current operation, though the adversarial ingredients are likely very limited. However, one can envision an operation of the grid where, in real time, an extremely large number of parallel simulations of future evolutions of the system are considered in order to suggest control actions, following a procedure reminiscent of that found in recent chess-playing computer programs. Such an endeavor would prove computationally expensive but could add further scientific rigor to grid operations. To the best of our knowledge, this approach is not used today.

4.3 ▪ Review of prior modeling work

In this section we review some of the prior work on analysis and simulation of cascading power system failures. The formal review of the literature starts in Section 4.3.1. To put the existing research into context, however, it is worth first outlining some of the central issues that arise from this work to understand not just the complexity that any attempt at optimization will have to deal with but the modeling challenges that need to be tackled before optimization is even considered. In this section we will discuss such modeling issues and discuss practicable solutions and corresponding critique. In later sections, primarily in Chapter 5, we will present attempts at resolving some of the difficult obstacles that we will discuss below.

Several features that stand out in the prior work are as follows:

(a) To a major extent, line tripping due to overloads is used as the primary modeling device for capturing the evolution of cascades.

(b) Modeling has primarily focused on comparatively slow cascade processes. Some cascades are very fast, and many start slowly but incorporate a very fast phase toward the end (see Figure 4.1).

(c) The standard linearized (DC) approximation to power flows is used.

(d) Short-term dynamics is largely not considered.

(a) *Focus on line tripping.* It is clear that line tripping is not the sole mechanism underlying cascades. Some cascades have taken place very rapidly, very likely due to fast transformer tripping or frequency issues which could cause generator tripping.

Transformer and generator tripping and other forms of system collapse

Transformers are common in transmission systems and play a critical role. Transformers receive power as an input and can output power with changed voltage magnitude and phase angle. They can be only partially modeled using the DC approximation, however, and from a simplified perspective they are typically modeled as having one input and one output, each connected to a bus. See [25, Chapter 5] for more background. Transformers can fail due to a number of reasons, including aging of components, but overload (current or voltage) is significant in the context of a cascade. Transformers are rated, in particular, in terms of power handling (KVA) capabilities and operating temperature. Transformer tripping appears to have played an important role in the progression of the September 2011 Arizona–Southern California cascade [71].

A generator may also trip and be taken out of service during a cascading failure. In this case the trip will be caused by protection systems installed at the generator to prevent physical damage. As discussed in Section 1.4, generators located in a (freshly created) island with an excess of load will experience a frequency drop; however, if the generation deficits are large enough (among other conditions), frequency response and secondary response may prove inadequate to stop frequency loss, and generators will likely trip. Finally, voltage decay can also prove a cause of system collapse.

Further, a cascade might be preceded by a comparatively long period of localized lower than desired voltages, caused for example by inadequate reactive power reserves together with large load levels. This condition can result in inefficient grid operation, and, possibly through remedial efforts to boost voltages (by providing additional reactive power to the loads) might produce higher line currents and thus higher line temperatures, which in turn place the grid at a higher level of cascading risk—higher likelihood that a contingency such as an exogenously caused line outage could result in a cascading process. To summarize the above, a model of cascading failures should incorporate a means to model such system-wide vulnerabilities. Refer to the discussion at the start of Section 3.3, where these points were presented in the context of the N-K problem.

In the initial phases of a cascade, however, line tripping appears to play a very significant role in the observed changes in grid behavior. The extent to which line outages tend to be stressed in the above accounts of cascades (Section 4.2) is remarkable. Even though the accounts also point out other types of events, line outages provide the major checkpoints in the narratives, signaling both changing conditions in grid behavior and opportunities for control action. From this perspective, feature (a) is justified.

(b) *Focus on modeling of slow-moving cascade conditions.* A weakness resulting from a reliance on (a) and (c) is that models based on these assumptions likely lose validity when a cascade enters a final, “exponential” stage characterized by very rapid system failure. Other phenomena, such as the loss of frequency synchronism among generators and voltage collapse, possibly play a larger role at that phase. However, as we discuss next, from a modeling and control perspective, such a final stage of a cascade is extremely complex and, arguably, research should focus on the earlier stages.

From a modeling perspective, there is a basic inadequacy of the steady-state physics model (the AC model) toward representing a grid in very rapid cascading conditions—the standard AC model is a steady-state model, and what is needed at this point is a global model of a grid in unstable mode,[24] with, as stated before, different phenomena taking place in parallel and, in particular, a large volume of line and generator tripping. As brought up in Section 1.4, numerical techniques can be employed to carry out simplified versions of this type of modeling (involving swing equations together with power flow modeling); however, the computational tasks can prove demanding. In fact, the development of a computationally practicable model of a grid in the final phase of a cascade that addresses all significant factors remains an open challenge. Among a myriad of technical difficulties, the nonconvexities in AC power flow computations will likely become more prominent as the grid decays and exhibits undesirable behavior (rendering the choice of starting points for iterative methods difficult). The model should also incorporate stochastics, both to account for model incompleteness and as a tie-breaking mechanism to handle various race conditions (see Section 5.1). Another practical challenge would arise in the calibration of the model, which would be extremely problematic as it would likely require extensive data captured during actual cascades.[25] These technical difficulties can already arise in the analysis of cascades in their early stages, but it appears fair to say that the task of performing extensive, accurate, complete simulations of synthetic cascades as they reach the exponential phase will likely exceed realistic computational capabilities *as opposed to set-piece simulations of past cascades, which rely on manual calibration and supervision.*

[24]From a power engineer’s perspective, the term “transient” might not be the immediate choice here because of the comparatively slow time frame.

[25]It is to be noted that significant interest is directed at very large scale instances of the problem, involving thousands of buses and lines.

(c) *DC versus AC power flow models.* A central question about cascading failure models concerns their reliance on DC power flow models. We can examine this question from various perspectives:

- DC models only account for active power flows, cannot directly describe conditions such as voltage decay, and cannot be used to model control actions such as the supplying of reactive power to loads.
- Standard DC models are only accurate when voltage levels are high and phase angle differences (across lines) are small. Both conditions may fail when a grid decays.

However, potential appeal from the improvement in the accuracy of a cascading model that would come from using AC power flows is somewhat diminished in light of several concerns. To begin with, many of the software tools for AC power flows are commercial, with many implementation details kept proprietary. This makes it difficult for widespread independent assessment by experts. In addition, it is worth revisiting some of the issues we discussed in Section 2.1:

- AC power flow computations are comparatively much slower than DC computations, making large-scale statistical analysis difficult.
- Power flow computations may not converge. Software packages will usually compensate by relaxing constraints (or, more properly put, by changing constraints in a manner judged minimal) to attain convergence. When simulating a cascade, multiple power flow computations are needed and the accumulated effect of such constraint adjustments is not clear.
- An AC power flow computation can be cast as an optimization problem (e.g., minimize the 2-norm of the feasibility errors). But this is a nonconvex problem and even when a computation converges, it may converge to a local minimum. In the case of a power grid in distress, there may be many local minima, and all may reflect some behavior that under normal circumstances would be viewed as undesirable; thus there is no "natural" local minimum to which any good software package "should" converge. Furthermore, different local minima may lead to different cascade trajectories. The choice of the local minimum to which a particular software converges may be idiosyncratic (i.e., may reflect underlying properties of the algorithm or even of its particular implementation). As in the prior point, the accumulated impact of convergence to local optima at several points during a simulated cascade becomes quite problematic.

Executive summary

The shortcomings of relying on the standard DC model are clear: it is a good partial approximation to the AC model only under very good operating conditions, which is not what we would expect in a cascade, possibly not even at a comparatively early point.

However, the basic driver for the reliance on the DC model is its computational ease and robustness, especially in the context of volume studies. When simulating cascading behavior of a grid, one might run tens of thousands of individual cascade simulations, each of which might call for many power flow calculations. Even in the slow phase of a cascade, a grid may already display unfamiliar behavior, and in this regime, the more accurate AC power flow solvers, in addition to being slower, may tend to fail to converge too often to produce useful statistics. When AC solvers do not fail, they nevertheless may succeed but at the cost of introducing artifacts into the simulation, such as the altering of problem data to attain "convergence."

This is an example of the conflation between *control* and *physics* brought up before. When simulating a cascade (in particular when evaluating a proposed control action), it is important to know, agnostically, what the underlying physics states about the power system. This goal may require many power flow computations that follow an evolving system in the midst of a cascade. If, at each computation, some subtle control is effectively introduced into the operation of the grid by the AC power flow solver, a troublesome situation arises if after a number of computations the simulation produces a grid that is inoperable, e.g., a large blackout condition. One is then left wondering: does the simulated blackout condition result from the initial grid condition, or is it a result of the sequence of controls unwittingly introduced by the solver?

(d) *Ignoring dynamics.* A final criticism of (a)–(c) is that each change in the topology of the grid, especially those that cause islanding, is likely to produce dynamics that are not adequately captured by the DC power flow model (or, for that matter, by the AC model). For example, in an island with severe generation shortage, the dynamics of secondary control, which play out in minutes, would materially affect power flows and thus could have an impact on line tripping. Capturing these dynamics and their impact on equipment tripping in a computationally tractable manner is a notable challenge. We will return to this issue below, in Section 4.3.1 and in more detail in Section 5.5. Initial work in this area can be found in [81], [80].

Ultimately, a competent software package that accounts for system behavior at multiple scales of time (and space) should be able to resolve all the points brought up in our discussion of (a)–(d). Of course, such a package would necessarily account for non-steady state behavior, that is to say, it would go beyond standard AC power flow modeling. This will likely require both methodological progress to render a practicable algorithm and an extremely good implementation.

Summary: The impact of incomplete cascade models

In examining existing work on cascade modeling it is clear that major compromises have routinely been made between the competing goals of attaining high fidelity on the one hand, and analytical clarity, methodological rigor, and computational practicability on the other. To put it differently, while it is certainly possible to assemble a representation of a power system that incorporates a wealth of complex details, the usability and usefulness of such a system becomes problematic if mathematically agnostic analysis tools are not available.

In our review of prior work, such compromises make criticism possible and even facile. When engaging in such criticism, our goal will be to exhibit the profound complexity that makes the compromises necessary to perform systematic analyses of cascading conditions.

4.3.1 ▪ Initial cascade models

We first review the work in [59], [60], [86], [61], which to a significant extent initiated this field.[26]

[26] A related, recent citation is [99].

One of the observations that emerges from the experiments in this work is that the frequency versus severity relationship of cascades shows evidence of a "power law" as opposed to an exponentially decaying behavior. The main point being made is not that inadequate investment on grid resources has taken place (in fact, significant investment has taken place) but that the pattern in the investments conspires with demand growth to create "critical behavior," whereby the grid is continuously operating just above a margin of safety, so that the probability of significant cascades does not decrease with time. An important point is that the above references were all presciently written prior to the 2003 Northeast blackout. The worldwide evidence since that event would tend to strengthen the validity of the theses in this work.

To approach this type of issue, [59], [60], [86], [61] study cascades in two different time scales: first, a short time scale which seeks to explain the dynamics of individual cascades and, second, a much longer time scale model which tries to explain why grids are susceptible to cascades, to begin with, and why this risk appears to be increasing.[27] Both viewpoints are clearly important. We consider the long-term time scale first.

The longer time scale model can be more completely described as follows:

(1) Time is discretized into periods. Denoting by $D_{i,t}$ the demand at bus i on period t, then

$$D_{i,t+1} = \lambda_t D_{i,t}, \tag{4.1}$$

where the λ_t are independent random variables taking values in an interval $[\lambda_{\min}, \lambda_{\max}]$ and with mean $\bar{\lambda}$ very slightly larger than 1. Likewise, denoting the maximum output of a generator i by $P^{\max}_{i,t}$,

$$P^{\max}_{i,t+1} = \bar{\lambda} P^{\max}_{i,t}. \tag{4.2}$$

(2) Let $U^{\max}_{j,t}$ denote the capacity (line rating) of line j on period t. Suppose a cascade takes place on day t, during which line j becomes overloaded. Then this line is "upgraded":

$$U^{\max}_{j,t+1} = \mu_t U^{\max}_{j,t}, \tag{4.3}$$

where the $\mu_t > 1$ are independent random variables taking values in an interval $[\mu_{\min}, \mu_{\max}]$ with $\lambda_{\max} < \mu_{\min}$.

(2) models the (reasonable) assumption that power system operators will spend resources on equipment that fails during a cascade. If many cascades take place during which a particular line j will overload, then that line will be upgraded to a comparatively higher degree.

At the opposite end in the model lies the shorter time scale, that is to say the modeling of individual cascades. The connection between the two time scales is provided by random line outaging events. On a given period t, let $f_{j,t}$ denote the (active) power flow on line j, and define the fractional load of line j on period t as

$$M_{jt} \doteq |f_{j,t}|/U^{\max}_{j,t}. \tag{4.4}$$

[27] For a correlated perspective on a similar issue, the increasing number of individual faults, see Figure 7.2 of [8] and the preceding discussion.

Then line j is outaged, independent of all other lines, and randomly:

$$\text{Prob(line } j \text{ becomes outaged)} = h^0(|f_{j,t}|/U_{j,t}^{\max}), \tag{4.5}$$

where h^0 is a nondecreasing, positive function bounded above by 1. In other words, lines with higher fractional load are more likely to fail. From a certain perspective, this is a very reasonable model, since lines with high overload will experience thermal stress, possibly leading to contacts or flashovers, as previously discussed (a critique of the model will follow later). In [86], $h^0(x) = 0.001$ (for all x) is used. However, the goal is to have line outages be very rare events (one in 30 days, for example), and for large grids a much smaller value would have to be chosen; equivalently, using a Poisson distribution with small rate would achieve the same effect.

If, in period t, some line or lines become outaged as just described, then a cascade may start. This is where the short time scale model steps in. In outline, the cascade model works as follows.

Model 4.3.1. Cascade model I [59], [60], [86], [61]

Step 0. Input: some lines have become randomly outaged using rule (4.5).

Step 1. Reset loads, generator outputs, and power flows to adjust to line outages.

Step 2. Using rule (4.6), test for new line outages.

Step 3. If no new lines are outaged, **STOP**. Otherwise, go to Step 1.

Steps 1–3 embody the dynamics of a cascade, in particular how line outages interact with new power flows (instantiated during the cascade) to continue the cascading process. We have already discussed how rule (4.5) is used to model random line outages. To handle outages of overloaded lines during the cascade a similar model is employed. Let $\tilde{f}_j$ denote the flow on line j at some point during the cascade. Then

$$\text{Prob(overloaded line } j \text{ becomes outaged)} = h^1(|\tilde{f}_j|/U_{j,t}^{\max}). \tag{4.6}$$

Note: This rule only applies to overloaded lines, i.e., lines j with $|\tilde{f}_j| > U_{j,t}^{\max}$. [86] uses $h^1(x) = 0.3$ (for all $x > 1$). As a result, a line which remains overloaded during several consecutive executions of Steps 1–3 will become outaged with high probability. In Section 5.3.2 we will provide a detailed stochastic model of line tripping that builds on these ideas.

The remaining item to discuss in Model 4.3.1 is the implementation of Step 1. This is critical because it relates both to the evolution of the cascades (namely, which lines are outaged in Step 2) and to the eventual severity of the cascade—the amount of demand lost once the cascade stops. There is an additional element at play here, which is that the implementation of Step 1 should amount to a faithful simulation of the way that the grid actually works, and not to an ideal control. A crucial issue is the possible islanding that might take place as the cascade evolves and, specifically, that some islands would likely experience an imbalance between available generation and load. A significant part of the problem here is that there is, currently, no automated methodology specifically designed for responding to an ongoing cascade, as opposed to automatic, local controls used to correct local deviations from normal behavior. Rather, the grid would respond using a combination of real-time mechanisms used to balance supply and demand; this is the frequency control described in Section 1.4. The situation is made even more complex by the many competing time scales at play, so it deserves some careful analysis.

In normal grid operation, a number of entities that provide power for customers will begin matching supply to predicted demand as far as one day in advance. This entity organizes markets operating at several time scales, involving operators which actually generate power and utilities that deliver power to customers. We refer to Section 1.4 for a discussion of the unit commitment and OPF problems; a central point is that algorithmic tools are used to control generator output in time increments with resolution as fine as a few minutes.

A related issue brings up a different question: what happens in real time as demand inevitably fluctuates? There are two perspectives to this question. As outlined in Section 1.4, a simple point is that an instantaneous increase in demand causes generator frequencies to drop, and following the disturbance, some generators will automatically attempt to adjust by producing more power (if possible), thereby increasing their speed of rotation. In addition, SCADA (supervisory control and data acquisition) and PMU (phasor measurement units) systems convey real-time data to central locations (see Chapter 1). This (partial) data, together with a state estimator system, is used to construct an approximate picture of how the grid is operating. In the case of significant disturbances (such as large changes in loads, generator outages, or line outages) this estimation can be used to effect central control to implement adjustments.

Now we return to the issue of islanding. The 2003 North American cascading blackout produced large islands with supply deficits; in the 2003 Italy event all of Italy effectively became an island with a shortage of generation. Simulations of severe cascades produce cases where many islands are created, sometimes in rapid succession. It seems clear that current SCADA systems would be unable to quickly determine the precise makeup of the islands—to do so, accurate knowledge of the set of tripped lines would be needed. Thus, central supervisory control would be unable to quickly respond, in a detailed manner, to islanding.

Island dynamics

Suppose an island with a generation deficit has just been created. In real time, generators in that island (if any) would experience a frequency drop. The power engineering background needed to analyze this situation has been discussed in Section 1.4 but is worth reviewing here. The corresponding mitigating action, i.e., primary and secondary response, would have some of the generators in the island attempt to boost their output, however limited by their mechanical capabilities—a generator would react at a finite speed and would also be constrained by its maximum total output. However, the precise dynamical evolution of generator frequencies and voltages would likely be quite complex as individual generator frequencies converge to an average island-wide frequency. In a normal situation the process will eventually stabilize, (perhaps) with acceptable generator frequencies, but in the case of a dangerous contingency, system behavior may become erratic and possibly lead to generator tripping events. Overall this necessitates the use of detailed models for generator dynamics. As discussed before, such models rely on systems of differential equations and model the effective interactions between different generator frequencies in narrow time scales using numerical techniques. In large-scale settings the solution of such systems can prove numerically challenging. Moreover, it is certain that there would be random effects that are difficult to model, due for example to lines becoming temporarily tripped. Nevertheless, from a nonspecialist standpoint it seems fair to say that if an island develops with a small demand deficit, generators in the island would likely be able to react and make up the deficit. In the case of a somewhat larger deficit, possibly, generation would be able to reduce the deficit (up to generation capacity); however, the eventual stable pattern of served loads is not precisely clear. And if the deficit is very large, the system could become so unstable that the entire island would shut down (blackout condition). This type of event appears to have taken place in the 2003 Italy blackout.

The case of an excess of generation in an island is similar—generators would slow down and reduce their output at a certain rate. While this situation seems more benign than the deficit case, if the generation surplus is large, instabilities could also arise.

To make matters worse, in all cases of islanding a parallel dynamic plays out. As generators adjust their output, new power flows are instantiated, which depend on the topology of the island and its line parameters. Should these power flows substantially exceed the ratings of some of the operating lines in the islands, those lines might also trip, changing the structure of the island before demand equilibrium has been reached. Accurately modeling this situation presents many challenges.

In Section 5.5 we will provide an elaborate model of island dynamics that appears suitable for simulation. Returning to the implementation of Step 1 in Model 4.3.1, in [59], [60], [86], [61] variations of the DC model are used. The following description amounts to an amalgam of features found in the references above:

(r.1) First, for each generator $g \in \mathcal{G}$ a variable P_g is used to model its output, and likewise for a load $j \in \mathcal{L}$ its demand is represented by a variable D_j.

(r.2) If a line (i,j) is outaged, then rather than removing the line from operation, two numerical adjustments are made. First, its reactance x_{ij} is scaled up by a large value $\varkappa_1$. Second, its limit u_{ij} is scaled down by a small value $0 < \varkappa_2$.

(r.3) To obtain the new power flows, loads, and generator outputs, a linear program is solved, with the following constraints. First,

$$P_g \le P_g^{\max} \; \forall g \in \mathcal{G}, \quad 0 \le D_j \, \forall j \in \mathcal{L}. \tag{4.7}$$

Here, for $g \in \mathcal{G}$, $P_g^{\max}$ is its maximum output. As a shorthand, write

$$\tilde{b}_i = \begin{cases} P_i & \text{if } i \in \mathcal{G}, \\ -D_i & \text{if } i \in \mathcal{L}, \\ 0 & \text{otherwise.} \end{cases} \tag{4.8}$$

The remaining constraints are per the standard DC model:

$$\begin{aligned} &\sum_i \tilde{b}_i = 0 \quad &&\text{and} \quad N f - \tilde{b} = 0, \\ &|f_{ij}| \le u_{ij} \quad &&\text{and} \quad x_{ij} f_{ij} = \theta_i - \theta_j \quad \forall (i,j) \end{aligned} \tag{4.9}$$

(where N is the node-arc incidence matrix of the grid). The objective to optimize is (in one of the versions)

$$\max \sum_{g \in \mathcal{G}} P_g + W \sum_{j \in \mathcal{L}} D_j, \tag{4.10}$$

where W is a large constant (e.g., $W = 100$). In [59] the following variant is used:

$$\max \sum_{g \in \mathcal{G}} |P_g - P_g^{\text{last}}| + W \sum_{j \in \mathcal{L}} (D_j - D_j^{\text{last}}), \tag{4.11}$$

where P_g^{last} is the output of generator g just prior to the line outage, and similarly with D_j^{last}.

This linear program models economical redispatch of the generators (with equal cost to all generators) but placing a high penalty on load shedding.[28] As such, it is probably most applicable to a slowly evolving cascade (or a slow stage of a cascade). We will return later to the choice of objective in the linear program; first we comment on the numerical adjustments in (r.2).

Note that as per rule (r.2) the u_{ij} (line ratings) of outaged lines (which appear in (4.9)) will remain positive. Thus the grid will always remain connected. This is a necessary step if Step 1 is to be implemented as a single optimization problem: if outaged lines were removed from the model, then the optimization problem would likely become infeasible if islands were generated. Preserving connectivity significantly simplifies the model and in particular hides the complexity of explicitly having to deal with demand imbalances in islands. As the u_{ij} of outaged lines go to zero, the solution provided by the above linear program will have the property that any islands that have been created will remain (nearly) balanced. The objective in (r.3) ensures that the necessary generator output adjustments and/or load shedding are minimal.

On the other hand, the linear program above will force flows on outaged lines to (near) zero, but will the flows behave as if the lines were actually out of service? That is to say, is the model realistic from a flow perspective? This is not an idle question, since

[28] As stated, this linear program allows loads to *increase* beyond their original values; we assume this was an oversight in [59].

the $x_{ij} f_{ij} = \theta_i - \theta_j$ equations in (4.9) are imposed on all lines. This brings in the scaling of reactances by the large value x_1. Here, we remind the reader of Theorem 3.3.1, proved in [42] and restated here for convenience.

Lemma 4.3.2. *Consider a DC loadflow problem with underlying network G. Let S be a set of lines whose removal does not disconnect G. Let* $\epsilon > 0$, *and suppose that for each line* $(s,t) \in S$ *we reset* $y_{st} = \epsilon$, *and for* $(i,j) \notin S$ *we leave* y_{ij} *unchanged. Let* $(f(\epsilon), \theta(\epsilon))$ *denote the resulting DC solution on G. Let* $(\bar{f}, \bar{\theta})$ *denote the DC solution on G—S, using the original line susceptances. Then we have the following:*

(a) $\lim_{\epsilon \to 0} f_{st}(\epsilon) = 0$ *for all* $(s,t) \in S$.

(b) *For any* $(u,v) \notin S$, $\lim_{\epsilon \to 0} f_{uv}(\epsilon) = \bar{f}_{uv}$.

(c) *For any* (u,v), $\lim_{\epsilon \to 0}(\theta_u(\epsilon) - \theta_v(\epsilon)) = \bar{\theta}_u - \bar{\theta}_v$.

The import of this result is that *if* the set of outaged lines in Step 1 does not form an island, the "trick" of increasing the reactances of outaged lines to large values will, indeed, succeed in modeling the behavior of the grid with those lines removed. Thus, the two adjustments in (r.2) do work in unison—this is, in fact, an ingenious device for avoiding complexity while maintaining a faithful model.[29]

Another point concerns the choice of objective in the linear program. As stated, it favors achieving balance in islands by adjusting generator output. In cases of large imbalances, this could result in large changes in some generators. In fact, since a linear program is being solved, an extreme point solution is most likely obtained. This would result in one generator per "island" accounting for most of the adjustment. From a real-time perspective this could be unrealistic—probably one would see several generators simultaneously changing their output. From a total power delivery standpoint this might not make a large difference, but from the perspective of power flows it could, and this might well impact the overall dynamic of line trippings in the model. In fact, note that both objectives (4.10) and (4.11) are global—they are a sum across all generators or loads in the network. Thus, potentially, qualitatively similar islands (similar load imbalances, and similar overall makeup) could be treated very differently by the solution to the linear program.

What the models say

Despite our technical quibbles, the models in [59], [60], [86], [61] are compelling and produce a very interesting overall picture. While one may criticize the accuracy of the detailed representation of individual cascades, it seems fairly certain that a cascade with severe outcome under the model would amount to a severe cascade from a real-world perspective, and vice versa.

The simulations involve very long term periods (over 100 years simulated one day at a time), resulting in thousands of individual blackouts. The principal dynamic under study is the interaction of long-term capacity expansion, given by (4.2), with growth in demand as in (4.1), punctuated by random outage events controlled by (4.5) and resulting in equipment upgrade as in (4.3).

[29]The "if" is important. Should the outaged lines create islands, we do not have a theoretical guarantee that the linear program in (r.1)–(r.3) will constitute a faithful facsimile of the DC loadflow linear program with the outaged lines removed, in particular in large-scale cases and with many islands.

An important feature related to the use of (4.3) is that while on the one hand it is an admittedly simple modeling device to describe the reaction to a blackout, at some level it does correspond to the way that grid operators would react—the proximate cause of the "last" blackout is addressed. A feature that arises from (many) simulations in [59], [60], [86], [61] is that, despite the repairs, subsequent blackouts still take place.

In fact, a related interesting point emerges from the simulations. In the experiments, the repair factors, that is to say, the μ_t in (4.3), are kept constant: $\mu_t = \mu$ for all t. What the experiments show is that under the given models, the size of blackouts (i.e., the amount of lost demand) can *increase* as a function of time if μ is not chosen large enough (see, e.g., Figure 1 in [86]). In other words, under the given model of (slow) demand growth and parallel generator capacity upgrades, an appropriately aggressive repair mode is called for, and this is a system parameter, rather than a quantity that could be determined by a simple rule of thumb.[30]

There are a number of ways to interpret this fact. From the practical perspective of a grid operator, it would be desirable to keep μ quantities small (to keep costs down). So the optimal engineering trade-off would be to make the μ just "large enough." But how large should that be? In fact, one could criticize equation (4.3) because, strictly speaking, its application does *not* prevent the cascade that just took place—it would *appear* to be guaranteed to do so if μ were larger than the overload factors experienced by lines that are outaged in Step 2 of Model 4.3.1. This is a random set of lines, as per the model, but when (4.3) is applied it is known.

However, there is a hidden and very interesting form of complexity here, which is (repeatedly) revealed when one performs cascading experiments. Namely, a version of Braess's paradox [49] can, and does, arise. To put it in a simple way, a given choice of μ may result in averting a given blackout, whereas a larger choice may *fail to do so*. This may appear surprising, but we will provide small examples, later on, that substantiate this view. The reader may then say that, surely, choosing μ large enough so that *every* outage in Step 2 is avoided would be good enough. There is a pitfall that arises now, namely, that cascade simulations produce outage sequences where some of the line loads are extremely high—this would necessitate a prohibitively large choice of μ.

To make matters worse, further thought shows that even this perspective is somewhat limited—we should not think about "averting a blackout" but rather think about the cost trade-off involved in reducing the size of the blackout. Simulations and algorithmic exploration show the resulting cost/benefit analysis to be extremely complex. In summary, the choice of an "optimal" μ_t is extraordinarily difficult. We conclude that the apparently simple approach followed in [59], [60], [86], [61] is reasonable, given the context, and that their experiments, involving alternative values for μ, expose the underlying complexity and the need for careful analysis. Similar issues, from an algorithmic context, have been considered in the cascade-averting models in [37]. In Chapter 6 we will further consider rigorous online control algorithms used to stop an ongoing cascade with minimum demand loss.

A second issue that emerges is what is termed a long-term "dynamic equilibrium." For a given choice of μ, the experiments seem to indicate the long-term average behavior of a blackout. For example, the size of the blackout is either constant or slowly growing once the equilibrium point is reached. This is the point we discussed above; in addition the dynamic equilibrium is better exposed by other blackout parameters, in particular, the

[30] Alternatively (not considered in the model), one would consider a different upgrade model, perhaps one where new power lines can be added, or new generation locations could be considered, as opposed to simply increasing generator capacity.

average relative line load on a given day t,

$$\bar{M}_t \doteq \frac{\sum_j M_{jt}}{m},$$

where m indicates the number of lines (refer to (4.4)). Clearly, this is a useful parameter from an operational perspective; larger (close to 1.00) values of M_t indicate potential vulnerability, and this is leveraged in rule (4.5) of the models. Later we will see that the $\bar{M}_t$ observed *during* a cascade prove useful as parameters for control application.

Simulations in [59], [60], [86], [61] show that the $\bar{M}_t$ attain a long-term equilibrium behavior (well-defined average and variance) for large enough t for a fixed value of μ. To be sure, the particular values $M_{j,t}$ can exhibit complex behavior for each given j as t grows. An interesting question concerns the behavior of the histogram of values $M_{j,t}$, and not just their average.

Summary

Put together, the pattern that emerges from the experiments is that, assuming a constant rate of demand growth, a constant rate of equipment investments (in particular, applied to remedy observed shortcomings) is not enough to avert blackouts—blackouts will continue, settling to a long-term behavior of relatively frequent blackouts, of significant severity, and with significant equipment overloads.

To close this section, we point out a potential criticism of this line of work. This concerns the stochastic nature of the cascade initiation events as per rule (4.5). Simulation experience, easily verified, indicates that random line outages produced in this manner will usually not devolve into cascades, even small ones. Cascades will be rare, and severe cascades quite rare. This is in line with the observations noted above. However, a common viewpoint on extreme events (earthquakes, large hurricanes, and so on) is that their relative frequency is not as important as their (conditional) severity. Taken from this perspective, the results above seem to be potentially flawed—they are all from the standpoint of "average" events. However, this criticism is itself incorrect. One of the goals of the models is to examine the relationship between demand growth, equipment investments, and blackout frequency. Investments, in particular, would be made in response to less than extreme blackouts. Thus, the random event dynamic is correct at least from this perspective. A useful way to complement the models is to parallel the random outage model with one that looks at rare events (possibly, driven by an adversarial or worst-case model), assuming a given history of equipment upgrades.

4.3.2 ▪ Subsequent models

The initial work of Dobson et al. has been followed by a number of papers that expanded the models and focused on a number of central points, reviewed here.

Anghel, Werley, Motter (2007)

Much work followed the initial models in Section 4.3.1. We next focus on [9], which introduced several interesting ideas and extensions to the work in Section 4.3.1:

(a) a stochastic model for exogenously caused line outages that incorporates line-specific information,

(b) a physical model for line outages due to thermal causes, and

(c) a model of the dynamics of line tripping, including a form of control on the part of the grid operator.

Cascade-initiating events Concerning (a), [9] models outages on a given line, k, as being exponentially distributed with rate $\lambda_{f,k}$, that is to say, the density function for the time until the next failure is $f(t) = \lambda_{f,k} e^{-\lambda_{f,k} t}$. Moreover, $\lambda_{f,k} = \lambda L_k$, where λ is small, and L_k is the length of line k (calibrating λ might pose some challenges, but it should be possible from existing data). This is a very reasonable model, so long as we are interested in "random" events, such as strong winds causing foreign objects to strike a line. Certainly, longer lines should be at comparatively higher risk. On the other hand, the propensity of a line to sag (due to thermal overload) and strike a foreign object can be enhanced by the load it carries, which is independent of line length, as well as other, exogenous factors that are more difficult to model. As pointed out in [9], some of the external factors, such as ambient temperature, could be incorporated into a line-specific p.u. length line failure rate parameter, though as pointed out before, external temperature could vary along the length of the line. Grid-specific factors, such as high line loads, cannot be modeled this way but could be used to fashion a (length-independent) line failure rate λ_k. Other low-probability events, such as human action, or unusual grid conditions, may be more difficult to capture by an exponential distribution. Nevertheless, this is an interesting feature that points to geographical effects in blackout dynamics.

Thermal line tripping Now we turn to (b). To address line failures due to overloading (essentially to implement Step 2 in Model 4.3.1), [9] uses an elegant model of line temperature. Grid operators attempt to limit a line's (internal) temperature, which is a function of current (amperage), to prevent excessive sagging and deformation, which could damage the line itself or cause it to contact a foreign object. See [101, pages 57–59].

For a given line, its temperature $T(x,t)$ at point x, at time t, is described in [9] by a model based on the heat equation:

$$\frac{\partial T(x,t)}{\partial t} = \varkappa \frac{\partial^2 T(x,t)}{\partial x^2} + \alpha I^2 - \nu(T(x,t) - T^{\text{ext}}), \tag{4.12}$$

where $\varkappa$, α, and ν are parameters that depend on the line material; I is the current; and T^{ext} is the ambient temperature. In [9] it is argued that the second-order term on the right-hand side of (4.12) can be dropped.[31] Following a number of approximations, one obtains the following equation to represent (location-independent) line temperature, at voltage V and power flow P:

$$T(t) = e^{-\nu t}(T^{\text{start}} - T_e(P)) + T_e(P), \tag{4.13}$$

where $T^{\text{start}} = T(0)$ is the starting temperature of the line (i.e., just before the current set of power flows was instantiated) and

$$T_e(P) = \lim_{t\to\infty} T(t) = \frac{\alpha I^2}{\nu} + T^{\text{ext}} = \frac{\alpha}{\nu}\frac{P^2}{V^2} + T^{\text{ext}} \tag{4.14}$$

[31] Equation (4.12) without the second-order term is consistent with that used in IEEE Standard 738 [126] to predict line temperature as a function of current; in the standard the constant ν depends on wind parameters. Also see Section 5.3.2.

is the equilibrium temperature of the line. Note that we can (equivalently) restate (4.13) as follows. For $\Delta > 0$,

$$\begin{aligned} T(t+\Delta) &= e^{-\nu(t+\Delta)}(T^{\text{start}} - T_e(P)) + T_e(P) \\ &= e^{-\nu\Delta}\, T(t) + (1-e^{-\nu\Delta})\, T_e(P), \end{aligned} \tag{4.15}$$

which highlights the hysteresis in the temperature process.

We will comment on several details in and implications of (4.13) and (4.14) below. However, this model is used in [9] to predict the precise time at which an overloaded line will fail. Let $P^{\max}$ be the maximum power that the line can carry. Assuming that the line is overloaded ($P > P^{\max}$), [9] stipulates that the time at which the line will fail is, deterministically, that time t^* at which its temperature reaches $T_e(P^{\max})$. Using (4.13), this is given by

$$e^{-\nu t^*}(T^{\text{start}} - T_e(P)) = T_e(P^{\max}) - T_e(P), \quad \text{and so}$$

$$t^* = -\frac{1}{\nu}\ln\frac{T_e(P^{\max}) - T_e(P)}{T^{\text{start}} - T_e(P)} = \frac{1}{\nu}\ln\frac{T_e(P) - T^{\text{start}}}{T_e(P) - T_e(P^{\max})}. \tag{4.16}$$

(Note: There is a misprint in [9], with the numerator and denominators in (4.16) reversed.) And, again according to (4.13), the time needed to reach temperature $\hat{T}$ equals

$$t(\hat{T}) = \frac{1}{\nu}\ln\frac{T_e(P) - T^{\text{start}}}{T_e(P) - \hat{T}} = \frac{1}{\nu}\ln\left(1 + \frac{\hat{T} - T^{\text{start}}}{T_e(P) - \hat{T}}\right). \tag{4.17}$$

This formula is qualitatively correct in several ways: First, $t(\hat{T}) \to +\infty$ as $\hat{T} \to T_e(P)$. Second, $t(\hat{T}) \to 0$ as $\hat{T} \to T^{\text{start}}$. Third, using (4.14), we see that for any fixed T^{start}, $t(\hat{T})$ is a decreasing function of T^{ext} (which only appears in the denominator of (4.17)).
At the same time, a valid criticism of this model arises by considering the implications of all the assumptions. [9] uses (4.14) to impute from the line limit $P^{\max}$ a corresponding equilibrium temperature $T_e(P^{\max})$. As a consequence, the value T^{ext} is not present in the denominator of the last expression in (4.16) but appears in the numerator. Thus, $t^* = t(T_e(P^{\max}))$ is an *increasing* function of T^{ext}, which appears surprising. The dependence on T^{ext} of the numerator in (4.16) disappears if we assume that T^{start} depends additively on T^{ext}; however, it is not clear why that should be the case, and even in that case, t^* simply becomes independent of T^{ext}—perhaps it should be a *decreasing* function of T^{ext}.

A more common approach would rely on a per-line maximum operating temperature obtained from material characteristics of the line. For example, a maximum temperature of 100°C is common. Using (4.14), we can then derive a corresponding ambient-temperature-dependent maximum power flow. This would leave the general methodological approach in [9] unchanged.

In any case, the model for line temperatures is interesting and could prove useful if properly calibrated. A potential difficulty in this regard is the need to include a model for weather and, in particular, wind speed. Wind speed would affect the parameter T^{ext} above; however, in the case of a medium to long line, wind speed would be both location dependent and, especially, time dependent. It is not obvious to us what impact these variations in external conditions would have on the temperature of an individual line (and if indeed the line temperature could depend on location within the line; this is the "x" that appears above in (4.12)). However the variations should have a system-wide impact in the sense that the modeling process that yields the T^{ext} parameters for different lines would have to include a location-dependent ingredient. These are *not* idle speculations. As [101] makes clear, an optimistic model for wind cooling was used in the case of some of the key lines that outaged in the 2003 event. (We will discuss related concepts in Section 5.3.2.)

We also refer the reader to [34] for recent related work, extending the above model to handle the case of *random* exogenous temperature T^{ext}, a relevant model given that this parameter is in general not precisely known, is dependent on location, and is affected by stochasticity of wind (which cools lines).

Another significant difficulty concerns the outage model itself. As presented, the outage process for a given line is deterministic, and it is solely due to the thermal process. However, a perusal of outage events reveals that whereas line outaging is correlated with high power flows, it is also due to other factors. In particular, as mentioned above, sagging of a line due to heating makes vegetation contacts more likely. This is hardly a deterministic process, and it is difficult to precisely calibrate. In the 2003 event, some of the key lines that tripped due to contacts were operating well within normal limits (as low as 44% of the limit; see [101, pages 58–59]) and with line temperatures in a nominal range. To make matters worse, those lines had no prior history (stretching back several years) of tree contacts. And, in fact, the limits to thermally safe operation of the local grid were attained by equipment other than the lines themselves.

The comments in the above paragraph should not be interpreted as criticism of the thermal model in [9]. The discussion highlights the overriding complexity of the line outaging process. It points out the need for incorporating robust (i.e., adversarial) modeling techniques. Line outaging should be viewed as correlated with large flows, but there is a need to capture exogenous effects that increase the probability of a line becoming tripped, even though a full accounting of those exogenous factors is impossible.

Event sequences and control [9] introduces a complex hierarchy of events that reflect the interaction between outages, power flows, and control actions. A central idea is that of *restoration*. When a line becomes tripped or damaged, an attempt will be made to restore it, albeit after a random time t_r elapses. This span of time is assumed to equal a known constant $c > 0$, plus a quantity that is exponentially distributed, with known rate λ_r. The second set of modeling ideas in [9] concerns the line outaging process, as well as parallel control actions taken with the possibly competing goals of restoring loads (shed during prior control actions) and stopping the cascade.

The overall process is fairly complex and in our view can only be partially gleaned from the description in [9], and so we will outline it here. When a line becomes overloaded, its temperature will reach criticality after an appropriate time span t_c elapses, as indicated above. Thus, when a line does become overloaded, the operator in [9] chooses an action as detailed in the following template, where p_1, p_2, and p_3 are given probabilities.

Procedure 4.3.3. Overloaded line control [9]

Application: A line has become overloaded.

(1) With probability p_1, trip the line to protect it.

(2) With probability p_2, allow the line to reach critical temperature and thus fail at a future time.

(3) With probability p_3, shed loads and adjust generator output to relieve the overload.

Here, $p_1 + p_2 + p_3 = 1$. We stress that with (2), if implemented, the time to failure t_c follows the thermal model discussed above and thus depends on a number of parameters, chiefly the flow on the given line. If case (1) is applied, or in case a line that was allowed by an application (2) to fail actually does fail, then an attempt to restore the line will be made after t_r units of time elapse, where t_r is also exponentially distributed. One early comment regarding the random choice model in Procedure 4.3.3 is that this approach *is* reasonable: it models decision-making under severe noise conditions.

We do not find in [9] an explicit description of how (3) is implemented; however, we assume that a simple linear programming model is run which seeks to minimize generator and load changes subject to alleviating (but perhaps not removing) all line overloads and thus is similar to the linear program (4.18)–(4.23) given below.

The line restoration model in [9] is important because it introduces the notion of overt control in the midst of a cascade, and, among other points, it highlights the fact that control can prolong a cascade. When a line is considered for restoration, an initial analysis takes place if the line joins two islands (created by prior line outages). If so, the controller will attempt to fully restore the loads in those islands to their initial values—load may have been shed in prior line restoration attempts, for example, or in prior applications of case (3) of Procedure 4.3.3. This will require generator redispatch and, therefore, will result in a global (or, at least, island-wide) restructuring of power flows. The computation of such a change would need to be implemented by solving a linear program (or at least a linear system); a precise description of how this is done is not provided in [9], but we assume, again, that it is done by solving an appropriate linear program. In any case, a new set of power flows is computed. If no new overloads are created, the line recovery is accepted, as is any load restored to islands. If, on the other hand, line overloads are created, the following template is applied:

Procedure 4.3.4. Line restoration with overloads [9]

Application: An outaged line joining two islands has been restored and overload(s) created.

(a) With probability q_a, the line restoration is accepted, despite the overloads.

(b) With probability q_b, load is shed and generators are redispatched, so as to alleviate the overloads.

Here, q_a and q_b are given probabilities with $q_a + q_b = 1$. We see that cases (3) (of Procedure 4.3.3) and (b) are similar. We now describe how these cases are implemented. To some extent this is similar to the load shedding and generator redispatch strategy in [59] that we discussed above in Section 4.3.1 (see (4.7)–(4.11)), but with some important new features. [9] solves a linear program of the form (borrowing notation from above)

$$\max \sum_{g \in \mathcal{G}} |P_g - P_g^{\text{last}}| \quad + \quad W \sum_{j \in \mathcal{L}} (D_j - D_j^{\text{last}}) \tag{4.18}$$

$$\text{s.t. } Nf - \tilde{b} = 0, \tag{4.19}$$

$$x_{ij} f_{ij} = \theta_i - \theta_j \quad \forall\, (i,j), \tag{4.20}$$

$$0 \le P_g \le P_g^{\max} \; \forall\, g \in \mathcal{G}, \; 0 \le D_j \; \forall\, j \in \mathcal{L}, \tag{4.21}$$

$$-\alpha u_{ij} \le f_{ij} \le \alpha u_{ij} \quad \forall \text{ operating lines } (i,j), \tag{4.22}$$

where the "last" superscript in the objective refers to the state of the grid prior to the restoration attempt and, repeating equation (4.8),

$$\tilde{b}_i = \begin{cases} P_i & \text{if } i \in \mathcal{G}, \\ -D_i & \text{if } i \in \mathcal{L}, \\ 0 & \text{otherwise.} \end{cases} \tag{4.23}$$

Note that this linear program makes sense only when flow variables f_{ij} and constraint (4.20) are restricted to lines (i, j) that are currently operating; [9] indicates that this feature is implemented as in the papers discussed in Section 4.3.1 (i.e., the reactance x_{ij} of an outaged line is increased by a large factor and its rating u_{ij} is decreased by a large factor). Also note that (4.19) implies that $\sum_i \tilde{b}_i = 0$.

Constraint (4.22) is an important new element in the implementation in [9]. Here $\alpha > 0$ is a *control parameter* (not a variable in the linear program). When $\alpha \leq 1$, all currently overloaded lines are returned to safe operating mode. That is because equation (4.16) applies to overloaded lines only, and in particular according to equation (4.13) the temperature that is not overloaded will converge to a limit below the line's critical level. Using $\alpha > 1$, on the other hand, may result in some lines remaining overloaded and the cascade therefore continuing. Even if $\alpha < 1$, the cascade may not stop—this is because previously tripped or failed lines may be due for future restoration attempts, and because of events (a) in Procedure 4.3.4 those attempts could create new overloads.

Comments and criticism The models in [9] make several important contributions, among which we highlight three: the thermal model of overloaded line failure, the notion of control in the midst of a cascade, and the explicit modeling of noise and errors in control actions, as captured by the probabilities in Procedures 4.3.3 and 4.3.4. It is the interaction of these realistic features, together with line outaging events, that brings huge complexity to cascade modeling. Next we will review what we consider to be some salient points.

1. *Optimal control.* An interesting feature that appears to emerge from the numerical experiments in [9] is that the optimal choice for the parameter α in linear program (4.18)–(4.23) may be $\alpha = 1$. That is to say (very roughly) that it is important to completely remove line overloads, as soon as such action is possible. (Later we will examine this issue in more detail.) To be sure, [9] is careful to state that this observation applies to the particular numerical cases that they consider, and that furthermore it might be best to choose different parameters α for different lines. An alternative paradigm that we will propose later would make α both *time* and *severity* dependent.
2. *Continuous time.* The model in [9] operates in continuous time, as opposed to those in Section 4.3.1. This makes it easier to incorporate under a single model several submodels that potentially work in different time scales. The feature also brings up other challenges, which we will discuss below.
3. *Randomness.* As discussed above, [9] introduces several sources of randomness to the model; the models of operator error and noise are of interest. On the other hand, the line outage model is *not* random: once a line is overloaded, its time to failure is precisely known and will not change unless power flows change. This is significant because simulated cascades can exhibit short spans of time with multiple line outage events. The precise sequencing of such events is important, because power flows would change after each failure, possibly changing the time to failure of the other overloaded lines. This particular feature *is* captured by the models in [9]; however, when many lines are close to

failure the deterministic behavior posited in [9] gives rise to only one of several outcomes that should be viewed as equally likely, and simulations show that choosing a different outcome can produce a vastly different cascade evolution.

An underlying complexity concerns the span of time between a line's failure (or trip) and the first attempt at restoration. When a line is outaged, power flows change, and instantly. Additionally, islanding could take place in this time span. Thus, there may be an intervening period of time where, under the model, several new islands could be created and power flows could change, but no power flow computation takes place. It is important to note that such power flow computations would be quite different from, the linear program discussed above. We have touched on this issue before; these computations would account for power flow changes that take place as a result of (automatic) primary and secondary control immediately after the structure of the grid changes. Thus, for example, the notion of control inherent in the objective (4.18) and in the constraints (4.22) would not apply. See Chapter 5 for various modeling ideas and implementations designed to deal with these issues.

4. *Noise in control.* As we discussed above, [9] introduced noise in the models in various interesting and realistic forms. The linear program (4.18)–(4.23) would also have benefited from incorporation of uncertainty modeling. In this regard, the main weakness of the linear program is that it requires exact knowledge of the grid, especially exact knowledge of the set of working lines. Experiments show that solving this linear program under a wrong assessment of the set of operating lines can lead to a generator redispatch/load shedding plan that is severely infeasible (producing unbalanced islands) and does not produce the control implicit in the choice of α. A model that explicitly considers noise in the implementation of controls is discussed in Section 6.2.

We close this section by stressing the daunting complexity of the modeling task at hand, in particular concerning islanding dynamics. It is essentially impossible to capture every detail in a completely realistic manner which is at the same time computationally practicable. And, as we hope the reader may appreciate, it is not clear that the underlying physics of rapidly evolving cascades is understood well enough to make such models possible.

Pfitzner, Turitsyn, Chertkov (2011)

Here we survey [182] (also see [183]). The main contribution of this work is an analysis of selective line tripping as a control tool to stop cascades. The models rely on DC power flows and have the following characteristics.

- *Random outage of overloaded lines.* At any point of the cascade simulation, one of the currently overloaded lines is chosen at random and outaged. Thus, the outage process is "uniform" in that the magnitude of overload does not play a role in the probability of becoming outaged. It is also "memoryless" in the sense that when the grid evolves, a line that was previously (heavily) overloaded may suddenly become not overloaded and will thus no longer be a candidate for outaging. A different model of random outaging of overloaded lines is considered in [28], [29], [30]. We stress that the model in [182] outages one line at a time.

- *More accurate modeling of islanding.* In [182] outaged lines are actually removed from service, as in [28], [29], [30], but in contrast to the models described above. This necessitates modeling of the dynamics of load versus generation when islands are created. Almost certainly, when a new island is created, there will be a mismatch between generation and load in the island, resulting in automatic (and immediately triggered) actions which are not best modeled using a generation redispatch optimization problem. To account for these changes, [182] proposes the following rule. Suppose $\mathscr{I}$ is a freshly created island. Define the following quantities:

$$\begin{aligned} D^{\text{old}}(\mathscr{I}) &= \text{total load among buses in } \mathscr{I}, \text{ just before } \mathscr{I} \text{ was created,} \\ G^{\text{old}}(\mathscr{I}) &= \text{total generation among buses in } \mathscr{I}, \text{ just before } \mathscr{I} \text{ was created, and} \\ g^{\text{old}}(i) &= \text{generation of bus } i \text{ in } \mathscr{I}, \text{ just before } \mathscr{I} \text{ was created (for each such} \\ &\quad \text{generator } i). \end{aligned}$$

Hence, if the ratio $D^{\text{old}}(\mathscr{I})/G^{\text{old}}(\mathscr{I})$ is larger than 1, the island has an excess of load, and if the ratio is less than 1 the opposite happens. Then for each generator $i \in \mathscr{I}$, [182] resets its output to

$$g^{\text{new}}(i) = g^{\text{old}}(i)\,\frac{D^{\text{old}}(\mathscr{I})}{G^{\text{old}}(\mathscr{I})}. \tag{4.24}$$

This action balances supply and demand in $\mathscr{I}$, without any changes in demand. An important case is when the ratio in (4.24) is larger than 1, which therefore results in generators increasing their output. If a generator reaches its limit (maximum output level), then [182] *fixes* its output at the maximum for the remainder of the cascade, thus removing it from any future applications of rule (4.24). However, it may be the case that (4.24) causes a generator to exceed, and not simply attain, its maximum. Even though [182] does not explain how to handle this situation, we assume that in such a case a sequence of rules of the form (4.24) is applied until island $\mathscr{I}$ is balanced (if possible) and perhaps several generators in $\mathscr{I}$ have reached their maximum.

In fact, it may be the case that $\mathscr{I}$ does not become balanced even after increasing generator output—there remains an excess of demand over supply. In that case, [182] reduces loads by 10% (in some fashion), and should this action reduce the total load of $\mathscr{I}$ below generation, then another application of (4.24) will result in balancing $\mathscr{I}$; otherwise the entire island $\mathscr{I}$ is "shut down"—all demand is lost. Overall, this constitutes a model for *droop control* governing the behavior of generators. Other, similar models for load balancing are considered in [28], [29], [30].

A control mechanism proposed in [182] involves deliberate tripping of overloaded lines (this constitutes a limited form of the *line switching* problem; see Section 6.3). Line tripping is to be applied repeatedly as the cascade unfolds, hopefully leading to an early cascade termination with minimal demand lost. The following versions of this idea are considered, where as before at a given time t the flow on a line j is denoted by $|f_{j,t}|$, and its rating by u_j, and we say the line is overloaded if $|f_{j,t}| > u_j$:

(A1) Trip the overloaded line j that attains *minimum* flow $|f_{j,t}|$.

(A2) Trip the overloaded line j that attains *maximum* flow $|f_{j,t}|$.

(A3) Trip the overloaded line j that attains *minimum relative* load $|f_{j,t}|/u_j$.

(A4) Trip the overloaded line j that attains *maximum relative* load $|f_{j,t}|/u_j$.

(A5) Trip a randomly chosen overloaded line (using the uniform distribution).

In the experiments presented in [182], criterion A1 partly emerges as a winner, followed by the (related) criterion A3. Criterion A5 also achieved some success. The reason offered in [182] to explain the strength of A1 is grounded in the analysis of tree-like networks and, more generally, hierarchical networks. A line tripped by criterion A1, in a hierarchical network, is likely to be placed relatively low in the hierarchy. In the case of a tree, such a tripping will result in a (small) island; furthermore, since most generation should be placed higher in the tree, the island would be unbalanced, with a large excess of demand over supply, and thus would shut down, according to the droop control discussion above. This, in turn, results in a "destressing" of the grid with global implications, because the demand of the island is shed, while most generation takes place higher in the tree. Another way to put this is that by shedding a relatively small amount of demand, many lines simultaneously experience a decrease in load.

Nevertheless, as [182] carefully points out, the experiments failed to select a fully reliable winner, and a strategy based on massive sampling of cascade trajectories might be best.

Also note that criterion A1, at least in the sense of the offered analysis for its performance, amounts, implicitly, to a careful load shedding scheme—small amounts of load are shed in carefully selected regions of the grid. This suggests that a strategy directly employing location-based load shedding would prove successful.

Eppstein, Hines (2012)

Eppstein and Hines [92] present a cascading failure model in a broader context, the N-k problem described before. They propose a random search heuristic that identifies sets of lines ("malignancies") whose simultaneous outage would lead to catastrophic system failure. Each prospective malignancy $\mathcal{M}$ is examined by simulating the cascade that would ensue by tripping all lines in $\mathcal{M}$. Here we focus on the cascading model in [92]; in Chapter 3 we discussed this paper in the context of the N-k problem proper.

This model relies on the DC approximation for power flows and is broadly similar to those described above but presents several interesting features in the way it precisely manages two of the crucial features: islanding and line tripping. The model is set up as a discrete-event simulation, with line trips constituting the events, and a submodel that produces the explicit points in time at which lines trip. Between successive line trip events, the simulation identifies islands and dispatches generators and/or sheds loads in each island to balance supply and demand. After this is attained, a new power flow computation is performed in each island. This process continues until either no line overloads are found or system failure results, defined as having a sufficiently large amount of demand lost.

Islanding is handled as follows. Immediately following some line trip(s), if in a freshly created island there is a generation shortfall, then load is shed (proportionally) to match generation. If, on the other hand, there is an excess of supply, then an attempt is made to decrease generator output, taking into account the generation upper and lower limits and ramp rates of generators and using an estimate of the amount of time available for this purpose.[32] If, after taking this action, a generation excess still exists, then generators are tripped, starting from the smallest, until the excess is removed. If a shortage is then obtained, then load is shed as indicated before.

The line tripping model in [92] is of special interest in that it approximates how transmission systems trip lines. The model relies on the concept of *accumulated line over-*

[32] As far as we can tell, the estimate used in [92] is the amount of time between the last power flow computation and the line trip(s) that engendered the new islands.

loads, defined next. For a line j (with limit u_j) its accumulated overload, $o_j(t)$, will be a piecewise-linear function of time with breakpoints occurring at the discrete events. Suppose a discrete event (line trip) takes place at time k, and the following event takes place at time $k+\delta$. Then for t in the (open) interval $(k, k+\delta)$ we operate using the following three rules, where $f_{j,k}$ is the flow on line j immediately after time k, that is to say, after load and generation levels have been adjusted in each island as per the above paragraph:

o.1. If $|f_{j,k}| \le u_j$ then we reset $o_j(k) = 0$. We also define $\omega_{j,k} = 0$.

o.2. If on the other hand $|f_{j,k}| > u_j$, then $\omega_{j,k} = |f_{j,k}| - u_j$.

o.3. For $t \in (k, k+\delta)$, $o_j(t) = o_j(k) + (t-k)\,\omega_{j,k}$.

Thus, in the interval $(k, k+\delta)$, the function $o_j(t)$ is linear with slope $\omega_{j,k}$ and is strictly increasing unless $|f_{j,k}| \le u_j$, in which case it equals zero. The model in [92] also includes per-line parameters $\bar{o}_j$ (described below). Given that a discrete event takes place at time k, rules **o.1–o.3** are used to deterministically decide the next discrete event as per the following:

o.4. The next line to trip will be $\check{j} \doteq \text{argmin}_j\{(\bar{o}_j - o_{j,k})/\omega_{j,k} \;:\; \omega_{j,k} > 0\}$, at time $k + (\bar{o}_{\check{j}} - o_{\check{j},k})/\omega_{\check{j},k}$.

Possibly, several lines tie in the minimum in **o.4** and they will all trip. If on the other hand all $\omega_{j,k} = 0$ then the cascade will stop—no more line trips. The quantities $\bar{o}_j$ are set to emulate existing procedures in transmission systems. A common rule may state (for example) that a line will trip after 10 seconds if it is overloaded by at least 20%. In such a case we would set $\bar{o}_j = 10 * 0.2 * u_j = 2u_j$. (In fact, several rules of the same kind would simultaneously apply: for example, the same line would also trip after 5 seconds with a 50% or higher overload. It would be simple to adapt the model in [92] to simultaneously accommodate all such rules, for example, by tripping if any rule applies.) In any case, the time to failure of a line is decreased (increased) in proportion to its current overload, and so a control action can lessen risk without completely removing all overloads.

To first order, this line tripping model is reminiscent of that in [9], discussed in this section, but with a muted representation of hysteresis because accumulated overloads are reset to zero as soon as lines are returned to nonoverloaded condition, as per **o.1**. On the other hand, the model in [9] only accounts for thermal processes, whereas model **o.1–o.4** simply correlates power overloads to tripping risk and thus may better capture some of the reasons that lines do trip in transmission systems.

Criticism The cascading model in [92] incorporates several realistic ingredients derived from transmission system practice, and as before the possibility exists of criticism due to the complexity arising from the interaction between line tripping and interisland load balancing dynamics. As we have noted before, this issue has only been dealt with under various assumptions and approximations in the literature.

As in previous work, an important issue concerns the behavior of the model when a sequence of two closely spaced line trips takes place. If the first trip results in the creation of two new islands, the generator dispatch submodel needs to be run to compute the power flows in the islands (because for example the next line to trip might be in one of the islands), but there is an implicit assumption in [92] concerning the time available as compared to the time until the next line trip. We will take up this issue in Section 5.5.

Another point concerns the line tripping model itself. As we indicated, the model incorporates elements reminiscent of power engineering practice. On the other hand, as soon as a line j is not overloaded, its "accumulated overload" o_j is reset to zero. If a line were to go through a sequence of somewhat long overload periods, punctuated by brief nonoverloads, the model would likely not trip the line, whereas in practice line temperature would probably remain high and thus make the line susceptible to tripping. In contrast, the model in [9] discussed above (see, e.g., (4.13)) does incorporate hysteresis. The models we will describe in Section 5.3.2 explicitly include hysteresis. Also see [30].

An important ingredient in the Eppstein-Hines paper is, of course, the incorporation of their model of cascades into N-K analysis. Their approach was discussed in Section 3.4.

Bernstein, Bienstock, Hay, Uzunoglu, Zussman (2011)

The cascading model considered in [32], [33] includes two features that we will discuss in greater detail in Chapter 5. The overall model is, very broadly, similar to those in Section 4.3.1 and [182]. When a line is considered tripped, it is removed from the grid (unlike the approach in Section 4.3.1), and when islands are created, any load/generation mismatch is removed by scaling down the larger of the two.

A primary difference in the modeling in [32], [33] is how line tripping is modeled. First, rather than using the (real) power flow f^t_{km} on a line (k,m) at time t to decide if line (k,m) trips at t, [32], [33] use a moving average of the form

$$\tilde{f}^t_{km} \doteq \alpha|f^t_{km}| + (1-\alpha)\tilde{f}^{t-1}_{km}$$

(where $0 \leq \alpha \leq 1$), which helps smooth out numerical artifacts and places the tripping process closer to the heat equation derived models described above. Second, tripping decisions in the model are not deterministic. Rather, a rule of the form

$$P(\text{line } (k,m) \text{ trips at time } t) = \begin{cases} 1 & \text{if } \tilde{f}^t_{km} > (1+\epsilon)f^{\max}_{km}, \\ 0 & \text{if } \tilde{f}^t_{km} < (1-\epsilon)f^{\max}_{km}, \\ p & \text{otherwise,} \end{cases} \tag{4.25}$$

is used. Here $0 < \epsilon < 1$ and $0 \leq p \leq 1$ are parameters and $f^{\max}km$ is the limit of line (k,m). The purpose of this submodel is to reduce the hair-trigger behavior of the overall model that can be exhibited by a purely deterministic rule. We will return to these points in Chapter 5.

A second focus of the work in [32], [33] is to investigate the impact of a severe geographically correlated event (e.g., electromagnetic pulse) on a specific interconnect, WECC. The data for this grid were partially obtained from publicly available information; e.g., the magnitude of loads was deduced from population densities, and a G.I.S. database was used to obtain geographical information. The paper estimated the impact of a cascade resulting from the simultaneous outage of all lines and generators within a 50 km circle and, using techniques from computational geometry, efficiently enumerated all distinct such events.

Remarkably, the San Diego area emerged as the most vulnerable. This was identified in simulations carried out *before* the San Diego event discussed above.

4.3.3 ▪ Models relying on AC power flow computation

A common feature in the work reviewed above is the reliance on DC power flow models. Clearly, richer and possibly more accurate cascade modeling would be attained if instead the AC power flow model were used. This is the approach followed in [165]. Software tools that rely on AC power flow modeling, and even richer modeling of controls and short-term dynamics, include ACCESS [181], TRELSS [93], [117], Eurostag [96], and others. The resulting cascade models are rich, at the cost of requiring extensive data calibration and heavier computational loads. Additionally, there are complications stemming from nonconvergence of AC power flow solvers.

For example, [117] outlines a power flow procedure (likely based on Newton–Raphson) that includes the following steps:

- If at an iteration of the procedure the magnitude of the voltage at a load bus becomes excessively low, that bus is tripped, thus starting a new stage for the cascade.
- If the power flow procedure does not converge, the system being simulated is assumed to have reached a collapse condition.

As previously discussed, this type of assumption, which is natural from the standpoint of having a software system that never "jams" (always provides an answer), creates the hazard that a different power flow algorithm, or even a different implementation of the same algorithm, could attain materially different results. Nevertheless, it is worth noting that systems such as those described above have found widespread use in power engineering practice.

Power systems in the midst of a cascade can experience transients, and predicting the course of the cascade becomes an arduous computational task. Interesting work in this direction is initiated in [81].

4.3.4 ▪ Abstract cascade models

As discussed above, cascade modeling presents difficult challenges—very detailed and accurate models can prove numerically intractable, while tractable models can fail to fully capture key aspects of the problem. An opposite approach is that of relying on abstract models of cascades. These are models that do not incorporate engineering details and instead use an analytical mechanism frequently borrowed from other disciplines, for example physics or biology. Even though there is a cost in neglecting realism, there is a benefit derived from the ability to make precise analytical statements regarding the simulated cascade mechanism and outcomes, in an attempt to capture key statistical and qualitative aspects of actual cascades. We will review two lines of work along these directions.

Dobson, Carreras, Newman (2005)

The article [88] presents the following model. We consider a system made up of n abstract components and described by five different parameters: $L^{\min} \leq L^{\max}$, L^{fail}, $D \geq 0$, and $P \geq 0$. The system operates as follows:

1. For $1 \leq j \leq n$, a random value L_j is uniformly chosen in the interval $[L^{\min}, L^{\max}]$.
2. For $1 \leq j \leq n$, component j is said to fail in Stage 0 if $L_j + D > L^{\text{fail}}$. Let M_0 be the number of components that fail in Stage 0.
3. For $t = 1, 2, \ldots$ write $S_{t-1} \doteq \sum_{i=0}^{t-1} M_i$. Then a component j fails in Stage t if it has not previously failed, and
$$L_j + D + S_{t-1}P > L^{\text{fail}}.$$
Write M_t = number of components that fail in stage t.
4. The process stops at the first stage t such that either $M_t = 0$ or $S_t = n$.

Interpretation The quantities L_j can be interpreted as component *loads* under normal operation (possibly "stress level" could have been a better term). A disturbance occurs, which increases the load of each component by a quantity D, and consequently some of the components fail (i.e., those whose loads exceed L^{fail}) in Stage 0. After that the proper cascade starts. At each stage, every component failure results in an increase by P units of the load of every (unfailed) component, resulting in additional failures in the next stage. For example, this means that the components j that are failed at the end of Stage 1 are those components such that

$$L^{\text{fail}} < L_j + D + M_0 P = L_j + D + S_0 P,$$

and similarly with failures in subsequent stages.

The approach in [88] focuses on the distribution of S, the total number of failed components at the end of the process. Note that if the process stops at stage t, then either $S = S_t = n$, or $M_t = 0$ and $S = S_t = S_{t-1} = \sum_{i=0}^{t-1} M_i$. Defining $d = (D + L^{\max} - L^{\text{fail}})/(L^{\max} - L^{\min})$ and $p = P/(L^{\max} - L^{\min})$, it is shown in [88] that

$$P(S = r) = \binom{n}{r} \phi(d)(d + rp)^{r-1} (\phi(1 - d - rp))^{n-r}, \qquad 0 \leq r \leq n-1, \tag{4.26}$$

where

$$\phi(x) = \begin{cases} 0 & \text{if } x < 0, \\ x & \text{if } 0 \leq x \leq 1, \\ 1 & \text{if } 1 \leq x. \end{cases} \tag{4.27}$$

Assuming D and P are small enough (i.e., as compared to the values $L^{\max} - L^{\min}$) and that L^{fail} is large enough, the process will eventually stop with $S < n$. This is guaranteed when $d + np < 1$. In this case the quantity $\phi(1 - d - rp)$ in (4.26) equals $1 - d - rp$, and (4.26) reduces to

$$P(S = r) = \binom{n}{r} d(d + rp)^{r-1}(1 - d - rp))^{n-r},$$

a *quasi-binomial* distribution (refer to [88] for further background), and it can further be shown that the joint distribution of the quantities $M_0, M_1, \ldots, M_{n-1}$ is *quasi-multinomial*. A question of central focus is the qualitative distribution of the random variable S under certain parameter regimes, in particular the "no saturation" regime $d + np \leq 1$. Assuming $n \to \infty$ and that there are positive constants λ and θ such that $np = \lambda$ and $nd = \theta$ (and thus $p \to 0$ and $d \to 0$, i.e., the disturbances P and D are vanishingly small as fractions of $L^{\max} - L^{\min}$), it is shown that

$$P(S = r) \approx \theta(r\lambda + \theta)^{r-1} \frac{e^{-(r\lambda + \theta)}}{r!}, \qquad 0 \leq r \leq n-1, \tag{4.28}$$

a *generalized Poisson* distribution. This expression is the distribution of the number of individuals produced in a so-called *branching process*, where the initial generation of individuals is Poisson with parameter θ and the following generations are Poisson with parameter λ.

An interesting feature of the model in [88] emerges under experimentation. For $n = 1000$ the authors examine the distribution of S when $L^{\text{fail}} = L^{\max}$ is fixed, but $L^{\min}$ increases, so that the average initial load of any component increases. More precisely, for any j, $L \doteq EL_j / L^{\max} = \frac{L^{\max} + L^{\min}}{2L^{\max}}$ approaches 1.0 from below as $L^{\min} \to L^{\max}$. In other words, we are studying a system that is progressively more loaded. Further, $D = P = 0.0004L^{\max}$; i.e., the initial disturbance as well as the subsequent load increases are quite small. The system as per (4.26) is shown to exhibit "thresholding" or "phase change" behavior:

- For $L \leq 0.6$ the tail of the distribution of S is approximately exponential and ES remains small.
- For $0.6 \leq L \leq 0.8$ the tail becomes heavier and ES grows as a function of L (approximate power law).
- Finally, for $0.8 \leq L$, ES rapidly grows, and moreover for $L \geq 0.9$, $P(S = n) \approx 1$.

Carrying an analogy to power systems, the model suggests that as demand levels increase, and resulting line loads increase, even a nontrivial "safety factor" (the gap between line ratings and average loads) may render a system vulnerable to a very serious cascading failure, even in the case of mild disturbances. The power law behavior observed for larger L approximately matches empirically observed blackout properties in North America and elsewhere. This is of interest since the cost of recovering from a blackout is correlated with the "size" of the blackout, i.e., the number of failed components, and assuming an exponential tail may produce an optimistic assessment of recovery costs. We refer the reader to [87] for additional discussion and references.

Xiao, Yeh (2011)

The paper [230] considers a setup where an infinite random graph is obtained by using as vertices randomly chosen points in $\mathbb{R}^2$, exponentially distributed with rate $\lambda > 0$, and such that two vertices v_i and v_j are connected by an edge if $||v_i - v_j|| \leq 1$. Even though this is an idealized model of a (large) grid, it is not entirely unreasonable, since of course grids can be viewed as contained in the plane, and, to some degree, grids have historically expanded by linking nearby points. Structural results are known for this setup; in particular there is a critical (finite) parameter λ_c such that if $\lambda > \lambda_c$, then with probability 1 the constructed graph will have a unique infinite component. This setup can be adapted to the finite case (only a finite number n of points created, plus the origin); the corresponding result states (abridged) that the largest connected component of the graph will have size proportional to n. This is an example of percolation-based analysis; see [230] and references therein. Interestingly, the quantity λ_c is not known. The regime $\lambda > \lambda_c$ is of interest because, in a sense, it corresponds to the natural setting of having a connected grid.

The second element in [230] is an idealized cascading failure setting that attempts to account for load redistribution (following a round of failures) in a manner that reflects graph topology. In the cascades considered in [230], for each edge e a random value $\psi_e \in [0, 1]$ is drawn from a common distribution with cumulative distribution function (CDF) F. The cascades are initiated by a single edge failure. This constitutes stage 0 of the cascade. In stage t, an operative edge e will fail if at least a fraction ψ_e of the edges that have a common endpoint with e have failed by stage $t - 1$. This notion of cascading damage spread attempts to mimic the redistribution of flow in a grid following a failure.

Note that at any stage of a cascade, if for a given vertex v of degree k (say) there is a single edge e incident with v in the failed state, the probability that a selected edge e' also incident to v will fail in the next stage is $F(1/k)$. We can interpret this value as an indicator of the likelihood that the cascade will spread "through" vertex v. Building on this point, the main result in [230] concerns cascades where λ is sufficiently larger than λ_c. Namely, if there is a fixed value $\lambda_1 > \lambda_c$ such that we can guarantee that $\lambda > \lambda_1$, then there is a parameter $\hat{k} = \hat{k}(\lambda_1, \lambda_c)$ with the property that if $F(1/\hat{k}) > \lambda_1/\lambda$ with probability 1 the cascade will produce an infinite set of failed links—this is termed a "cascading link failure." At the same time, [230] shows that if we can only guarantee $\lambda > \lambda_c$ (i.e., not bounded away by a positive quantity), then we can have cases where with probability 1 the number of failures will be finite.

Postscript: quasi-randomness

Models such as those in [88] and [230] are of course tantalizing because they are underpinned by the type of strong analytics one hopes can be brought to bear on cascade studies, which otherwise run the risk of becoming mired in huge simulation complexities and extensive data calibration. Of course abstract models can be difficult to motivate (especially to a power engineering community) because of the seemingly arbitrary nature of their assumptions.

Possibly, an alternative course is available. The reader may be aware of the classical theory of random graphs, introduced by Erdös and Rényi in [94], which later grew into a theory of random combinatorics [95]. In one of the constructions a graph with n vertices is obtained by setting each pair of vertices as an edge, independently, with probability $p = p(n)$. One of the celebrated results in this body of work is that if p is chosen so that the expected number of edges grows at least as fast as $cn \log n$ (for large enough $c > 0$), the graph will be connected with probability converging to 1 as $n \rightarrow \infty$, whereas a slower rate of growth guarantees that the graph will not be connected.

In fact, a number of structural graph attributes can be shown to hold with high probability if p is large enough, all of them, seemingly, attributes of the random construction. In fact [68], it can be shown that a large family of such structural attributes are *equivalent* in the sense that any graph (random or not) that satisfies any of them will satisfy all of them. Such a graph is termed *quasi-random*.

Carrying this analogy to the setting of cascades of power grids, an interesting possibility exists that once certain structural attributes are observed (such as distribution and timing of line faults), other attributes will necessarily follow (i.e., amount of lost load, number of islands). Were this to be true, it would make it possible to carry out realistic (enough) simulations of cascades without having to exactly model every engineering detail. Possibly this was the hope of the authors in [88] and [230].

Chapter 5
Toward more comprehensive cascade models

In this chapter we will discuss models of cascades that build on and extend those reviewed in Chapter 4 and those that appeared in [28], [29], [30]. The work in this chapter can be used to justify the methods that will be used in Chapter 6 to develop online controls for arresting an ongoing cascade, in particular controls that operate while under a noisy regime (i.e., errors and delays in observations) and with imperfect execution (e.g., incorrect and/or delayed application of control).

The focus of this chapter is twofold. First, we wish to produce models of cascades that, while including a number of realistic details, remain computationally practicable, albeit possibly involving significant computational resources. A second goal of this chapter will be to stress the idea that cascading processes necessarily need to be viewed as stochastic processes with key events in the timeline best modeled through the use of randomness.

5.1 ▪ Key modeling issues

Chapter 4 highlighted the complexity and potential pitfalls arising in the modeling of cascades. An examination of the models pinpoints several key problematic issues that have to be tackled but which perhaps will not admit a completely satisfactory resolution.

I. **Race conditions.** A cascade is punctuated by discrete events: the tripping of lines or other equipment, such as generators or transformers.[33] Each such trip event impacts power flows and voltages and, possibly, generator output levels. If a trip causes a new island to form, generator outputs in that island might change by a large enough amount that a nontrivial amount of time is required until the change is fully effected, and while this is taking place power flows will likely be changing in the island—perhaps precipitating other line trips in the island before the generators have completed their output change. More precisely, secondary frequency control takes place in a span of seconds to minutes, which overlaps with the time span of thermal line trips, especially in the case of heavily overloaded lines. This situation constitutes a classical example of a *race condition*: whether the generators complete the change before any new line trips, or otherwise, could vastly impact stability within the island, possibly having a major effect on the evolution of the cascade.

[33]Note that each tripping event may not necessarily amount to a simple process: for example, a line may (permanently) trip after "opening" and "reclosing" a number of times, possibly quite rapidly. A rich enough model of a cascade might attempt to capture such dynamics; however even in that case each opening or reclosing itself would amount to a discrete event.

High-fidelity versus statistically valid simulations

In this chapter we focus on developing techniques that can produce statistically valid information concerning the impact and evolution of a cascade. The term "statistical" will be valid in our context since we will rely on randomness to model poorly understood (or unknown) factors that can affect the course of a cascade. Our goal stands in contrast to that of developing very accurate emulations of specific cascading failures. This type of emulation can require a vast amount of data for calibration purposes, as well as very accurate models of equipment behavior. Additionally, in practice, such emulations of past events are often scripted or supervised, that is to say, a partial timeline is provided and the simulation is itself calibrated through additional real-data injections at key points of the simulation. For example, there are no simulations that predict the entire four-hour course of the August 2003 event using the state of the system at time zero (e.g., noon) as the sole data input from which the evolution of the system will be predicted.

As we will argue below, noise and uncertain behavior should constitute a critical characteristic of realistic models of cascades. This idea is not yet accepted in the power engineering community. A key point is that an observed cascade may simply represent one possible evolution out of many starting from a given initiating event—perhaps even a low-probability evolution. Hence a reader may wonder about the utility of cascade simulation tools—what use are they if they cannot reproduce past events? The best answer is that a simulation tool that is computationally practicable may discover gross patterns of system behavior where current operational practice is inadequate.

From a fundamental perspective, we would argue that it should be possible to derive statistically valid models for the behavior of cascades, especially in the absence of some data and while relying on clearly approximate and incomplete models of complex equipment behavior, in other words models that fail from a mechanistic perspective and yet provide useful statistical information.

It is worth bringing up the example provided by *statistical mechanics*, a branch of physics developed in the late 1800s. Classical statistical mechanics is still widely used today. It provides useful and accurate large-scale laws of system behavior that stem from probabilistic observations. Yet classical statistical mechanics, from a modern, rigorous perspective, is not quite correct—it posits small-scale behavior which, according to quantum mechanics (which was not developed until the 1920s), does hold.

For a conceptually simpler example, consider a situation where several lines are, simultaneously, highly overloaded. Quite likely, many of these lines will trip, possibly with very brief time spans between consecutive trips. A challenge lies in predicting the particular *sequence* of trips that will be attained. In fact, it seems clear that no satisfactory (scientifically validated) model exists that would select one particular sequence, except in very special circumstances. This is partly due to the particular nature of line tripping—it responds to exogenous conditions that are difficult to model and calibrate.

II. Power flow model and system dynamics in approximation. The vast majority of the work surveyed above relies on the standard DC approximation to the AC power flow model. While this approximation is valid under stable operating conditions, it becomes progressively less so as a cascade intensifies. In essence, the approximation becomes less accurate just when precision is most needed. However, switching to an AC power flow model comes with its own challenges and risks—in fact, accuracy is not guaranteed during a fast phase of a cascade, though of course an AC power flow representation may be effective at diagnosing the *start* of stability loss (e.g., the start of voltage collapse). Ultimately, what is needed is a model that accounts for system dynamics at a very fine temporal scale, and at the global (spatial) scale of an entire grid. As discussed before, the AC power flow model is simply an approximation to the steady-state solution of a system of partial differential equations that embody the appropriate laws of physics (electromagnetism). It is this system that is of interest—although we should probably say systems, since the timeline is punctuated by discrete events (equipment trips) which can have dramatic impact. And there are exogenous factors (such as ambient temperature) and noise, both of which are extremely difficult to model and are best represented using randomness.

Items I and II are clearly interrelated, and they impinge upon a pressing need for computational practicability. We need models whose analysis is fast enough that massive off-line statistical studies are possible. In real-time operation we need accurate tools that can run extremely quickly to, for example, permit diagnosis of a freshly started cascade (dangerous or not?) and suggest real-time control actions. Additionally, increasing model accuracy comes with the added cost of an increased need for model calibration against detailed real data (which may not exist) and for accurate data gathering, which may be extremely problematic in a real-time setting.

In this context, one would need to assume that, in real time, data arrives at control centers asynchronously, incompletely, and with errors.[34] In fact, whereas events during a cascade certainly take place in real time, actual observation and consideration of the data do not. Technically, data (such as PMU data) arrive at a centralized control center quite rapidly—but at a *finite* rate, which may be outstripped by the speed of tripping events. The speed of dissemination of digital information in a power grid during unstable operation is an intriguing research topic that has not yet been explored.

5.2 ▪ Numerical integration

We will describe a hierarchy of simulation models that progressively approximate system behavior on a temporal scale. A central concept we will rely on is a *duration* parameter, $\Delta t > 0$. Our simulation will be checkpointed at time points that differ by Δt—that is to say, we will checkpoint at time $0, \Delta t, 2\Delta t$, and so on. The period between two consecutive simulations will be termed a *time-step*. The simulations will not attempt to model in a detailed manner what happens within a time-step, effectively assuming that system behavior is constant (or, more precisely, linearized) during a time-step, with possibly random changes taking place just before (or after) the end of a time-step. When Δt is relatively large, we obtain a coarse description of a cascade, at the cost of decreased accuracy, but computationally more accessible. As $\Delta t \to 0$ the simulation becomes increasingly more accurate, and finer details, such as the resolution of race conditions, become better approximated. This comes with increasing computational cost, but possibly decreasing robustness. By convention, time-step 0 will cover the interval between time 0 and Δt and will handle the period immediately following the contingency that starts the cascade.

[34] All of this in particular during an unstable, fast phase of a cascade.

Dynamical systems and numerical integration

Many systems arising in science and engineering are described using *dynamical systems*. We will discuss their use in the context of power grid stability in Section 6.0.1. Often a dynamical system embodies a system of equations (usually, differential equations) that is used to describe the time evolution of a *state space* or *phase space*. The system may additionally incorporate stochastics, frequently used to model imperfectly observable behavior. Dynamical systems can sometimes be solved in closed form. When the system is sufficiently complex, this option may not be available and numerical techniques are preferred. A common technique is that of *numerical integration*. When applying numerical integration, time is divided into discrete steps of length Δt, termed *time-steps*. In each time-step, system behavior is simplified, typically through linearization of the system equations, possibly by holding certain quantities constant. The length Δt is chosen small enough that the simplifications are (approximately) valid.

In other words, our general approach to modeling cascades is akin to numerical integration. For example, as discussed above, one can describe the dynamics of frequency response using systems of differential equations. Equipment tripping processes (e.g., line tripping) can likewise be modeled. Our use of numerical integration will also rely on randomness, both as a tool for accounting for poorly understood exogenous factors that contribute to the evolution of a cascades and as a means for agnostically resolving several kinds of race conditions that arise in cascades, as discussed above.

Modeling time scale, islanding, and event separation

The models cited in Chapter 4 (as well as those in [28], [29], [30]) amount to event-driven, discrete-time simulations of a cascade. Some of these attempt to describe, with high precision, the precise timing of events, while others effectively proceed in large time-steps. The advantages and problems arising from either choice are well known in simulation; models with coarse time resolution can be simpler, and faster, but may not capture all details of relevance, while models with fine time resolution can be more accurate but also slower and carry the implicit cost of assuming that there is a system description that is in fact accurate to extremely small time resolution.

A key point is that all the models described before separate, in the time domain, line tripping from islanding dynamics.

Models with this feature have difficulty capturing race conditions, in particular race conditions that arise when an equipment trip closely follows the creation of a prior island (due to previous trips) or a situation where many lines (or other equipment types) are, simultaneously, highly overloaded. In either case the inter-event dynamics could play an important role in the evolution of the cascade. Additionally, in the case of many overloaded lines, we can expect that some or all of them will trip in some sequence, and experimentation (see, e.g., [182]) shows that the particular sequence that is attained can have significant impact on the cascade.

Models that separate island dynamics from line tripping are, necessarily, coarse time scale models—in a model with fine enough time resolution the separation cannot be effected because both types of events overlap in the time scale considered. As discussed above and in more detail below, this feature has some modeling shortcomings but also provides computational expediency and bypasses complex power engineering issues.

In Section 5.3 we will describe our model in the case of "large" Δt, allowing for separation of islanding and tripping, which provides a venue for resolving race conditions while maintaining comparative computational ease. We will then perform numerical experiments to illustrate the hazards of such simplifying assumptions, as well as techniques for reducing the impact of coarseness. In Section 5.5 we will describe how to adapt our models as Δt becomes smaller, thereby allowing us to model (albeit approximately) complex issues such as the loss of generator frequency. Using smaller values for Δt also allows for better modeling of such race conditions; however, as we noted, this will come with several challenges. Whether Δt is large or small, we will apply randomness (coin flipping) to resolve inter-time-step behavior. In particular, our model for line tripping will include randomness; see Section 5.3.2; also see [30], [182]. In Section 5.3.3 we discuss the incorporation of AC power flow models into the simulation.

5.3 ▪ Separated time scale models

Here we will describe a class of models that are obtained by using a large value for the duration parameter Δt. To some degree these models can be considered a synthesis of those cited above, in particular in that they separate generator dynamics from line tripping dynamics in the time scale. However, we will also introduce some modifications, in particular in the development of stochastic models for line tripping. We will use the term *separated time scale* to refer to these cascade models because we will separate island dynamics from tripping. Our goals in introducing these models are first to highlight the underlying assumptions that are needed to obtain logically consistent structures, and second, through a number of experiments, to explore the capabilities of the models and also expose difficulties that suggest the need for more sophistication.

5.3.1 ▪ Islanding

As lines trip during a cascade, islands can form—this has been observed in actual cascading blackouts and is confirmed by simulations. In the course of a cascade, a single line trip (likely preceded by other trips) can cause the creation of two islands, and quite likely at that instant there will exist a mismatch between supply and demand for power in each of the two islands. This situation is, literally, created in a fraction of a second. What will actually happen in each island? This apparently simple question is laden with complexity, because there is no single correct answer or, more properly, as we will see, the answer could depend on the context.

The phenomenon of islanding, of course, is not one that would be encountered in normal operation of a power grid. In particular, grids are supposed to be (N-1)-secure, meaning that the loss of a single line should not constitute a critical event. Thus, if islanding is in fact created by the loss of a single line, it should be the case that each island is automatically self-sufficient with little loss of load and no further instabilities.[35] Nevertheless, if the loss of several lines caused islands to be created, the grid would then respond to correct any load/generation mismatches, and how this is done reflects how a grid responds to rapid changes in (effective) demand. That a grid must be able to respond to such changes is clear, since demand is constantly fluctuating and also changes relatively slowly as a function of time. Furthermore, events such as a sudden generator outage could cause a generation deficit without islanding and must be seamlessly handled to accommodate the N-1 criterion. In fact, several mechanisms come into play to handle load changes.

[35] Our simulations involving real grids reflect this behavior.

See Section 1.4. A survey of the literature shows somewhat vague criteria and nomenclature and differences in practices across not just countries but individual power delivery entities. See [228], [136], [191], [192], [193].

It is worth repeating some of the elements reviewed in Section 1.4. A key element is that in a power transmission system, generators operate in synchronized form, producing AC power at a common target frequency (e.g., 60 Hz in the United States). It is essential for the sake of stability and efficiency of operation that this frequency be kept as close to constant as possible, and sophisticated automatic control mechanisms are in place as an integral component of each generator to maintain frequency in an extremely narrow band. Should loads increase (or should a generator go off-line), frequency at operating generators will immediately lag. A grid first responds to a load generation mismatch to correct the observed frequency drift, and only in a slower time scale to balance supply and demand. The overall process is characterized by the three stages of primary, secondary, and tertiary control described in Section 1.4.

We now turn to the issue of how to model islanding during a cascading failure. An obvious point is that human operator or algorithmic actions during the course of a cascade can make the cascade worse. We wish to distinguish purely automatic control from intelligent or computed control.

Consider first the case of an island with a generation shortfall. Automatic control, in the form of primary and then secondary control, would immediately take hold. If the shortfall were small enough, this process would eliminate it. An issue that immediately arises, however, is the role that secondary control available on the island would play, starting at some point after the creation of the island. The commitment of such reserves would typically involve control actions taken from a centralized location, but this assumes that accurate information on the island is quickly available at that centralized location and that communication capabilities between the island and the control location exist. An alternative would be a form of automatically developed decentralized control that allows an island to rapidly constitute itself into an independent control entity, with sufficiently accurate and complete information to apply control. To the best of our knowledge, such capabilities do not exist today.

There is another factor arising from the application of secondary control: in the midst of a cascading blackout, the deployment of additional generation within a given island would help avoid frequency instabilities in that island, which is certainly an important goal; a large frequency drop would likely lead to the loss of all loads in the island. However, a potentially unfamiliar combination of island structure and generator output levels[36] may result in power flows within the island exceeding the rating of some lines. And, in fact, there may exist overloads in the island before additional generation reserves are used. In other words, the application of automatic control may hasten the progress of the cascade, and, likely, there may be cases where partially withholding reserves (and shedding a small amount of load) is preferable. A precise metering of generation reserves would require an understanding of the likely progress of the cascade in the island, as a function of generator output. This clearly constitutes a case of intelligent control and, except in extreme cases (large generation shortfall or very slowly progressing cascades with small overloads), may not be currently available. We note that grid operators have long applied models of frequency control, line tripping, and islanding; see, e.g., [45]. Our models below are specifically geared to account for stochasticity (i.e., race conditions) and rapid computation.

[36]Unfamiliar because the precise island makeup, including the subset of operating lines, may be unfamiliar.

Basic islanding model under separated time scale

We model islanding dynamics in an attempt to balance the issues described above, but leaning toward models that account for automatic control but only allow limited intelligent control. We first consider the case of an island with a generation shortfall. For a given generator g we denote by $P_g^{\max}$ its maximum output. Additionally, we will use two other parameters: $0 \leq \rho_g \leq 1$, the generator's *maximum ramp-up fraction*, and $0 < v_g$, its *ramp-up rate*. Finally, we will make use of an additional parameter, $0 < \Gamma < 1$, the *maximum shortfall ratio*. Intuitively, when a generation shortfall is instantiated in an island, each generator g will increase its output up to at most P_g, by at most $\rho_g P_g$, and at speed v_g. If this generation ramp-up is not sufficient to remove the generation gap, then the island sheds up to a fraction Γ of each load. And if that step also proves insufficient, then all loads are lost in the island. To model an island with generation shortfall, we proceed as in Model 5.3.2 below, which gradually reduces the shortfall while taking into account generator ramp-up rates and other limits.

Lemma 5.3.1. *If Step* 1(a) *of Model 5.3.2 does not apply, i.e.,* $S \leq \sum_{g=1}^{k} \delta_g$, *then* j^* *and* t^* *as in Step* (b) *exist.*

Proof. For $t \geq 0$ small enough, the amount of additional generation produced by time t equals $\sum_g v_g t$. As t increases, generator 1 will first reach its effective limit, δ_1, at precisely time δ_1/v_1 (that generator 1 is the first to reach its limit is guaranteed by condition 4 in Model 5.3.2). From that point on, the output of generator 1 will be fixed at δ_1, and as time increases, other generators will reach their limits, in the order prescribed by condition 4. For a given $t \geq 0$, therefore, the total additional output achieved by time t equals

$$F(t) \doteq \sum_{g: v_g t > \delta_g} \delta_g + \sum_{v_g t \leq \delta_g} v_g t.$$

We have that $F(0) = 0$; $F(t)$ is strictly increasing; and, since Step 1(a) does not apply, for t large enough $F(t) \geq S$. So there is a smallest value t^* where $F(t^*) = S$. Writing $j^* = \max\{g : v_g t^* > \delta_g\}$, we are done. ∎

Model 5.3.2. Island with generation shortfall

Notation:

1. S = generation shortfall in the island; k = number of generators in the island.
2. The generators in the island are labeled $1, 2, \ldots, k$.
3. For each generator g, $1 \leq g \leq k$, write

$$P_g^0 = \text{generator output just before the creation of the island,}$$
$$\delta_g = \min\{P_g^{\max} - P_g^0,\ \rho_g P_g^{\max}\}.$$

4. For notational purposes, assume $\frac{\delta_1}{v_1} \leq \frac{\delta_2}{v_2} \leq \cdots \leq \frac{\delta_k}{v_k}$.
5. For each generator g, write r_g = amount by which generator g will increase its output.

Step 1.

(a) If $S > \sum_{g=1}^{k} \delta_g$, then set $r_g = \delta_g$ for each generator g.

(b) Otherwise, let the index $j^* < k$ and value $t^* > 0$ be such that $\sum_{g \leq j^*} \delta_g + \sum_{g > j^*} v_g t^* = S$.

(Comment: we will show below that such a j^* and t^* are well defined.)

For $1 \leq g \leq j^*$, set $r_g = \delta_g$, and for $j^* < g \leq k$, set $r_g = v_g t^*$.

Step 2.

If Step 1(b) applies, then the generation shortfall has been overcome and the procedure stops. Otherwise, let G and L denote, respectively, the total generation in the island after Step 1 and the total load in the island. (Comment: $L - G = S - \sum_g \delta_g$.)

(a) If $L - G \leq \Gamma L$, then a fraction $(L-G)/L$ is shed from each load in the island.

(b) Otherwise, all loads are lost in the island.

Even though Model 5.3.2 could be applied in a variety of parameter settings, in a separated time scale model the overall process described by the model is assumed to take place quickly (instantly, in fact, even though the term "time" appears in the proof of Lemma 5.3.1). With this point in mind, the value t^* is less important than the final disposition of the generators and loads in the island. Thus, it is not the precise values of the rates v_g that are important, and instead what matters is their relative comparison. For example, if we assume that all ramp-up rates are approximately equal, we can simply assume that they are all equal to 1, and other than having an incorrect estimate for t^* the final condition of the island will be roughly the same.

A similar but simpler process models an island with a generation excess. In this case we will assume, for simplicity, that all generators have equal *ramp-down* rate; on the other hand we will assume we make a significant assumption.

Assumption 5.3.3. In an island with a generation excess, no load is increased as part of an automatic control, even if that load was previously decreased during an application of Model 5.3.2.

Model 5.3.4 describes our model; whenever possible we use similar notation to that of Model 5.3.2.

Model 5.3.4. Island with generation excess

Notation:

1. E = generation excess in the island.
2. For notational purposes, assume $P_1^0 - P_1^{\min} \leq P_2^0 - P_2^{\min} \leq \cdots \leq P_k^0 - P_k^{\min}$.

(a) If $E > \sum_{g=1}^{k} P_g^0 - P_g^{\min}$, then all generators in the island are shut off and all loads lost.

(b) Otherwise, let the index k^* and value $x^* \geq 0$ be such that $\sum_{g \leq k^*}(P_g^0 - P_g^{\min}) + (k - k^*)x^* = E$, $P_k^0 - P_k^{\min} \leq x^*$ for $k \leq k^*$, and $P_k^0 - P_k^{\min} \geq x^*$ for $k > k^*$.

For $1 \leq g \leq k^*$, the output of generator g is set to $P_g^{\min}$, and for $k^* < g \leq k$ it is set to $P_g^0 - x^*$.

It is seen (as in Lemma 5.3.1) that if case (a) does not apply, then the parameters j^* and t^* in case (b) exist—namely,

$$j^* = \max\left\{ h : \sum_{g=1}^{h}(P_g^0 - P_g^{\min}) + (k-h)(P_h^0 - P_h^{\min}) \leq E \right\} \quad \text{and}$$

$$t^* = \frac{E - \sum_{g=1}^{j^*}(P_g^0 - P_g^{\min})}{k - j^*}.$$

The procedure models the situation where all generators in the island are winding down at a common speed, and each time a generator reaches its minimum output it remains at that value. Case (a) indicates a collapse of the island. Note that we do not model the careful shutting down of generators. The choice of which generators to shut down constitutes a combinatorial problem and should be viewed as a form of intelligent (not automatic) control.

Accurate modeling of islanding dynamics remains a challenging task. The precise behavior of a freshly created island, in the midst of a cascade, would reflect procedures designed for normal operation of the grid (such as frequency response) but would also interact with line tripping, a nontrivial process discussed below. As argued above, line tripping has an impact in the islanding models because generator response could make line tripping more likely, thereby changing the structure of the island even before generator response is complete, and this is captured by a separated time scale model.

In Section 5.5 we will return to this issue; however it should be clear to the reader that arriving at a satisfactory model for island behavior and thereby capturing race conditions between a number of competing processes (line and generator tripping, response from several generators, load shedding) would be complex task. The task is made more challenging by arguably inadequate understanding of the aggregate physics and poor data availability. The use of randomness in our islanding models and in our line tripping models should help produce a practicable approach.

Islanding models under AC power flows

Models 5.3.2 and 5.3.4 produce generator output vectors resulting in total output that precisely matches total island load. They are thus feasible for the DC model of power flows, where zero transmission losses are assumed. Under the AC power flow model, this, in general, is not the case. Thus, an application of Models 5.3.2 or 5.3.4 should be followed by an AC flow computation designed to compute a nearest feasible operating point (to the generator outputs computed by the models), if one exists. In fact, it may turn out that *no* close enough feasible operating point exists. Such an outcome should be interpreted as stating that the given island is infeasible. See Section 5.3.3 and Experiments 5.4.9 and 5.4.10.

5.3.2 ▪ Tripping models

Transmission lines are designed to quickly trip (or open) in response to unusual conditions, such as large power flows or voltage levels, to protect the line. Under normal conditions (e.g., lightning strikes) the overload condition will be brief and the line will reclose to resume operation.

However, if the power flow on the line remains above its rating or limits the line will not reclose, thus placing it out of operation. There are in fact three types of limits that can arise (see, e.g., [224]):

(a) In the case of very long lines (hundreds of miles) a large power flow will result in a large phase angle difference between the ends of the line. Such a state is undesirable in terms of grid stability and is the primary reason for limiting the flow on the line.

(b) For short lines the main limitation is of a thermal nature.

(c) In the case of intermediate-length lines, large power flows can result in large voltage magnitude drops, again an undesirable condition.

Recent cascading blackouts have primarily involved thermal trips. As the power transmitted on a line increases, its temperature will rise due to resistivity effects. This causes the line to gradually sag. In an extreme case, this sagging may affect the physical characteristics of the line (its annealing properties), resulting in permanent softening; such a line must be replaced. Even in a less damaging case, however, the sag may cause an actual or near contact with a foreign object (such as vegetation) resulting in an arc that will trip the line.

Here we focus on the thermal limit, which was already discussed in Section 4.3.2. The need for safe operation of a power line is usually parameterized through a temperature limit, a power limit, or a current limit (or a subset of these criteria). Power flow is easiest to control, and power limits appear most common. Furthermore, there is typically an "emergency" limit, which is usually stated as a given percentage higher than the "normal" limit (for example, 10% higher); during an emergency, power can exceed the normal limit (but not the emergency limit), but only during a limited time period (e.g., 15 minutes).[37] To make matters even more complex, ambient conditions play a role—higher ambient

[37] In fact, utilities may rely on short-term and long-term emergency limits.

temperatures raise line limits; wind speed and direction may reduce them. In addition, the limiting thermal factor in the operation of the line may be the terminal equipment (gear located at the end of the line). Transformer operation is similar. Transformers trip due to overload conditions, which are implemented as relay settings, controlled by both normal and emergency ratings. For example, NERC Reliability Standard PRC-023-1 states that transformer relays should be set to an overload level at least as large as the maximum of 150% of normal and 115% of emergency ratings.

As the reader may appreciate, the resulting situation is one of extraordinary complexity and it is not surprising that operating practices amount to careful attempts at balancing conservatism in the face of an overwhelming avalanche of details that appear difficult to accommodate in a robust manner. See, for example, the 2006 version of IEEE Standard 738 [126]. Also see [9], [34]. Even if sufficient sensorization were available to obtain real-time line temperature data at multiple points of a line, the difficulty would still be one of forecasting *future* line temperature and sag levels, and corresponding risk exposure. An was noted in Section 4.3.2, during the 2003 Northeast U.S.–Canada event, some of the lines that initially tripped were operating at levels substantially below their (emergency) rating. A failure to take ambient conditions into account may have played a role here.

We now describe the family of models we rely on in our simulations. These are guided by a number of criteria:

- Model calibration should include a significant data-driven component. What we mean here is that a particular functional model (for example, a particular model for the underlying physics) should only be a starting point for the analysis; this starting point should then be subjected to intensive *simulations* of the behavior of the line, as a function of environmental parameters (such as weather and proximity to foreign objects), to obtain estimates for numerical parameters.

- The model should be probabilistic. This is a useful device to accommodate the likely fact that multiple obscure details will escape the modeling of a particular device, such as a power line. Using probabilistic models should lead to more robust control algorithms and more realistic simulations. A probabilistic model is used in [226], which suggests a specific stochastic model for line temperature as a function of current and ambient conditions and uses this model to derive a method for setting line limits as a function of the tolerable risk that can be assumed.

- The model should incorporate memory or hysteresis. To some extent this is achieved by the use of the heat equation in IEEE Standard 738 (and the refinement of the heat equation in [9], [34]). What we mean here is that both simulations and control will rely on estimates (or readings) of the flow on a line taken at discrete points in time (even if these are closely spaced). As power flow levels change (or if they change) between one checkpoint and the next, the "effective" power flow should change smoothly.

- The model should have a simple functional form. Simplicity adds robustness, in the sense that a model with overly complex functional form may prove very difficult to justify and, especially, to calibrate. Additionally, simplicity helps speed up simulations, an important consideration if simulations are to be used as a real-time tool for developing controls.

Following these notions, the first ingredient of our models is the notion of an *effective* load on a line.

Line load parameters

For a line e, let $0 \leq \eta_e \leq 1$ be a *hysteresis* parameter.

Definition 5.3.5. *Write* f_e^t *= (active) flow on line e at time t. The* **effective load on line** e **at time** t, $\tilde{f}_e^t$, *is defined as follows:*

$$\tilde{f}_e^0 = |f_e^0|, \tag{5.1}$$

$$\tilde{f}_e^t = \eta_e |f_e^t| + (1-\eta_e)\tilde{f}_e^{t-\Delta t}, \quad t = \Delta t, 2\Delta t, \ldots. \tag{5.2}$$

Also, $|f_e^t|/u_e$ = **relative load of line** e **at time** t,

$\tilde{f}_e^t/u_e$ = **effective relative load of line** e **at time** t.

We see that for $\eta_e = 1$ the model has no memory (effective load = absolute value of flow), whereas for $\eta_e = 0$ the effective load is constant and equal to $|f_e^0|$. For values $0 < \eta_e < 1$, should grid conditions remain fixed after time t, then flow on line e will remain constant at f_e^t and so $\lim_{j\to+\infty} \tilde{f}_e^j = |f_e^t|$. In this light, (5.2) describes a hysteresis-driven model that computes a value between the moving average, $\tilde{f}_e^{t-1}$, and the long-term limit, $\tilde{f}_e^t$. A thermal model for line tripping can be justified along these lines (compare with (4.15)), with the quantity η_1 close to 1 as the time elapsed between checkpoints becomes comparatively large. Moreover, a central reason for the hysteresis model involves *robustness*: whether simulating line conditions or relying on observations, the quantities f_e^t are bound to incorporate noise and errors, and (5.2) smooths the data stream. From this perspective, using $\eta_e = 1/2$ might suffice.

Our second ingredient is a *tripping probability function*, which for a given effective relative load function gives the probability that a line will trip between two consecutive checkpoints. Starting from [60], [86], [61] (see (4.6)), the general model we rely on is described by a function $\Pi_e^{(\Delta t)}(x)$ such that

$$\lim_{x\to 0^+} \Pi_e^{(\Delta t)}(x) = 0, \quad \lim_{x\to+\infty} \Pi_e^{(\Delta t)}(x) = 1.$$

Our model is as follows: let u_e denote the rating for line e. Then we have Model 5.3.6.

Line tripping model

Model 5.3.6. *Line e trips between checkpoints t and* $t + \Delta t$ *with probability* $\Pi_e^{(\Delta t)}\left(\tilde{f}_e^t/u_e\right)$.

This is similar to the model in (4.5), however, using the effective relative load rather than the relative load. Our notation stresses the dependence of the tripping function on the duration Δt; we will return to this topic later. When the duration is known, we may suppress the superscript for clarity. Many functions $\Pi_e^{(\Delta t)}$ are possible, and we will discuss the specific functional form we use in experiments later. However, we stress that our model will allow tripping (with suitably low probability) even in the case of less-than-overloaded lines $\tilde{f}_e^t < u_e$, to capture random or exogenous effects that are difficult to model or calibrate, and to achieve a measure of robustness against data or numerical errors.

Suppose that trips of a line are caused by the following stochastic mechanism:

(a) The line is subdivided into N segments.

(b) Under effective fractional overload x, segment i $(1 \leq i \leq N)$ experiences "stress" $s_i = x\boldsymbol{v}_i$, where we use boldface to indicate random variables. We assume that the $\boldsymbol{v}_i$ are i.i.d., with mean μ and variance σ^2.

(c) The line trips if $\sum_{i=1}^{N} s_i > K$, where K is a given parameter.

Thus it follows that the line trips with probability

$$P\left(\sum_{i=1}^{N} s_i > K\right) = P\left(\sum_{i=1}^{N} \boldsymbol{u}_i > K/x\right)$$
$$\approx 1 - \Phi\left(\frac{K - N\mu}{x\sqrt{N}\sigma}\right) \tag{5.3}$$

for N large, where Φ is the CDF for a standard normal distribution. The right-hand side of (5.3) is a sigmoid function;[38] the parameters K and σ can be calibrated so that the inflection point of the curve (as a function of x) takes place at $x = 1$. Or, using a variant of the reasoning, one can obtain two sigmoid curves (one for $x < 1$ and the other for $x > 1$), to model different behavior under the two regimes. A slightly more general case of interest to which the analysis extends is that where $\boldsymbol{v}_i = a_i + b_i \boldsymbol{u}_i$, where a_i and b_i are deterministic and $\boldsymbol{u}_i$ is randomly distributed in $[0, 1]$ (e.g., uniformly distributed).

Even though the model appears arbitrary, it is not entirely implausible. A line will generally trip due to a fault experienced at some location of the line (e.g., a tree contact). If the condition of the line is evaluated at a set of uniformly spaced discrete points, the independence assumption is justified. Our cumulative "stress" criterion (c) can be interpreted as a (continuous) approximation to what would otherwise be an extremely complex stochastic process, because tripping does not happen instantaneously. A model incorporating total stress should capture the idea that when many segments are simultaneously stressed the line is more likely to trip. We note that the concept of splitting a line into separate segments for the purpose of analyzing tripping is loosely related to the well-known method of *distance relay tripping* ([8, Appendix B], [25, Section 13.3], [106, Section 10.9]).

Motivated by the above discussion, the specific model we use for the tripping probability function relies on

- parameters $0 \leq \epsilon_e^L \leq 1$, $0 \leq \epsilon_e^U$, and $0 < p_e < 1$;

- an increasing function $L(x)$ with $\lim_{x \to 0^+} L(x) = 0$ and $\lim_{x \to (1-\epsilon_e^L)^-} L(x) = p_e$; in particular, we want $L(x) \approx 0$ for $x < 1 - \epsilon_e^L$; and

- an increasing function $U(x)$ with $\lim_{x \to (1+\epsilon_e^U)^+} U(x) = p_e$ and $\lim_{x \to +\infty} U(x) = 1$; in particular, we want $U(x) \approx 1$ for $x > 1 + \epsilon_e^U$.

[38] Not surprisingly since it is a tail probability.

Then

$$\Pi_e^{(\Delta t)}(x) = \begin{cases} L_e(x) & \text{if } 0 \leq x < 1-\epsilon_e^L, \\ p_e & \text{if } 1-\epsilon_e^L \leq x \leq 1+\epsilon_e^U, \\ U_e(x) & \text{if } 1+\epsilon_e^U < x. \end{cases} \tag{5.4}$$

The function $\Pi_e^{(\Delta t)}$ as defined will essentially have the appearance of two sigmoid functions (one for $x \leq 1$ and the other for $x \geq 1$). The primary reason for relying on the "dead band" $[1-\epsilon_e^L, 1+\epsilon_e^U]$ is greater numerical robustness in the critical regime $x \approx 1$. In fact, note that $0 < \Pi_e(1) = p_e < 1$—a typical choice we will make below is $p_e = 1/2$. The reader may at first view (5.4) as either excessively benign (a line at 100% loading should immediately trip) or the other way around (at power levels below the line rating there should be no risk of tripping). But we note, first, that the quantities ϵ_e^L and ϵ_e^U can be chosen arbitrarily small, and second, as stated $U_e(x)$ should be selected so that $U_e(x) \approx 1$ already for values x just slightly larger than 1, and $L_e(x)$ should be selected so that $L_e(x) \approx 0$ already for values x just slightly smaller than 1. Moreover, and most important, we can appropriately shift the "rating" u_e to fit any perception of risk.

There are many functions $L_e(x)$ and $U_e(x)$ that fit the general form (5.4), and we expect that for a given choice of p_e, ϵ_e^L, and ϵ_e^U they should all produce similar results. However, a specific choice we will make for most of our experiments is as follows, where $m_e > 0$ is a parameter (a discussion of our choice is given later, especially concerning calibration):

$$L_e(x) = p_e \frac{e^{1-m_e\frac{1-\epsilon_e^L}{x}+\frac{1}{m_e(1-\epsilon_e^L-x)}}}{1+e^{1-m_e\frac{1-\epsilon_e^L}{x}+\frac{1}{m_e(1-\epsilon_e^L-x)}}}, \tag{5.5a}$$

$$U_e(x) = p_e + (1-p_e)\frac{e^{m_e x-\frac{1}{m_e(x-1-\epsilon_e^U)}}}{1+e^{m_e x-\frac{1}{m_e(x-1-\epsilon_e^U)}}}. \tag{5.5b}$$

Figure 5.1 shows the behavior of $\Pi_e(x)$ for various values of m_e, assuming $p_e = 0.5$ and $\epsilon_e^L = \epsilon_e^U = 0.01$. We observe that for m_e large, the tripping probability rapidly approaches 0 (resp., 1) when the effective relative load is below $1-\epsilon_e^L$ (above $1+\epsilon_e^U$). Our choices for $L_e(x)$ and $U_e(x)$ are smooth, which should lead to model smoothness under simulation.

Discussion The reader may wonder why we chose the specific functions L_e and U_e described above. Part of the reason is simply to provide a smooth version of (5.4). However, note the general sigmoid appearance of the curves in Figure 5.1 (in fact, two sigmoids per curve).

Other examples of sigmoid-like behavior can be applied to derive alternative, plausible models for tripping. Of course, whether we use the type of model we just described or impose a functional relationship such as (5.5a)–(5.5b), significant **calibration** of individual lines is necessary. A number of IEEE standards are in place to manage line tripping, in particular the above-mentioned IEEE Standard 738 (used to compute temperature of overhead transmission lines under various conditions). Also see [69]. A typical operating constraint is of the following generic form: a given line may attain the emergency limit for up to a fixed length of time, L (15 minutes, for example).

We can use such a constraint to help us calibrate the tripping probability function for a line e, under different durations of the form rL, for $0 < r \leq 1$. First we set u_e to be the emergency limit; ϵ^L and ϵ^U small (one percent, for example); and p_e close to 1, say

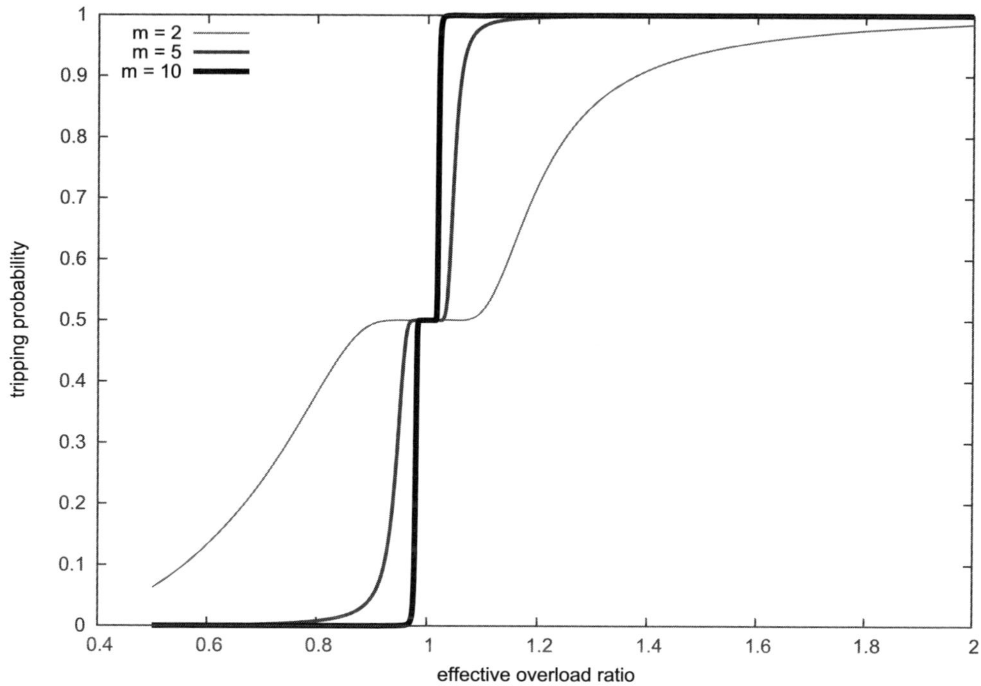

Figure 5.1. *Tripping probability curves.*

$p_e = 0.99$. Using (5.4), (5.5a), and (5.5b), these choices define the line tripping distribution $\Pi_e^{(L)}(x)$, i.e., the distribution valid for duration $\Delta T = L$. Next, for $0 < r < 1$ we set

$$\Pi_e^{(rL)}(x) = 1 - \left(1 - \Pi_e^{(L)}(x)\right)^r. \tag{5.6}$$

Essentially, this models the tripping process as an exponential process. Note that when $r < 1$ and $x \neq x'$ then in general $\frac{\Pi_e^{(rL)}(x)}{\Pi_e^{(rL)}(x')} \neq \frac{\Pi_e^{(L)}(x)}{\Pi_e^{(L)}(x')}$, that is to say, the proration (5.6) is in general not proportional. Another important point is that (5.6) simply identifies a correct way to model (stochastic) line tripping when the Δt parameter is decreased. However, using a smaller Δt will in general also introduce additional richness in the simulation—in other words, simulations carried out using different values of r will, in general, not produce equivalent results.

5.3.3 ▪ AC power flow model

In many of the computational experiments below we will use the DC approximation for power flows. This is done simply for reasons of speed and robustness of the computations. However, we will also perform experiments using several AC models, to better capture nuances of the power flows and conditions such as voltage collapse and large phase angle differences.

The first version of the AC flow model that we consider was chosen to reduce numerical difficulties, in particular regarding nearly infeasible cases. See the discussion in Section 2.4 concerning the experiments performed in [56]. In the formulation below we will also use the following notation and assumptions:

- To simplify the description, we assume that each bus that houses a generator has zero load and is a (P,V) bus, and that these are all the (P,V) buses.[39]

- Let k be a generator bus. Denote by $|V_k^0|$ its voltage magnitude at time 0 (known) and by $\bar{P}_k^t$ its generation at time k, set as per the islanding criteria described above.

- Let k be a load bus. Denote by $P_k^{d,0}$ and $Q_k^{d,0}$ the active and reactive demands at bus k at the start of the cascade (refer to Section 1.2.2). Denote by $\bar{P}_k^{d,t}$ the demand of bus k at time t of the cascade, again set as per the islanding rules.

The islanding criteria given above set *active* power loads and generation. To run an AC power flow computation, we also need to set reactive loads. For any load bus k we will set

$$\bar{Q}_k^{d,t} = \frac{\bar{P}_k^{d,t}}{P_k^{d,0}} Q_k^{d,0}. \tag{5.7}$$

In other words, we will maintain a fixed *power factor* [25], [8], [106] at each load.

Now we turn to the formulation we use. For each load k we treat the quantities $\bar{P}_k^{d,t}$ and $\bar{Q}_k^{d,t}$ as given inputs; for each generator k we allow P_k^t to vary but we penalize for deviations from target output values. For each bus k we allow $|V_k|$ to vary in a range; however when k is a generator the range is very tight. Following notation as in Chapter 1, the voltage at bus k is represented as $e_k + jf_k$ (where e_k and f_k will be variables in our formulation); we write $v = (e_1, e_2, \ldots, e_n, f_1, f_2, \ldots, f_n)^T \in \mathbb{R}^{2n}$ (n = number of buses); and the active and reactive power injections at a bus k are $v^T M_k v$ and $v^T N_k v$, respectively (refer to eqs. (1.22)–(1.26)). Finally, $\mathcal{G}$ is the set of generator buses. Our formulation is

$$\min \sum_{k \in \mathcal{G}} (P_k^t - \bar{P}_k^t)^2 \tag{5.8}$$

$$\text{s.t.} \quad \left.\begin{aligned} v^T M_k v - P_k^t &= 0 \quad (5.9)\\ v^T N_k v - Q_k^t &= 0 \quad (5.10) \end{aligned}\right\} \text{for each bus } k,$$

$$\left.\begin{aligned} P_k^{\min} \le P_k^t &\le P_k^{\max} \quad (5.11)\\ Q_k^{\min} \le Q_k^t &\le Q_k^{\max} \quad (5.12) \end{aligned}\right\} \text{for each generator } k,$$

$$\left.\begin{aligned} P_k^t &= -\bar{P}_k^{d,t} \quad (5.13)\\ Q_k^t &= -\bar{Q}_k^{d,t} \quad (5.14) \end{aligned}\right\} \text{for each nongenerator } k,$$

$$\left.\begin{aligned} e_k^2 + f_k^2 - |V_k|^2 &= 0 \quad (5.15)\\ V_k^{\min} \le |V_k| &\le V_k^{\max} \quad (5.16) \end{aligned}\right\} \text{for each bus } k.$$

[39] The experiments we perform do not rely on the assumption.

This formulation can be seen as minimizing the L_2-norm of active power mismatches at generators. Here, equations (5.9) and (5.10) are used to define the active and reactive power *injections* P_k^t and Q_k^t at generator bus k. Constraints (5.11) and (5.12) constrain such injections at generators, whereas constraints (5.13) and (5.14) fix the injections at nongenerator buses. The objective (5.8) minimizes the 2-norm of the vector of generation "errors" $P_k^t - \bar{P}_k^t$. In (5.16), $V_k^{\max}$ and $V_k^{\min}$ are used to constrain voltage magnitude.

This formulation implements a standard approach similar to that of Overbye [178], [179] described in Section 1.2.2. The rationale for this model is as follows: First, recall that the generator outputs are chosen in our islanding model to result in a load/generation balance in each island. This assumes zero transmission losses, which is only the case when resistances are zero. Thus, the $\bar{P}_k^t$ may be (slightly) incorrect; the objective function (5.8) attempts to minimize the generation mismatch. The formulation also attempts to model the fact that generator voltages are, in general, tightly constrained—we will typically choose the $V_k^{\min}$ and $V_k^{\max}$ quantities at each generator k close to the prescribed voltage value for that generator. For example, we will use $V_k^{\max} = \bar{V}_k(1+\delta_k)$ and $V_k^{\min} = \bar{V}_k(1-\delta_k)$, where $\bar{V}_k$ is the precontingency voltage magnitude at bus k. Here, the quantities $\delta_k > 0$ provide some slack to facilitate feasibility of the optimization problem.

To use this model together with the line tripping model we need to extract a single active power flow estimate for each line. Given a line (k,m), the solution to the above model will produce two values—p_{km} and p_{mk}—which are negatives of each other when line (k,m) has zero resistance (i.e., no loss). If the resistance is positive, but small (the typical setting), p_{km} and $-p_{mk}$ will be slightly different; our estimate for "flow" from k to m is $\frac{p_{km}-p_{mk}}{2}$.

Using the AC power flow computations provides a richer model where several forms of tripping can be captured, including under/overvoltage tripping and line tripping due to large phase angle difference. Additionally, our optimization problem (5.8)–(5.14) may produce large errors $P_k^{(t)} - \bar{P}_k^t$ for generator buses k. If we assume that the underlying OPF solver has correctly solved the optimization problem, such large errors indicate that the pure loadflow problem with generator outputs $\bar{P}_k^t$ is infeasible. We will discuss later how to parameterize the errors to avoid excessive conservatism in this infeasibility assessment; in general when we decide on infeasibility we will then trip the entire island. Likewise, the OPF solver may fail to converge. This could happen for a number of (solver-dependent) numerical reasons, or because problem (5.8)–(5.14) is indeed infeasible. Later we will discuss some recourse actions that can be taken to better diagnose the underlying difficulty; however, should nonconvergence take place in our simulations we will also trip the entire island.

5.3.4 ▪ Final details

As we have noted before, power flow computations can prove problematic. The generic problem is that a power flow computation may fail to converge or may erroneously be declared infeasible. While we can try to adjust for such difficulties (e.g., using the method of Overbye described in Section 1.2.2, or the formulation given by equations (5.8)–(5.14)), the cost is that we may introduce artifacts into the simulation. Ultimately, the simulation relies on the idiosyncrasies of a numerical engine, which creates a problem since discrete events (such as line tripping) are being simulated on the basis of numerical criteria.

The introduction of stochastic line tripping helps in this regard by smoothing simulation outcomes, but it does not help with power flow computations. We have experimented with two additional techniques: in *stochastic smoothing*, the loads (demands) are perturbed by small random amounts at the start of each cascade simulation. In *stochastic load shedding*, each load is decreased by a small random amount at the start of each time-step of the simulation—decreasing the load tends to favor feasibility. A more drastic technique (not implemented) would run each power flow computation a number of times, each using a random and small perturbation of the loads, and report infeasibility only if a majority of the cases claim so. Note that in each of these cases an adjustment of loads necessitates a corresponding change in generation; this is relatively straightforward in the DC approximation but in the AC case would require a formulation such as the one we gave in Section 5.3.3.

Cascade termination

The previous sections described models for islanding, line tripping, and power flows. A final issue to take up concerns the termination of a cascade. The classical cascade template, Model 4.3.1, terminates a cascade as soon as there are no line trips. However, our line tripping model is stochastic, with a positive tripping probability no matter how small the flow on a line is, and furthermore the model applies to the effective relative load, a quantity with memory.

There is a subtle modeling issue at play, in the sense that our model may trip a line with effective relative load (much) smaller than 1, for reasons that are unrelated to the actual cascade. The risks of such an error are somewhat mitigated by our use of a large parameter m and a small parameter ϵ_e^L. More significantly, if we choose any finite limit on the total simulated time then the cascade may "end" with some lines that remain in an overloaded condition.[40] There is a need to balance the correct accounting of low-probability cascades with the speed of simulation. To that effect, we impose the following rules to decide termination.

- A line e is termed *risky* if its relative load exceeds $1-\epsilon_e^L$. The cascade can terminate at the end of a time-step if no lines are risky.[41]
- The cascade can terminate if no line trips have taken place during the preceding $\lceil 1/\Delta t \rceil$ time-steps.
- Let N be the total number of simulated time-steps. This quantity is chosen so that $N\Delta t$ is "large." At the end of the last time-step, if any lines are still overloaded, load is shed uniformly in any island where overloads exist.

The use of "can" in the first two criteria means that both conditions must be satisfied for the cascade to terminate. The use of $\lceil 1/\Delta t \rceil$ in the second condition is meant to model a minimum-length period of time without trips. Finally, the total simulated time, $N\Delta t$, should be calibrated to match a reasonable period of time during which control action can be taken; the action indicated in our model (load shedding) approximates a common control step.

Thus, if our model terminates prematurely, any future line trips would only arise as a result of the stochastic ingredient of our line trip model, as applied to lines with effective relative loads smaller than 1. Further, from a practical perspective, many interesting (read: high-impact) cascades include a stage or stages with high line overloads, and thus our termination criterion would indicate that a relatively quiet period has been reached.

[40]Note: Assuming $p_e > 0$, then the expected number of consecutive time-steps during which line e can remain overloaded is finite.

[41]Later we will amend this condition to include generator trips and generator ramp-up models.

Nevertheless, one can produce realistic cascade simulations where our criterion terminates a cascade with several lines at power flow levels near their rating. If this is the emergency line rating, then arguably the grid is still in an unstable condition. Bringing the grid back to a fully stable status would require a *control* action, the analysis of which is outside the scope of a pure "simulation." However, control becomes easier to apply when a cascade is progressing slowly, making our criterion somewhat reasonable. Below, we will consider control algorithms that attempt to arrest a cascade. In the context of such algorithms we will employ a stricter termination criterion.

5.4 ▪ Numerical experiments using the separated time scale model

The next sections are devoted to experiments that will explore various features of the model, with particular focus on exhibiting the impact of race conditions. In Section 5.4.4 we rely on the AC power flow model, and otherwise all experiments use the DC model.

5.4.1 ▪ Artificial examples

Here we describe some simple, instructional examples. We stress that these examples represent cases with extreme parameter configurations which would not arise in practice; however the behaviors they illustrate also arise in realistic networks, albeit in a more muted fashion.

A nonmonotone, high-variance case

Consider the system in Figure 5.2(a). Bus 1 is a generator with maximum output 100, and buses 2 and 3 are loads, each with demand 50. All lines except $(4,5)$ (which has limit 50) have limit 100 (limits are shown in lighter shade). All lines have reactance of value 1. The resulting flows are as follows: line $(4,5)$ carries zero flow and all other lines carry flow 50 (flow values are shown in black, next to the arrows). The arrows indicate flow directions.

In Figure 5.2(b) we show the same grid after the contingency where line $(1,2)$ is removed from operation. The quantities next to the lines show the resulting flows. Note that by choice of the line limits, lines $(1,4)$ and $(4,2)$ are at full load and therefore stressed, but line $(4,3)$ is only at 50% of capacity.

Since two lines are under stress, a cascade may ensue. We assume $\eta_{i,j} = 1$ for all lines (i,j), i.e., there is no memory in the line tripping model (see Section 5.3.2). Assuming a model of the form given by (5.4), denote by p_{ij} the tripping probability for each line (i,j) under the flows shown in (b); we assume that $p_{4,3} (= \Pi_{4,3}(1/2)) = 0$. For the two other lines (i,j) in (b) we have $p_{i,j} = \Pi_{i,j}(1)$ and we assume $0 < p_{i,j} < 1$. Finally, we also assume that $\Pi_{1,4}(1/2) = 0$. These assumptions are justified under the general model given by (5.4).

We will now consider the evolution of a cascade. Suppose that a given time-step of the cascade starts with all lines shown in Figure 5.2(b) in operation. Then the following is an exhaustive list of events:

(a) No line trips. This event has probability $(1-p_{1,4})(1-p_{4,2})$.

(b) Line $(4,2)$ trips but line $(1,4)$ does not. This event has probability $(1-p_{1,4})p_{4,2}$.

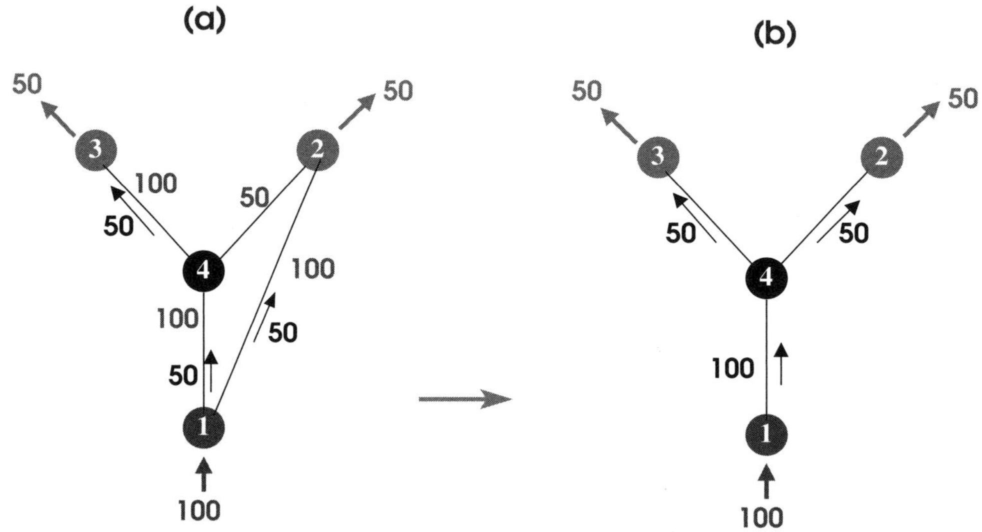

Figure 5.2. *A simple race condition.*

(c) Line $(4,2)$ does not trip but line $(1,4)$ does. This event has probability $p_{1,4}(1-p_{4,2})$.

(d) Both lines $(4,2)$ and $(1,4)$ trip. This event has probability $p_{1,4}p_{4,2}$.

In case (a), the next time-step of the cascade will start under identical conditions. In case (b) two islands are created. One island has bus 2 alone and thus 50 units of load are lost. The other island has an excess of generation of 50 units, and by Model 5.3.4 generator 1 will reduce its output by 50 units. Thus, flow on line $(1,4)$ will drop to 50, which is half the line limit. Since by assumption $p_{4,3} = \Pi_{1,4}(1/2) = 0$, the cascade will terminate in one more time-step (refer to Section 5.3.4 for termination criteria) with a total of 50 units of demand being delivered (to bus 3). On the other hand, in cases (c) and (d) the generator becomes disconnected and the cascade terminates with zero demand being delivered.

Consequently, for any $k \geq 1$, we have the following:

(1.k) The probability that the system reaches time-step k with all lines operating, and that no line trips in time-step k, equals $[(1-p_{1,4})(1-p_{4,2})]^k$. In this case, at the start of time-step $k+1$ the system will be in identical condition as at the start of time-step k.

(2.k) The probability that the system reaches time-step k with all lines operating, and that in time-step k line $(4,2)$ trips but line $(1,4)$ does not, equals $[(1-p_{1,4})(1-p_{4,2})]^{k-1}(1-p_{1,4})p_{4,2}$. In this event the cascade will terminate at the end of time-step $k+1$ with 50 units of demand being served.

(3.k) The probability that the system reaches time-step k with all lines operating, and that in time-step k line $(1,4)$ trips but line $(4,2)$ does not, equals $[(1-p_{1,4})(1-p_{4,2})]^{k-1}p_{1,4}$. In this event the cascade will terminate immediately, with zero demand being served.

Thus, the behavior of the system incorporates a simple race condition involving outcomes (2) and (3). Denoting by y the *yield* of the cascade, defined as the amount of demand being served at termination, and writing

$$\pi = (1 - p_{1,4})(1 - p_{4,2}),$$

we have that

$$E(y) = 50(1 - p_{1,4})p_{4,2} \sum_{k=1}^{+\infty} \pi^{k-1} = 50\frac{(1 - p_{1,4})p_{4,2}}{1 - \pi} = 50\frac{(1 - p_{1,4})p_{4,2}}{p_{1,4} + p_{4,2} - p_{1,4}p_{4,2}}. \tag{5.17}$$

(Effectively yield behaves like a geometric random variable.) Suppose that the contingency in Figure 5.2(b) is considered likely enough that it is important to improve the outcome of the cascade. One way to do so would be to invest in capacity—we invest in a line (i, j) to increase its limit; this action *decreases* the corresponding value $p_{i,j}$. And since $(1, 4)$ and $(4, 2)$ are the two stressed lines, we would wish to invest in these lines. An alternative to static investment is to apply real-time control: in real time, we can decrease the flow on a line (i, j) for example by switching on a standby, parallel line, again reducing $p_{i,j}$. In either case, we have

$$\frac{\partial E(y)}{\partial p_{1,4}} = \frac{-50 p_{4,2}}{(p_{1,4} + p_{4,2} - p_{1,4}p_{4,2})^2} < 0, \tag{5.18}$$

and so investing in line $(1, 4)$ indeed improves the expected yield. However,

$$\frac{\partial E(y)}{\partial p_{4,2}} = \frac{50(1 - p_{1,4})p_{1,4}}{(p_{1,4} + p_{4,2} - p_{1,4}p_{4,2})^2} > 0, \tag{5.19}$$

and therefore investing in line $(4, 2)$ paradoxically *decreases* expected yield. There is a simple explanation for this apparently baffling effect: as $p_{4,2}$ decreases, line $(1, 4)$ becomes comparatively more likely to trip before $(4, 2)$ does, resulting in zero yield. In fact, it is advantageous to *decrease* the limit for line $(4, 2)$—or, should the contingency be realized, an effective solution (if possible) would be to purposely trip line $(4, 2)$.

This simple example highlights the possibility that all manner of counterintuitive, nonmonotone or nonconvex behaviors may be observed in cascades. As before, these behaviors are reminiscent of Braess's paradox [49]. In the particular case above, direct inspection provides a straightforward explanation; in large-scale cases the behavior may be observed but it could be very difficult to arrive at an analytical explanation. Instead, a data-driven process may be more effective at evincing appropriate corrective measures.

Another interesting point regarding the above example arises from the standard deviation of yield; using the properties of geometric random variables we obtain

$$\sigma(y) = 50(1 - p_{1,4})p_{4.2}\frac{\sqrt{\pi}}{(1 - \pi)^2}. \tag{5.20}$$

Suppose that $p_{1,4} = p_{4,2} = Q$ (say). Then

$$E(y) = 50\frac{Q - Q^2}{2Q - Q^2} = 50\frac{1 - Q}{2 - Q} \text{ and } \sigma(y) = 50\frac{(1 - Q)^2}{Q(2 - Q)^2}. \tag{5.21}$$

Thus it is possible for the standard deviation to be significantly larger than the mean. High variance of yield is a phenomenon that we will observe on simulations of cascades on realistic networks.

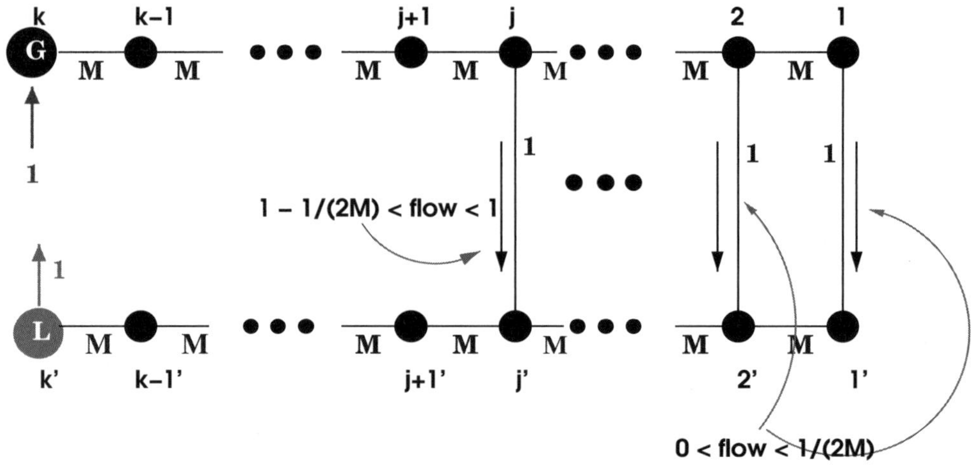

Figure 5.3. *(k,j)-ladder.*

As a final note on this example we remind the reader that the concrete line tripping model (5.5a)–(5.5b) described above requires all tripping probabilities to be positive (except at zero load). However, by choosing the parameters m_e appropriately large, we will have that $\Pi_{2,4}(\frac{1}{2})$ and $\Pi_{1,2}(\frac{1}{2})$ can be made arbitrarily small, and the analysis above goes through largely unchanged.

An example with multiple time scales

The example shown in Figure 5.3 is parameterized by two positive integers, k and j. Bus G (indexed as k) represents a generator and bus L (indexed as k'), a load. The corresponding generation (resp., load) is 1 in both cases. The figure shows line reactances (but not the line limits). For each value i, $2 \leq i \leq k$, there are a line $(i, i-1)$ and a line $(i', (i-1)')$, both with reactance $M \gg 1$ and limit 2. Finally, for each $1 \leq i \leq j$ there is a line (i, i') with reactance 1 and line limit $u_i > 0$. We assume that $\Pi_{i,i-1}(1/2) = \Pi_{i',(i-1)'}(1/2) = 0$ for $2 \leq i \leq k$, and that $\Pi_{i,i'}((u_i 2M)^{-1}) = 0$ for $1 \leq i \leq j$. We call this configuration a *(k,j)-ladder*.

A little analysis shows that the flow on line (j, j') will be between $1 - 1/(2M)$ and 1 as shown in the picture.[42] On all lines of the form (i, i') for $i < j$ as well as all lines of the form $(i, i-1)$ and $(i', (i-1)')$ for $2 \leq i \leq j$ the flow is therefore less than $1/(2M)$. Finally, lines $(i, i-1)$ and $(i', i-1')$ for $j < i \leq k$ carry 1 unit of flow.

We conclude that all lines other than (j, j') are unstressed; more precisely their respective tripping probabilities are zero. On the other hand, line (j, j') will trip with probability approximately $\Pi_{j,j'}(1/u_j)$, assuming, as we did, that M is very large.

Suppose that after an unspecified contingency that removes equipment from operation, a grid is in a (k, k)-ladder configuration. Suppose we use again a line tripping model without hysteresis. Then, (k, k') will trip after a period whose expected length is, approximately, $1/\Pi_{k,k'}(1/u_k)$. At that point the grid will be in configuration $(k, k-1)$. After a period of expected length approximately $1/\Pi_{k-1,k-1'}(1/u_{k-1})$, line $(j-1, (j-1)')$ will trip. Inductively, therefore, we will have individual line trips separated by (approximately) the

[42]This can be obtained by using classical series-parallel reductions.

following time spans:

$$\frac{1}{\Pi_{k,k'}(1/u_k)}, \quad \frac{1}{\Pi_{k-1,(k-1)'}(1/u_{k-1})}, \ldots, \frac{1}{\Pi_{2,2'}(1/u_2)}, \quad \frac{1}{\Pi_{1,1'}(1/u_1)}. \tag{5.22}$$

The situation described by this equation is completely general. Suppose, for example, that u_i is small for i odd (e.g., equal to 1/2), but relatively large for i even (e.g., 1.2). Then the cascade will progress by alternating between brief periods where a severely overloaded line trips quickly, and longer periods where a somewhat overloaded line eventually trips.

In fact, any arbitrary sequence of varying time scales can be achieved by appropriately choosing the u_i quantities. This is significant from the perspective of *control*. In a real-time setting that includes noise and partial observation, periods of high overloads would be perceived as more critical and therefore a grid operator might feel more compelled to take action right away. However, because the grid might change state rapidly, such action could prove ineffective or even harmful.

Also note that the resulting cascade cannot be stopped by deliberate line tripping. At any point in time, tripping any subset of lines will fail to stop the cascade, or, worse, it may result in a faster cascade that ends with zero demand being met. A control strategy that does work is that of *load shedding* (deliberately removing demand). Using a reasonable line tripping model, when the grid is in configuration (k, j) only a fixed percentage of load needs to be removed for the tripping probability of line (j, j') to drop to a small value. Of course, for such an action to be successful, it must be taken before the grid changes state—if line (j, j') trips before the action is complete, then the control may prove unfruitful. The point here is that a control agent must be aware of the changing time scales and of the potential for incorrect control action. A robust load shedding control, when the grid is in configuration (k, j), might consider the state of the grid at some future point in time, for example by anticipating that the grid will be in configuration $(i, 1)$ for some $i < j$ and shedding load so that, *at that future point in time*, line (i, i') is not overloaded. In these examples, perfect clairvoyance is possible because there is essentially one cascade that can take place. Thus an ideal control is always possible. In more complex examples a robust control would have to hedge.

Finally we note that with the line data as stated, all lines will eventually trip. However, this is easily modified so that the cascade will stop, with probability 1, at some time-step of the cascade.

An example with islanding

The grid in Figure 5.4 is obtained by first assembling a number of ladder configurations into chains, or spokes, and then arranging the spokes with a common generator bus at the core.

As before, we assume that the grid is in this state postcontingency. We assume that the generator at the core is the only one with positive output; the solid arrows indicate overall direction of flows. The ladders making up this grid are (k,k)-ladders but for different values of k and may have different physical attributes (reactances, limits, load values). In particular we assume that in each ladder the line carrying the majority of flow is stressed, as in the previous example. Thus, lines will start tripping, independently, in every ladder in the configuration.

At some point in the cascade the first split into two islands will take place; see Figure 5.5. One island, containing the generator at the core, will have a generation excess, while the other island will have a generation shortfall. According to our islanding models (Models 5.3.2 and 5.3.4), the core generator will ramp down, while the generator in the

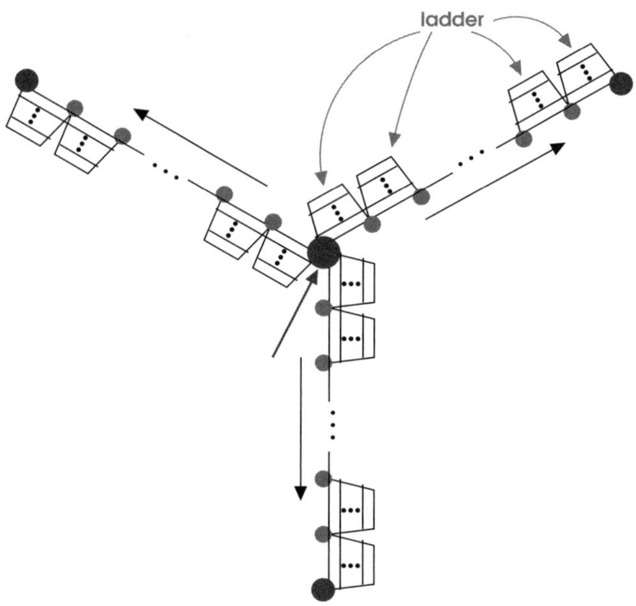

Figure 5.4. *An example with multiple islanding–postcontingency configuration.*

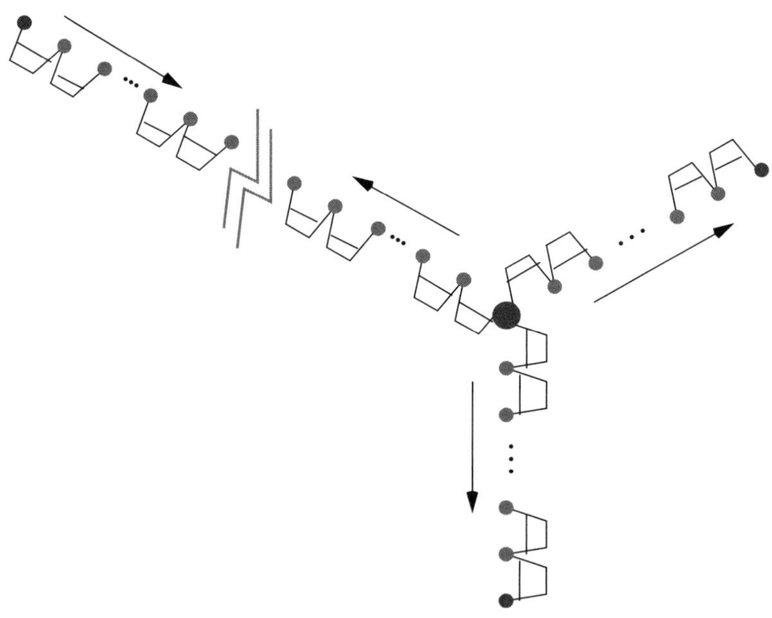

Figure 5.5. *First instance of islanding takes place.*

other island will ramp up. Possibly, in both cases, some load loss may take place. As the arrows indicate, power flows are reversed in the island with generation shortfall. Further still in the cascade, other islands may be formed, including islands with no generator, in which case load will be lost. The final disposition of the cascade therefore depends in a complex manner on various race conditions involving line tripping, with, potentially, a wide variety of final yield values.

The cascades represented by these examples are complex in that they would superimpose line trips in multiple time scales. In addition, islands may be rapidly forming. From the point of view of a centralized control agent, the cascade would be characterized by a large amount of asynchronous, rapidly changing information without a clear pattern. Two possible cascade strategies reappear: controlled line tripping and load shedding. Unlike the situation in the previous example, controlled line tripping *can* be effective here to leverage the generators at the ends of the spokes; however this would require a careful evaluation of the status of lines in each spoke (or, more precisely, in each island), in other words, global information. Load shedding is simpler, in principle, because its execution only requires knowledge of the largest line overload in any island. In both cases a robust control algorithm would have to negotiate race conditions, errors in data gathering, and delays in control application.

Summary

The examples presented in this section are highly artificial and unlikely to occur, verbatim, in a real scenario. However, the qualitative nature of the phenomena arising in these scenarios is not unnatural and has been observed in real-world cascades. For example, in the 2003 Northeast U.S. and Canada cascade, power flows on a major line were reversed at an intermediate stage of the process, and likewise with the 2011 San Diego event. In fact, the cascades we described in Chapter 4 were not orderly processes; instead, they can be described using the admittedly simplistic term "chaos." Whereas the events were initiated by simple, easy to understand causes, systems quickly evolved into profoundly unfamiliar configurations that evaded control attempts through continuing structural change; the process incorporated a noticeable dearth of accurate information. To put it differently, even though *some* elementary or atomic causes for the start of a cascading failure may be well understood (such as tree contacts), there is an exponential variety of postcontingency states, and such states may exhibit unfamiliar, unusual, and counterintuitive behavior.

5.4.2 ▪ Basic experiments

In Section 5.4.3 we will describe experiments involving realistic grids with thousands of buses and lines. Here, we begin with experiments using the classical 118-bus IEEE Test Case [172], also available through the MATPOWER package [234], [235]. The 118-bus data set represents an (old) actual power grid and is widely used by researchers because of its simplicity. This test case has 51 generators, 57 loads, and 186 lines, with a total (active power) generation of 4377.4 MW (out of a maximum of 9966.2 MW) and a total active power load of 4242 MW. However, much of this demand is located at buses that also house generators. To magnify some of the behaviors that we discuss in this section we will first work with a modified version of the 118-bus case. This modification was obtained by netting out the load and *maximum* generation at any bus.[43] With this modification, the total (active power) generator output amounts to 3785.4 MW (out of a maximum of 7326.2 MW), while the sum of loads becomes 1602.0 MW. To render the example usable under the DC power flow model, which is lossless, we scaled down the generator outputs by a common ratio (approximately 0.42321) to make generation and supply equal at the start of the simulation.

Additionally, the 118-bus case does not have line limits. To obtain line limits, we scaled the flows obtained from the generation and loads described above by the "safety margin" 1.3 (when the flow was zero, we used the minimum nonzero absolute flow, in-

[43]Note that in several cases, generators at buses with positive load initially produce zero power.

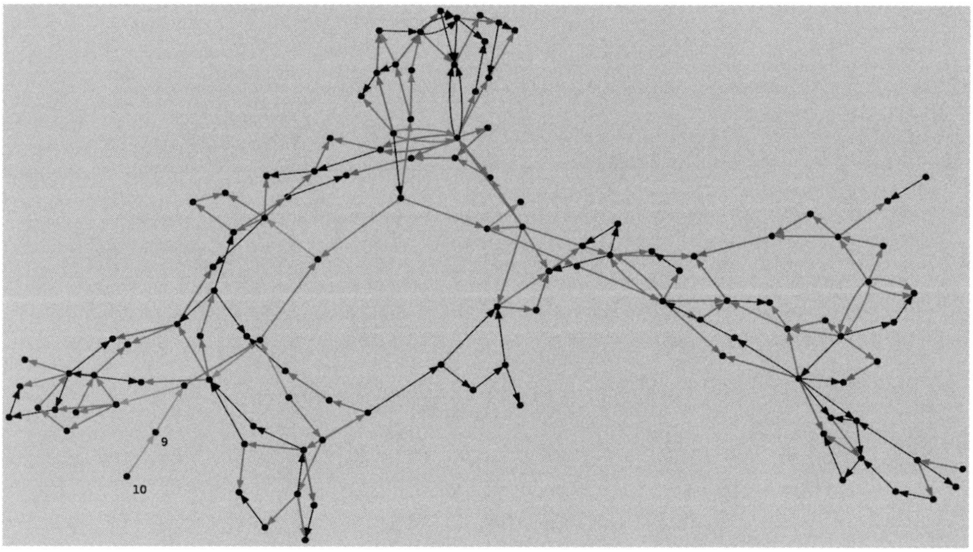

Figure 5.6. *Flow pattern for modified* 118-*bus case.*

stead). Figure 5.6 displays the grid with lines rendered to show flow direction and shaded to reflect flow magnitude, according to the following convention:

- Dark gray = flow < 10.
- Progressively lighter shades of gray are used to denote increasing flow magnitude, 10 ≤ flow.
- Near-white = 140 ≤ flow.

We note the light-colored lines near line (9, 10) (bottom left-hand corner of the picture). In particular, the bus with a single line emanating from it is bus 10, the fourth-largest generator, and that line is (9, 10), which in the solution depicted in Figure 5.6 carries a flow of approximately 190 MW.

Experiment 5.4.1.

In Figure 5.7 we show the corresponding case that arises in the contingency where line (9, 10) has been disabled (shown in white). (Note: Disabling this line is equivalent to disabling the generator at bus 10.) We can point out some differences with Figure 5.6, in particular changes in flow magnitude (for example, the other line incident with bus 9), and some flow reversals (examples indicated by white arrows). What this picture fails to convey is the true impact of the contingency, because different lines have different flow limits.

To better address this issue, in Figure 5.8 we display the postcontingency (relative) line loads, but no flow directions. Here the coloring convention is as follows:

- Dark gray = relative load < 0.001.
- Increasingly lighter shades of gray indicate larger relative load.
- Very light gray = relative load > 1.10.

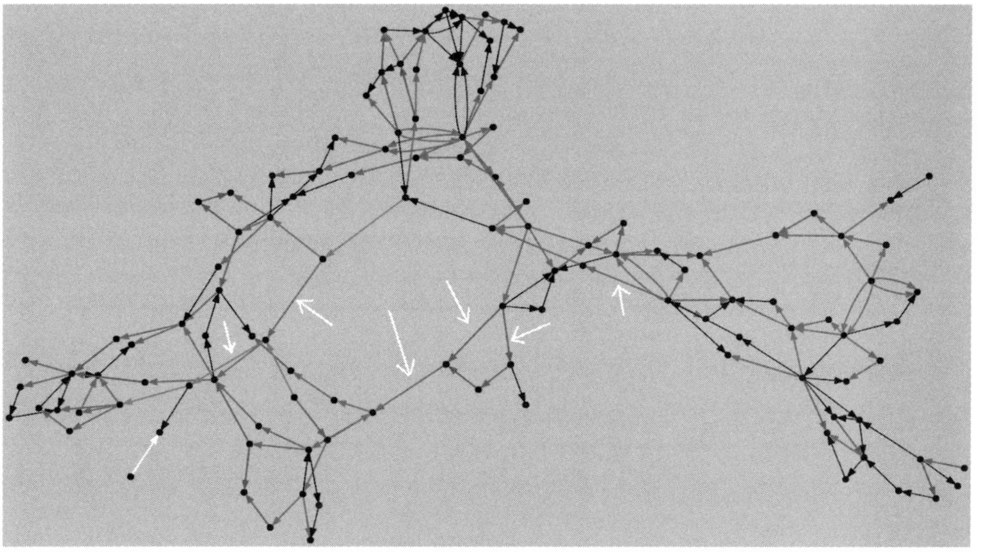

Figure 5.7. *Experiment* 5.4.1. *Flow pattern, postcontingency; some flow reversals indicated.*

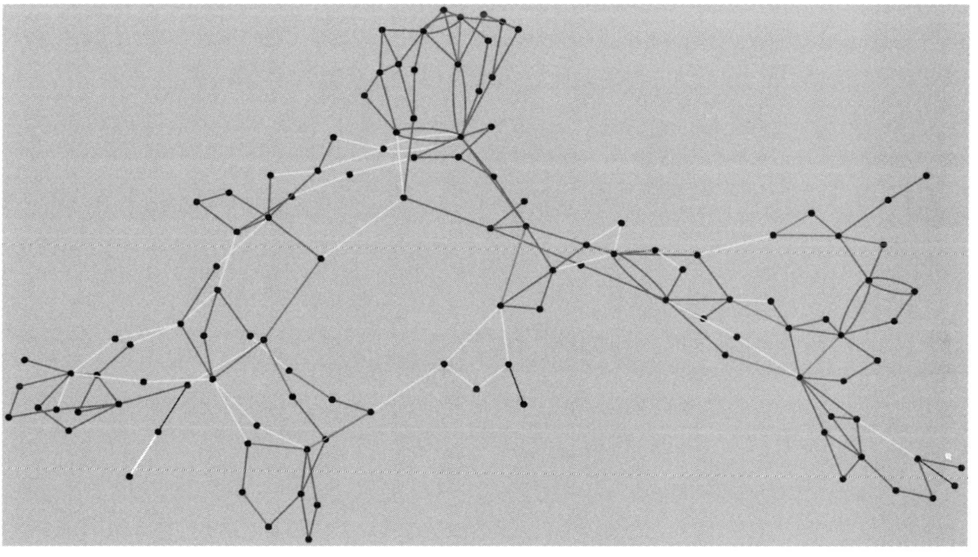

Figure 5.8. *Experiment* 5.4.1. *Relative line loads, postcontingency (white line = contingency).*

- White = contingency.

We note the large number of lines that are strictly over the line limit. We stress that, in contrast, all lines in the precontingency state have relative load of at most $1/1.3 \approx 0.77$. Since several lines are overloaded, a cascade can now start. As a first experiment, we used the following parameters to simulate a cascade. First, in terms of our Model 5.3.2 (used to handle islands with generation shortfall), we used a maximum ramp-up fraction $\rho_g = 0$ for every generator g. This means that in an electrical island with generation shortfall, generators will not be able to ramp up at all, and instead load will be shed, proportionally, to remove the shortfall. And in terms of our line tripping model we used $\eta_e = 1$ for every

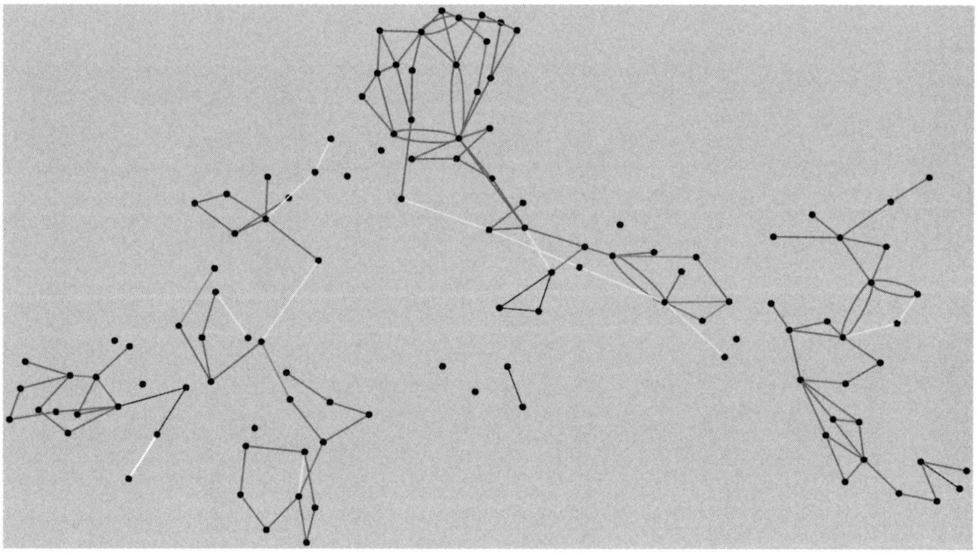

Figure 5.9. *Experiment* 5.4.1. *Relative line loads, end of time-step* 1 *(tripped lines not shown).*

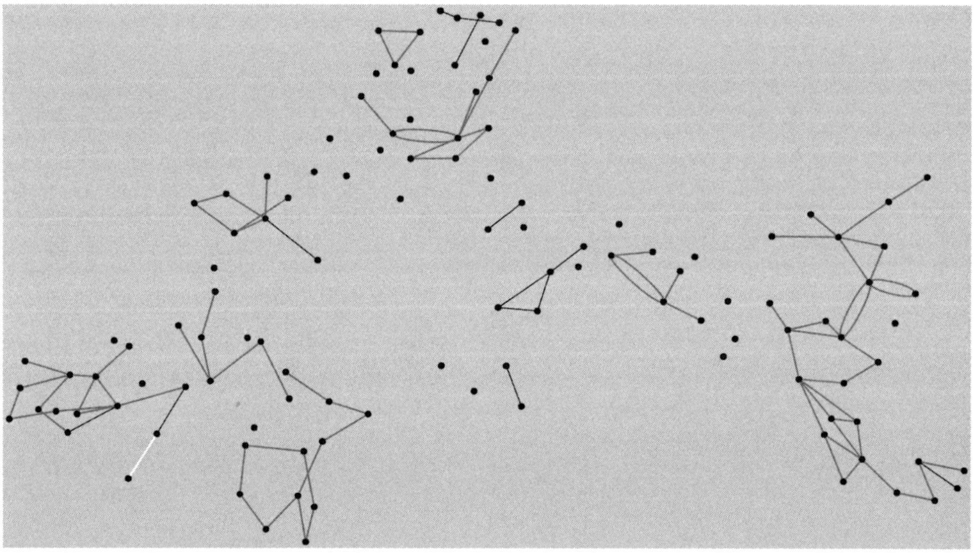

Figure 5.10. *Experiment* 5.4.1. *Relative line loads, end of time-step* 5 *(cascade ends).*

line e (no memory; see (5.1)–(5.2)) and a deterministic model that trips a line as soon as it is overloaded.

This particular cascade ends in five time-steps (see Figures 5.9 and 5.10) with a total of 91 tripped lines, yield (approximately) 37.36%, and 42 islands (including single-bus islands). The color convention is the same as in Figure 5.8, with the added convention that tripped lines are left out. We see that there are several islands, in particular one near the left of the image. Clearly, the cascade is a serious one, which is surprising, especially since in the precontingency setting all lines are operating well below their limit.

The reader may find this fact striking given that the contingency appears to simply remove from operation a generator (albeit large) located in the periphery of the grid. The flow reversals and flow magnitude changes highlighted in Figure 5.7 help explain the story: when line $(9,10)$ is disabled the grid compensates by shifting generation from bus 10 to other units. However, precisely because bus 10 is located at the extreme left of the grid, this causes a shift of flows—postcontingency we will see an increase of flows from the right to the left of the picture to meet loads on the left side. In this unusual flow pattern many of the lines are overloaded, which sets off the cascade. Here we recall our modeling assumption that all maximum ramp-up fractions ρ_g are set to zero, thus preventing generators from ramping up. As a result, since the initial generator output was scaled down from 3785.4 MW to 1602.2 MW (to match loads), the remainder (2183.2 MW) is unavailable to offset the loss of generator 10. And to further compound the handicap, in the initial configuration only 15 out of 51 generators are actually operating; the true total generation available is 7362 MW.

The maximum ramp-up fraction represents, as a fraction of a generator's maximum output, the maximum output increase that the generator is capable of in the amount of time corresponding to a time-step. Together with the maximum ramp-up range, we use this parameter to model a generator's participation in so-called secondary response to load changes—as such, this is a fully automatic control, not easily configurable in real time to deal with a cascade. It is therefore of interest to examine the impact of having a positive maximum ramp-up fraction at all generators; this is examined in the next experiment.

Experiment 5.4.2.

In Table 5.1 the column labeled ρ indicates the common value for all ρ_g, "Yield" is the yield at cascade termination, "Trips" is the total number of tripped lines during the cascade, "Islands" is the number of islands at termination, and "Time-steps" is the total number of time-steps in the cascade. For all values $0.65 \leq \rho$ the values for these four quantities stay fixed. The table displays several interesting characteristics, among them the large number of line trips and islands and the poor yield, independent of ρ. The table does show some expected behavior, namely yield (roughly) increases as a function of ρ. But, on the other hand, the increase stops at $\rho = 0.65$, and further we also note the nonsmooth and nonmonotone behavior of yield as a function of ρ.

The first issue has a straightforward explanation. First, immediately after the contingency, and for all $\rho \geq 0.65$, all but possibly the smallest of the generators in the large island will ramp up by approximately 3.81 MW. For each line in the large island the resulting flow is close enough (across all $\rho \geq 0.65$) that in time-step 1 of the cascade exactly the same set of trips will occur, for $\rho \geq 0.65$, thus creating the same set of 22 islands, and with only slightly different shortfall (or excess) amounts. The result of this is that the ramp-up (or ramp-down) process in each island ends up producing exactly the same outcome for every $\rho \geq 0.65$. The cascades are then identical for the remaining time-steps. In summary, this behavior does not appear counterintuitive.

Next we take up the nonmonotone and (possibly puzzling) nonsmooth behavior of yield as a function of ρ. Part of the answer is that, as islands with generation shortfall are created (and our contingency creates one such island at the very start), positive ramp-up fractions imply (nonlinearly) changing power flow patterns within the island; such changes may result in line trips that follow irregular sequences, which in turn help bring about the behavior in the table. Roughly speaking, this argument can be used to explain the non*monotone* behavior near $\rho = 0.30$ (and other values). Consider the setting immediately after the contingency, as shown in Figure 5.7. The disabling of line $(9,10)$ creates two islands, one containing bus 10 only. The other island (henceforth the "large" island)

Table 5.1. *Experiment 5.4.2. Cascade behavior as a function of maximum ramp-up fraction.*

ρ	Yield (%)	Trips	Islands	Time-steps
0.00	37.36	91	42	5
0.02	38.01	91	43	7
0.03	40.15	97	45	7
0.05	48.80	104	53	5
0.10	45.24	110	55	8
0.15	48.17	107	53	7
0.20	50.95	111	56	6
0.25	52.54	111	56	6
0.30	54.13	111	56	6
0.305	52.37	112	57	6
0.35	52.12	115	59	6
0.40	53.32	115	59	6
0.45	52.81	117	61	6
0.50	53.04	117	61	6
0.55	54.14	116	60	6
0.60	54.50	116	60	6
0.65	54.56	116	60	6
0.70	54.56	116	60	6
1.00	54.56	116	60	6

contains all loads, as well as the remaining 50 generators. The output of the generator at bus 10, just prior to the contingency, equals approximately 190.442—thus, the generation shortfall in the large island is precisely by that amount. Following Model 5.3.2, and recalling that all ramp-up *rates* v_g are assumed equal in these experiments, some or all of this shortfall will be apportioned among the 50 generators in the large island. For most of these generators, the amount available for ramping up (that is to say, the difference between their maximum output and their current output) is sufficiently large that wide ranges of values of ρ elicit nearly the same qualitative behavior.

In particular, as ρ increases beyond 0.30, the flows of several lines in the large island will increase, following the ramping-up that takes place immediately after the contingency. This causes additional lines to trip (as compared to the case $\rho = 0.30$), causing further islanding and effectively giving rise to more severe cascades (e.g., with lower yield), again compared to that for $\rho = 0.30$. As ρ continues to rise, these secondary islands are able to recover more load through the additional ramping up that is possible, generally giving rise to increasing yield values, as shown in Table 5.1.

This explanation, although qualitatively correct, is however incomplete because it fails to address the nonsmooth behavior in Table 5.1. This observation becomes apparent when considering the magnitudes of the ramping-up amounts immediately after the contingency.

- In the case where $\rho = 1$ all generators ramp up by the common value $3.80884 = 190.442/50$, and with $\rho = 0.5$, all but the smallest generator ramp up by 3.84576 units, and the smallest by 2 units—thus, for values of ρ between 0.5 and 1 we will see roughly the same behavior. And, since (precontingency) 14 generators in the large island are responsible for generation (i.e., 1602 MW), on average the output of these generators changes by a very small amount.

- Even with $\rho = 0.25$, all generators but the two smallest ramp up by 3.89463 units (and the two smallest by 1 and 2.5 units, respectively).
- At $\rho \approx 0.0282$ there is a change—for values of ρ smaller than this threshold quantity the amount of generation that is ramped up is insufficient to match the shortfall of 190.442 units, and some load will be shed as per Model 5.3.2.

Thus, even though changes in ρ should be expected to bring about changes in the power flow patterns, they do not solely explain the pattern in Table 5.1, in particular the "hair-trigger" behavior near $\rho = 0.30$. To explore this issue we compare the cascades obtained under $\rho = 0.30$ and $\rho = 0.305$.

- In both cases the behavior at the end of the first time-step is similar, as expected. Under $\rho = 0.30$, at the end of time-step 1 the yield is 91.11%, whereas, at $\rho = 0.305$, the yield is 91.31%; in both cases there are 56 tripped lines and 22 islands.
- The first difference emerges during time-step 2, with 22 additional lines tripping under $\rho = 0.305$ but only 20 tripping under $\rho = 0.30$. One of these lines is $(7, 12)$, with limit 25.925, which under $\rho = 0.30$ carries flow -25.696 (and thus does not trip) but which under $\rho = 0.305$ carries flow 25.960 and thus *does* trip (the other tripping line behaves similarly).
- At the end of time-step 2 the yield under $\rho = 0.30$ equals 74.65% but only 73.24% under $\rho = 0.305$, and this difference is maintained through the remainder of the cascades.

In other words, the change observed going from $\rho = 0.30$ to $\rho = 0.305$ is due to a set of lines that trip in the second case, because their flow is *slightly* over the limit, but do not trip in the first case, even though in that case flow is *just* below the limit. Now we recall that these experiments have been carried out under a completely deterministic line tripping model—the nonsmooth behavior forcefully points out the inadequacy of such a model. It is to be expected (and it is observable) that a deterministic model will bring about much more counterintuitive, hair-trigger (and, incidentally, much more difficult to diagnose) behavior.

Experiment 5.4.3.
To address this issue we repeated Experiment 5.4.2 but this time using the stochastic line tripping Model 5.3.6, with (as per equations (5.5a), (5.5b)) $p_e = 1/2$, $m_e = 10$, and $\epsilon_e^L = \epsilon_e^U = 0.05$ for every line e. To refresh the memory of the reader, this means that when a line has relative load between 0.95 and 1.05, a line will trip with probability 1/2, and if the relative load is above 1.05 (or below 0.95), the tripping probability rapidly, but smoothly, approaches 1.0 (resp., 0). Table 5.2 presents the results of 100 runs; for each value of ρ the table describes the maximum, minimum, average, and standard deviation of yield.

The experiment displays nonmonotonicity in average yield, as we qualitatively would expect as per the discussion above. However, average yield is a fairly smooth function of ρ, in particular considering the magnitudes of the standard deviation. Another interesting observation is the small standard deviation as compared to the dispersion of the distribution.

Experiment 5.4.4.
To investigate this issue, we repeated the experiment with $\rho = 0.50$ but using 10000 observations. See Figure 5.11. The average and standard deviation of yield were 50.94 and 5.08,

Table 5.2. *Experiment* 5.4.3. *Yield under stochastic tripping as a function of maximum ramp-up fraction (%).*

ρ	Maximum	Minimum	Average	Std. deviation
0.00	39.40	31.90	35.19	1.89
0.02	44.03	23.21	35.91	3.43
0.03	46.44	17.68	38.51	4.29
0.05	49.21	14.69	38.38	4.87
0.10	55.24	30.52	44.71	4.29
0.15	56.13	24.38	47.97	4.73
0.20	59.32	22.53	50.76	4.61
0.25	59.17	35.75	51.83	4.76
0.30	57.84	6.22	49.98	5.94
0.305	58.90	21.87	49.60	4.84
0.40	59.97	22.85	49.98	5.04
0.50	60.45	24.41	51.81	5.11
0.55	61.35	24.30	51.73	6.01
0.65	64.73	24.35	51.99	5.92
0.70	61.42	23.92	51.71	6.49
0.75	63.42	36.42	53.32	4.24
0.80	63.42	36.42	53.10	4.42
0.85	64.54	24.09	52.77	5.31
1.00	64.54	24.09	52.77	5.31

respectively, closely matching the results in Table 5.2; on the other hand, the maximum yield was 64.70 and the minimum, 4.95.

To construct Figure 5.11 we split the range between minimum and maximum yield into 100 equal-size intervals and then counted the number of observations in each interval. The nearly symmetrical hump is approximately centered at the mean and is roughly two standard deviations wide in each direction. We also observe a significant tail involving very small yield values—there are 233 outcomes (2.33% of the outcomes) with yield smaller than 41, i.e., more than two standard deviations smaller than the mean, and seemingly a few local maxima in this tail. Even though this effect can be explained by the existence of some low-probability cascade paths, one would wonder if any of the model features could contribute to this behavior. Now we recall that all the simulations in this section have been conducted using the "no-memory" version of our line tripping model (5.1)–(5.2), i.e., using $\eta_e = 1$ for all lines e.

Experiment 5.4.5.

To study this issue, we repeated some of the experiments in Table 5.2 (Experiment 5.4.4) but now using $\eta_e = 0.5$ for all lines e; the results are reported in Table 5.3. We see a decreased (but still statistically significant) dependence of average yield on ρ; as noted above this is a cascade where very large line overloads are observed; a model with memory will thus "remember" this feature regardless of the amount of ramping-up that takes place. More important, standard deviations are twice as large as before. The impact of this fact is better displayed in Figure 5.12, similar to Figure 5.11 but now using $\eta = 0.5$. Even though the average yield resulting from these 10000 runs is now 50.89 and the standard deviation is 12.10 (and thus broadly in agreement with Table 5.3), Figure 5.12 clearly shows three likely cascade modes, which are only hinted at in Figure 5.11.

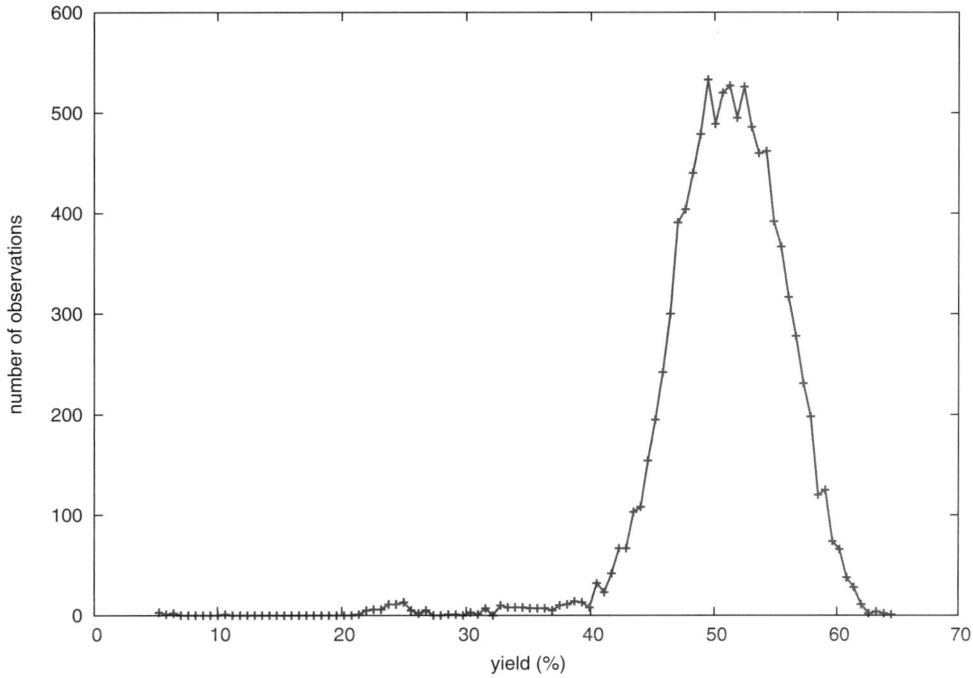

Figure 5.11. *Experiment* 5.4.4. *Yield histogram in* 10000 *cascade simulations with memoryless line tripping.*

Table 5.3. *Experiment* 5.4.5. *Yield under stochastic tripping as a function of maximum ramp-up fraction (%) ($\eta = 0.5$).*

ρ	Maximum	Minimum	Average	Std. deviation
0.00	53.2842	17.6831	41.7211	8.55038
0.05	58.0003	27.9392	45.4948	6.37709
0.10	62.8647	16.6515	44.702	11.2579
0.15	62.9371	23.4443	48.2766	9.77732
0.20	62.5299	17.0336	48.7986	10.8535
0.25	62.5299	17.0336	48.7986	10.8535
0.30	68.4369	24.7031	49.8716	10.2679
0.35	64.1163	24.1285	50.756	11.0545
0.40	69.2275	26.019	51.3989	11.229
0.45	66.1755	25.3317	51.8731	11.7186
0.50	66.0424	23.1776	49.7228	12.2841
0.55	67.6008	25.314	50.7448	12.0882
0.60	67.0412	22.0373	49.8311	13.1876
0.65	65.0437	23.2668	51.1984	11.9349
0.70	68.3521	27.0511	53.1251	10.6651
0.75	68.2896	25.3989	52.2017	12.0226
0.80	66.4794	23.3399	51.0131	11.6641
0.85	64.9134	24.7458	49.5856	11.5415
0.90	65.1708	25.4918	49.4748	12.8753
1.00	65.5431	26.3969	50.4303	11.9863

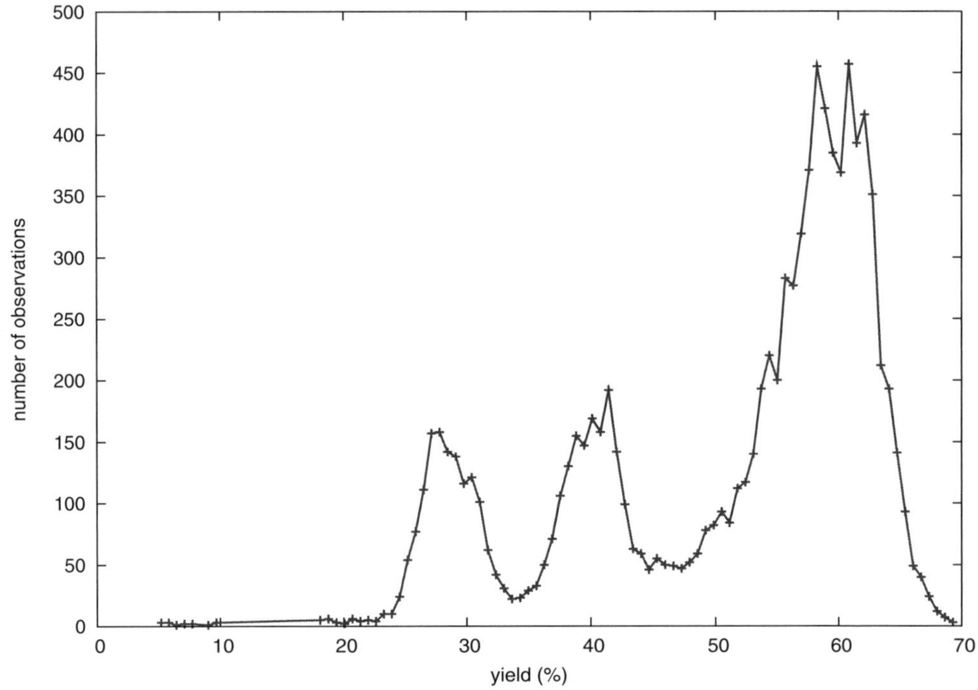

Figure 5.12. *Yield histogram as in Figure* 5.11 *(Experiment* 5.4.4*) but with* $\eta = 0.5$.

Table 5.4. *Case* 2746wp *(Polish grid) statistics.*

Buses	Branches	Generators	Generator buses	Total load
2746	3514	520	388	24.87 GW

5.4.3 ▪ Experiments using larger grids

We now perform experiments using realistic power grids, beginning with the so-called case2746wp of the Polish power grid, available through MATPOWER. Statistics for this grid are provided in Table 5.4. Some of the generators are housed at a common bus; the entry headed "Generator buses" indicates the number of buses that hold at least one generator.

Under the DC power flow model no active power losses take place; to match active power injections to total load we scaled all active power injections by the ratio of the total load to the sum of the generator injections provided in the case file. This ratio equaled approximately 0.98. This resulted in an average generator injection of approximately 82% of maximum.

Using the supplied line ratings, this choice of generator outputs and of loads results in a DC power flow that nearly exceeds line ratings. In particular, one of the lines reaches nearly 98% of its rating. We therefore anticipate stability issues; in another experiment we will increase line ratings. Some of the line flows are on the order of 600 MW, several others are somewhat smaller (on the order of 300 MW), and many are below 100 MW.

First we will perform some basic experiments (similar to the above) using this grid; however, below (Experiment 5.4.8), we will perform other tests that address a particular point of interest: the geographical pattern of line trips as a cascade develops—is it the case

Table 5.5. *Experiment 5.4.6. Overall cascade behavior of Polish grid.*

Time-step	Yield (%)	Trips	Islands	Largest island	Max relative load
1	100.00	3	1	2746	1.053
2	100.00	2	1	2746	1.327
3	99.62	3	1	2746	1.261
4	99.62	2	2	2741	1.180
5	99.62	2	2	2741	1.140
6	99.62	2	2	2741	1.251
7	99.54	9	2	2741	1.277
8	95.89	30	4	2739	2.284
9	95.89	0	26	2657	0.997

that the cascade spreads geographically? That is to say, can one argue that line trips in a given time window Δt are geographically near those line trips that took place in (recent) prior time windows?

Experiment 5.4.6.

Our first experiment with the Polish grid uses the deterministic line trip model with no memory ($\eta = 1.0$). In this experiment we study the interesting cascade initiated by deliberately tripping line $(7,8)$ and line $(40,42)$. The cascade stops after 9 time steps, losing only approximately 4.2% of the total load, however resulting in 54 line trips and 26 islands. Table 5.5 displays some statistics. The column headed "Islands" indicates the number of islands, and the column headed "Max relative load" indicates the maximum relative line load, both at the start of each time-step. "Trips" is the number of line trips during the time-step. "Largest island" is the number of buses in the largest island at the start of the time-step.

Two features stand out from Table 5.5. First, even though a large number of islands is created in time-step 8, all but one are quite small—on average, these islands have approximately 4 buses (in fact the sizes of the remaining islands are 35, 14, 7, 5, 5, 2, 2, 2, with the remaining all of size 1). Naturally, the largest island comprises most of the load and generation. This seems to be typical behavior in our simulated cascades and, superficially at least, seems to mirror the behavior of random graphs [94]. The second feature is that although the cascade does terminate according to the rules of our deterministic model, at least one line is nearly at risk at termination. In fact a closer look at the raw data shows that more than 10 lines have relative load greater than 0.9. This suggests that an assertion that the cascade is benign may be premature, and it would prove beneficial to rerun the simulation using a stochastic line outage rule.

Experiment 5.4.7.

Here we will repeat the previous experiment, using hysteresis parameter $\eta_e = 0.5$, for each line e, and the stochastic line tripping rule (5.4) (using (5.5a)–(5.5b)) with $\epsilon_e^L = \epsilon_e^U = 0.05$ for each line e, $p = 0.5$, and $m = 10$. To refresh the memory of the reader, this model will trip a line with probability 0.5 if its effective relative line load is between 0.95 and 1.05; when the effective relative load is less than 0.95 the probability is near zero, and when the effective relative load is more than 1.05 the probability is near one.

Table 5.6. *Experiment* 5.4.7. 20000 *runs of Polish grid under stochastic line tripping.*

	Max	Min	Average	Std. dev.
Yield	0.996	0.482	0.864	0.081
Time-steps	95	7	24.41	10.81
Num. islands	373	2	74.44	47.27

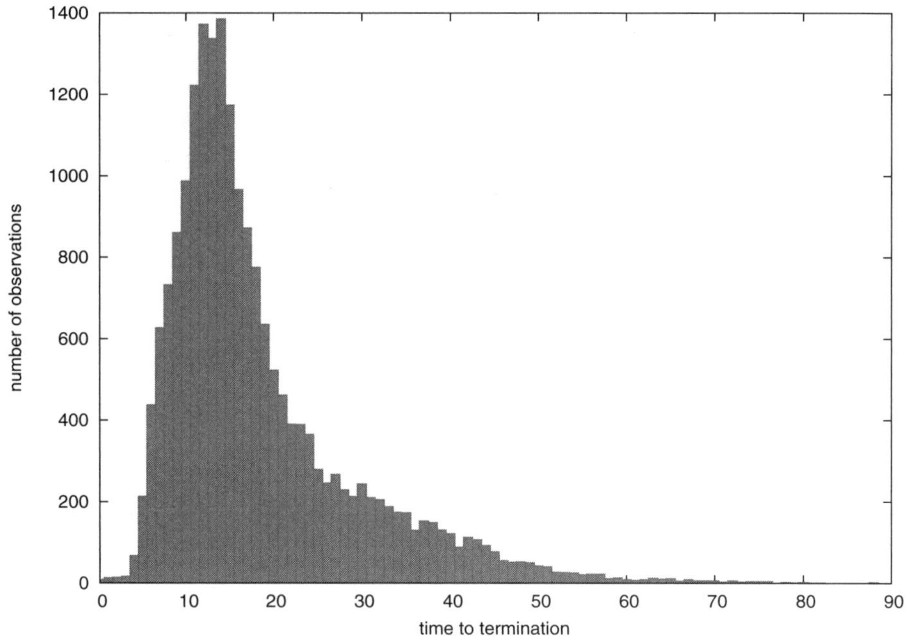

Figure 5.13. *Experiment* 5.4.7. *Distribution of time to cascade termination in* 20000 *simulated cascades with a common start for the Polish grid.*

Table 5.6 presents a summary of 20000 runs of this model for three parameters: yield at termination, time-steps to termination, and number of islands at termination. The table shows a much more nuanced view of the risk embodied when lines (7,8) and (40,42) are simultaneously tripped. Also note the large sample variance for the last two parameters, despite the fairly narrow range of randomness of the line tripping rule.

Figures 5.13, 5.14, and 5.15 show (respectively) the distribution of the sampled time to termination and number of islands. Considering the structure of the observed distributions, we can make the plausible assumption that the length of a cascade is positively correlated with the number of islands at termination and with decreased yield. In fact, in this experiment the (sample) correlation between time (to termination) and number of islands (at termination) is 0.80, that between time and yield is −0.85, and that between yield and islands is −0.96.

This experiment thus raises the (as before) plausible conclusion that control to terminate a cascade should be attempted sooner rather than later, at least in the sense that delaying control actions for too long increases the probability that a longer (and more damaging) cascade may ensue. We will see a modified version of this principle in Chapter 6.

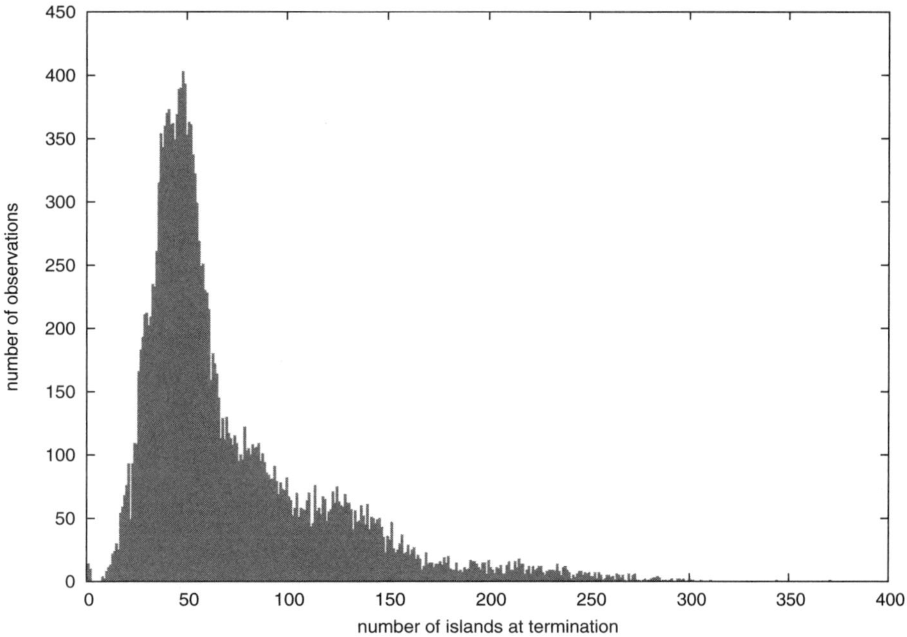

Figure 5.14. *Experiment* 5.4.7. *Distribution of number of islands at cascade termination in* 20000 *simulated cascades with a common start for the Polish grid.*

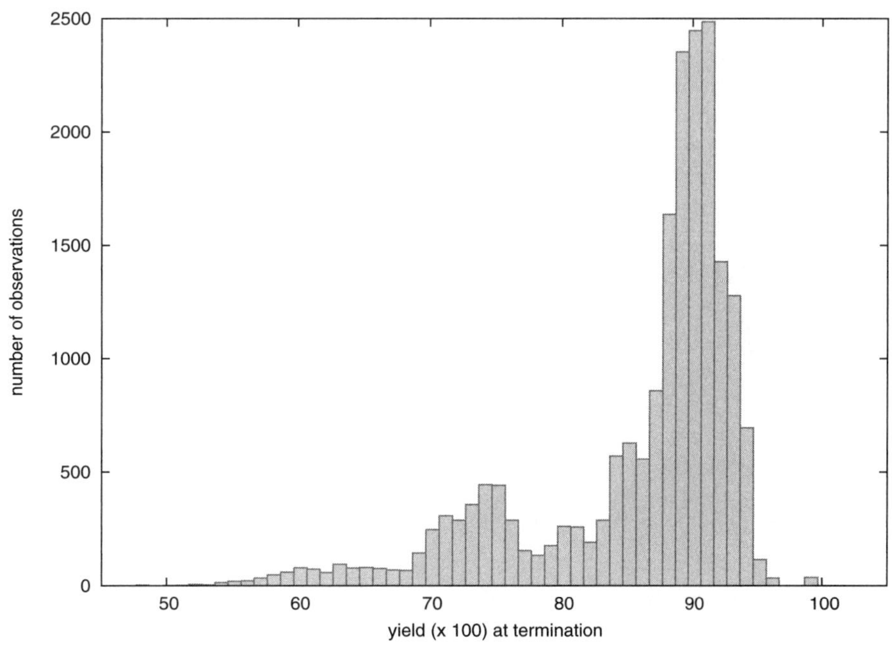

Figure 5.15. *Experiment* 5.4.7. *Distribution of yield at cascade termination in* 20000 *simulated cascades with a common start for the Polish grid.*

Table 5.7. *Statistics for model of Eastern Interconnection.*

Buses	Branches	Generators	Generator buses	Total load
15023	23769	2255	7696	579168 GW

Experiment 5.4.8.

Our next experiment uses a model of the Eastern Interconnection (one of three large U.S. transmission systems).[44]

Data for the model we consider here are given in Table 5.7. Below we will carry out many experiments using this system and modifications thereof; however for this experiment we consider a relatively simple case. We will examine a hypothetical but very challenging (multiple) contingency, where 55 (randomly chosen) lines are simultaneously tripped. To keep the simulation simple we use the deterministic line tripping rule without memory ($\eta_e = 1.0$ for every line e). The resulting cascade terminates after 22 time-steps, with a yield of 58.30%, 6199 tripped lines, and 2143 islands.

Figures 5.16, 5.17, and 5.18 provide a snapshot of the cascade at the end of time-steps $0, 4, 8$ and $12, 16, 20$ (respectively). These figures show a common subset of lines, in each case rendered so that more recent trips are shown darker (and some nontripped lines are rendered in gray). These pictures appear to show a mixed process, with (apparently) a clear progression of the cascade toward the center of the map, but also with (again, apparently) some spontaneous trips developing not near previous trips (such as the formation toward the middle right of the map in time-steps 16 and 20; see Figure 5.18). However, these observations are to be questioned because the figures only show a small subset of all lines tripped at any point in time, which is often large in this cascade. This discussion concerns the issue of geographical "spread" of a cascade: does a power grid cascade extend from tripped lines to nearby tripped lines? Many abstract models of cascades, such as those in [230] (see Section 4.3.4), postulate precisely such a behavior while bypassing detailed modeling of a grid (in particular power flow physics), but will simulations based on realistic models and using real data support such a model? This is indeed a very difficult issue to settle. Further experiments that we conduct below and later in this work will seek to address this question.

Another interesting facet of this (very significant) cascade is its behavior as a function of time. Table 5.8 displays the number of trips in each time-step ("trips"); the yield at the end of the time-step; and the maximum relative line load ("overl") also at the end of the time-step. We observe that early on in the cascade little yield is lost; however very quickly the number of tripped lines and the maximum relative line load become very high. The cascade remains very intense until (roughly) time-step 18, at which point it begins to abate—however at that point the yield has essentially reached its final value. An interesting question motivated by this behavior concerns the timing of control: suppose we wanted to arrest the cascade by (for example) shedding a small amount of load. What would be the optimal way of doing so? Would it be optimal to shed that load all at once—and if so, when? The real question is what policy one would implement, in real time, on the basis of the currently available information. Nevertheless, a theoretically interesting point concerns the optimal policy even assuming perfect information of the entire cascade.

These points will be considered in Chapter 6, from a mathematical standpoint. From a basic standpoint, we recall that any fair load shedding policy (i.e., one that sheds load in

[44]Kindly provided to us by Prof. Ian Dobson.

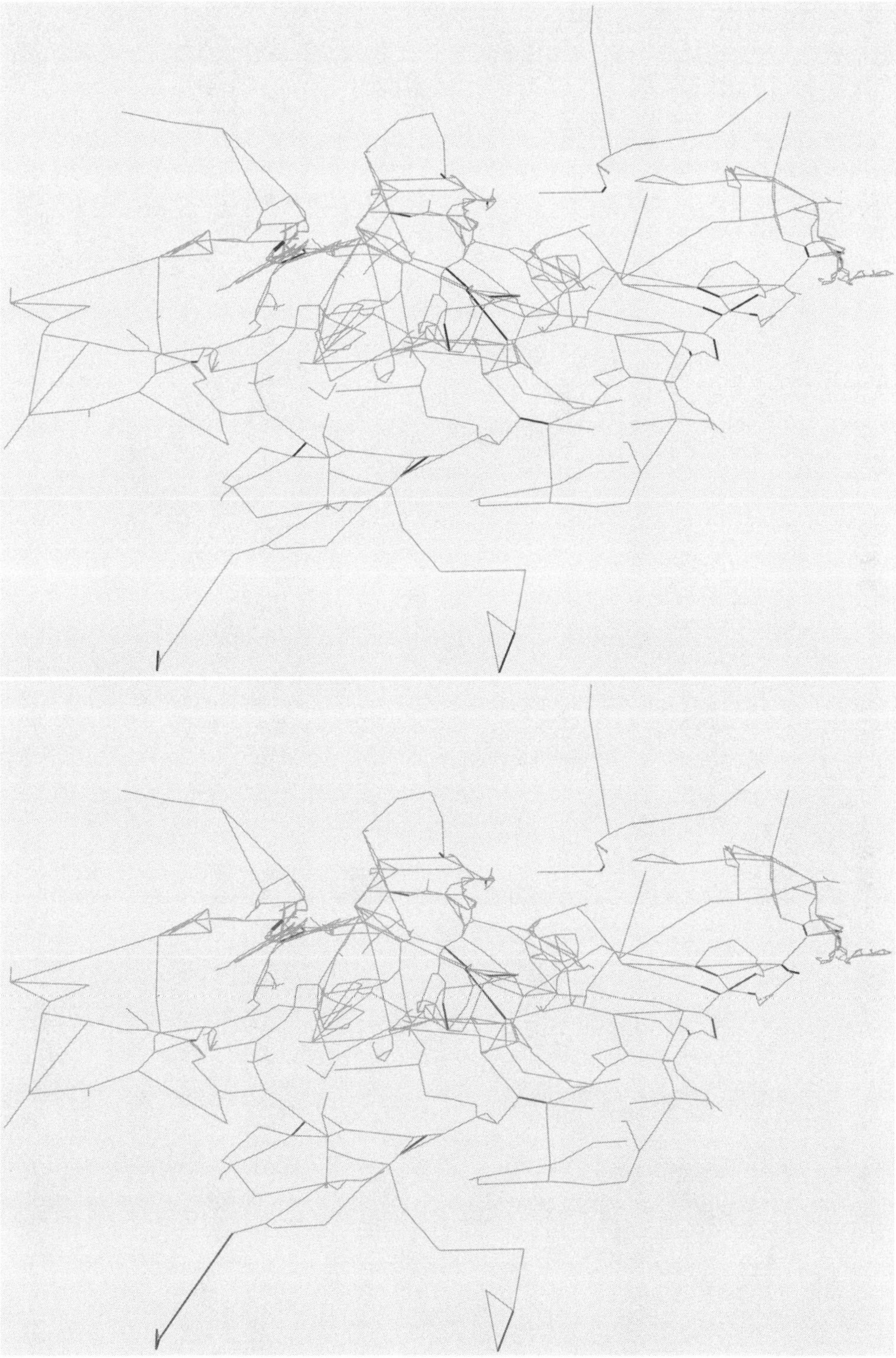

Figure 5.16. *Experiment* 5.4.8. *Time-steps* 0, 4.

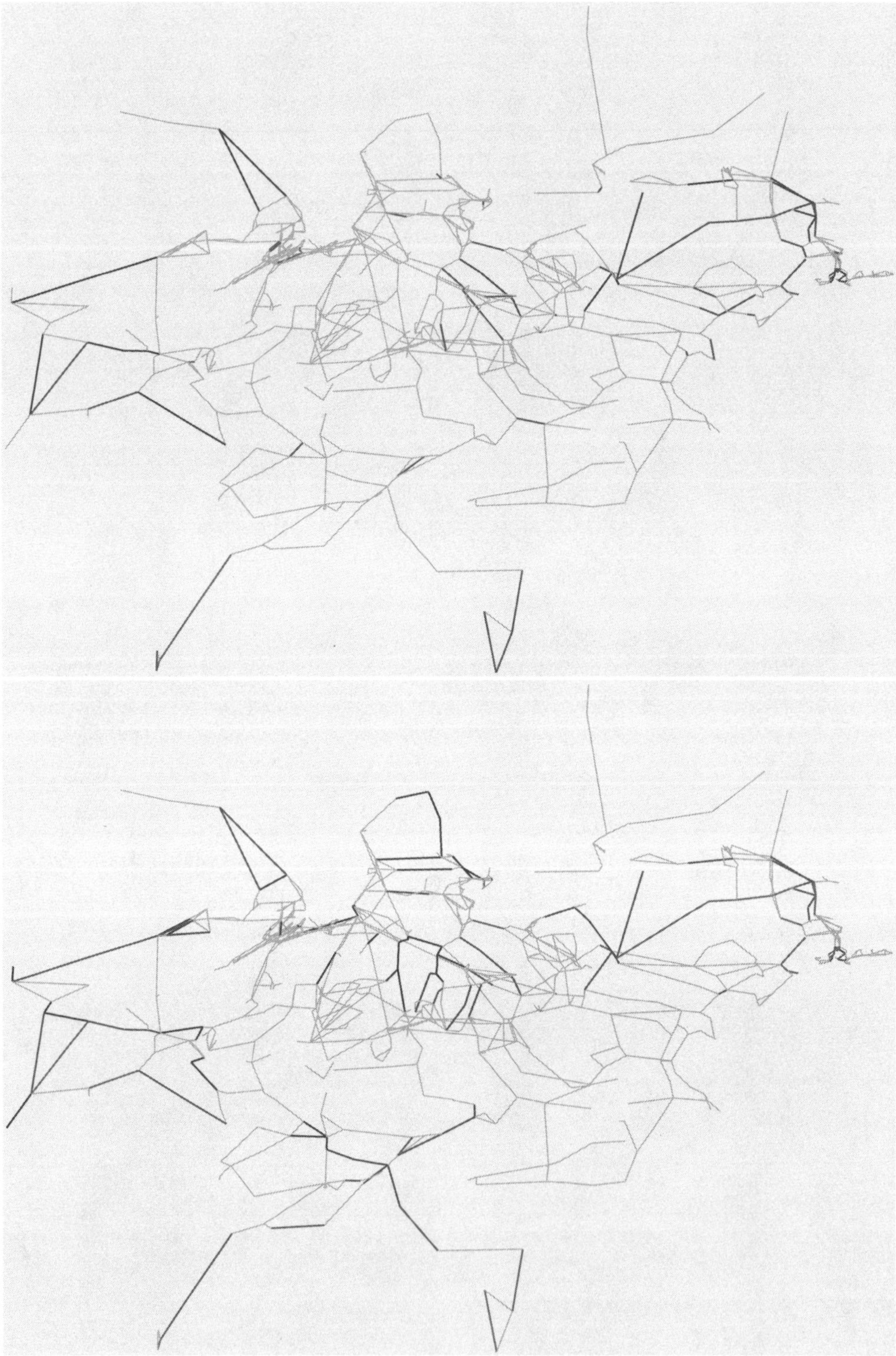

Figure 5.17. *Experiment* 5.4.8. *Time-steps* 8, 12.

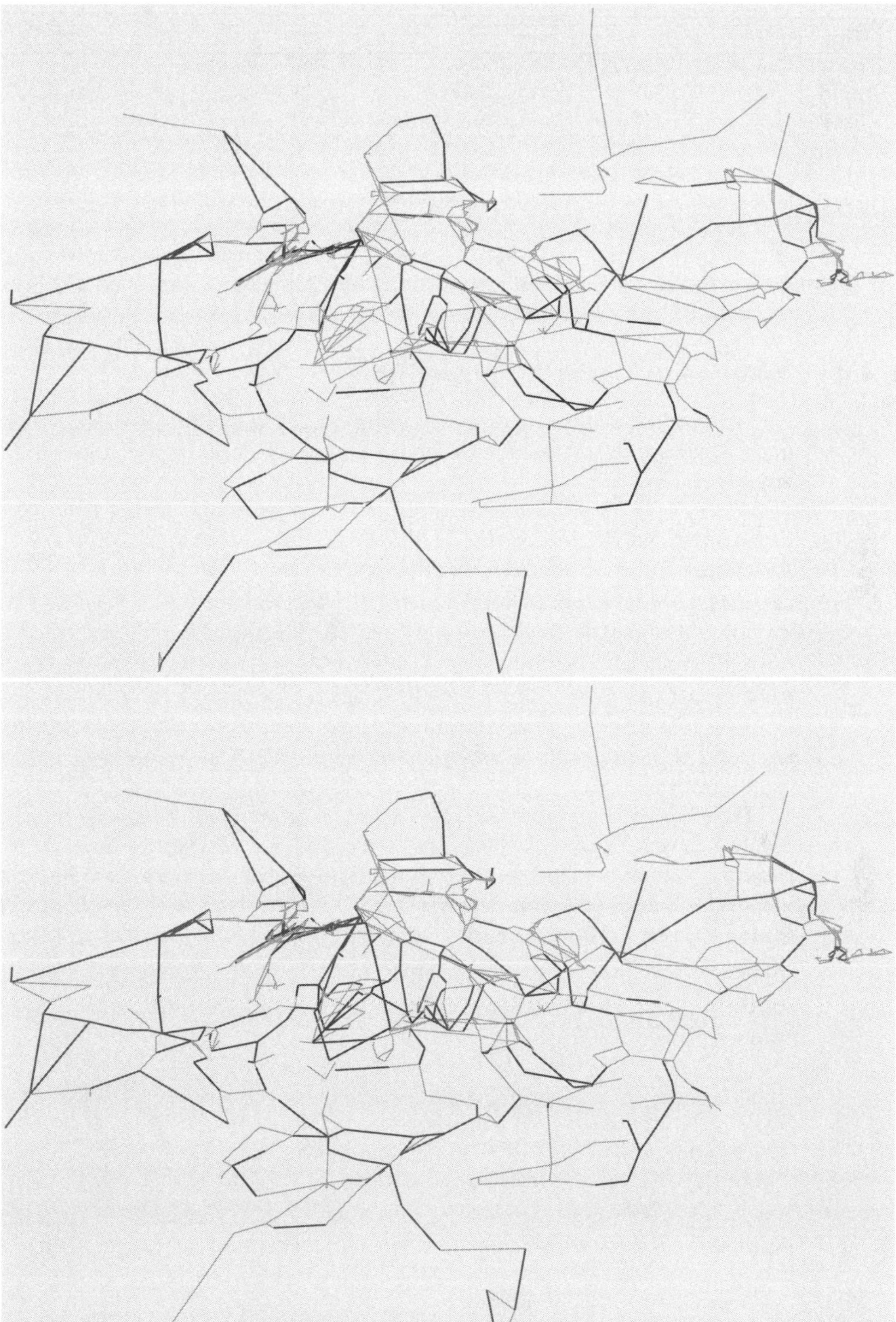

Figure 5.18. *Experiment* 5.4.8. *Time-steps* 16, 20.

Table 5.8. *Experiment 5.4.8. Cascade progression on Eastern Interconnection.*

Time	1	2	4	6	8	10	12	14	16	18	20	22
trips	24	56	132	467	485	378	326	266	536	103	22	0
yield	1.00	1.00	0.98	0.94	0.89	0.85	0.79	0.74	0.66	0.60	0.58	0.58
overl	1.50	8.57	10.13	47.15	13.21	28.97	62.42	34.39	115.62	10.85	3.18	0.99

equal proportion throughout the grid) will, broadly speaking, reduce power flows in proportion to the decrease in load. It therefore makes sense to shed load when the maximum relative overload is at a minimum (so that all lines become stable). Considering Table 5.8 we see that the optimal timing for democratic, "all-at-once" load shedding is very early in the cascade. Is this a general observation for all cascades?

5.4.4 ▪ Experiments using AC power flow models

Here we describe some experiments using our AC power flow–driven cascades model under separated time scale. For the power flow computation, we first use the formulation given in Section 5.3.3.

Experiment 5.4.9.
For the first experiment, we use the standard 118-bus model, and as above we consider the cascade that ensues after tripping line $(9,10)$. We used $\rho = 0$ (no generator ramp-up) and $\eta = 1$ (no memory in the line tripping model). The quantities $V_k^{\min}$ and $V_k^{\max}$, which in our OPF model (5.8)–(5.14) are used to constrain voltage magnitude changes, were set as follows. For a generator bus k we set $V_k^{\max} = \bar{V}_k(1+\delta_k)$ and $V_k^{\min} = \bar{V}_k(1-\delta_k)$, where $\bar{V}_k$ is the precontingency voltage magnitude at bus k and $\delta_k = 0.01$. For nongenerator buses we set $\delta_k = 0.5$. Since $\bar{V}_k \approx 1$ for all buses k, this choice allows nongenerator voltages to significantly change.

Table 5.9 reports results obtained when we computed the cascade under both AC- and DC-driven versions. In both cases the cascade ended after five time-steps. Each row describes numerical values at the end of a time-step (the values for the fifth time-step are identical to those for the fourth time-step); "Max. rel. load" is the maximum relative line load. The last two columns in the table were obtained as follows. Let f_{ij}^k denote the value of the DC flow on line (i,j) at the end of time-step r and likewise g_{ij}^k denote our estimate for the value of the active flow on line (i,j) at the end of time-step $k = 1,2,3,4$. Then define

$$\delta_{ij}^k \;=\; \frac{|f_{ij}^k - g_{ij}^k|}{\max\{|f_{ij}^k|,|g_{ij}^k|\}} \quad \text{if } |f_{ij}^k - g_{ij}^k| > 0.001, \quad \delta_{ij}^k = 0 \text{ otherwise.}$$

The column headed "Rel. diff." reports the average value of δ_{ij}^t (over all lines). In other words, this column reports the average discrepancy between flow values. This error measure, while useful from a general perspective, could be misleading if large discrepancies occur on lines (i,j) with small $|f_{ij}^k| + |g_{ij}^k|$. To guard against this eventuality, we define

$$\omega_{ij}^k \;=\; \frac{|f_{ij}^k - g_{ij}^k|}{\frac{1}{\hat{m}}\sum_{h,k}(|f_{hk}^k| + |g_{hk}^k|)}, \quad \text{where } \hat{m} = \text{no. of observations.}$$

The column headed "Scaled $\bar{\omega}$" reports the average value of the ω_{ij}^k. This metric thus provides a sense of the average discrepancy as a fraction of the average flow value.

Table 5.9. *Experiment* 5.4.9. *AC- versus DC-driven cascades,* 118-*bus case.*

	Trips		Islands		Yield		Max. rel. load		Ave. discrepancy	
Time-step	**AC**	**DC**	**AC**	**DC**	**AC**	**DC**	**AC**	**DC**	**Rel. diff.**	**Scaled $\bar{\omega}$**
1	45	45	18	18	72.22	72.22	4.82	4.82	0.050	0.020
2	60	63	25	25	62.56	62.56	3.90	4.12	0.055	0.031
3	82	86	35	38	51.60	50.91	1.67	1.67	0.048	0.032
4	84	88	37	40	47.46	46.77	0.91	0.92	0.051	0.045

Table 5.10. *Experiment* 5.4.9. *Error metric (see* (5.23)*) under AC model.*

t	0	1	2	3	4
$\mathcal{E}(t)$	7.8e-4	4.3e-3	4.5e-3	1.4e-2	4.3e-2

We note, in this example, a close similarity between the overall behavior of both cascades, and even a relatively close match between flow values. It is known, however, that the 118-bus case is well behaved, and it is clearly questionable whether this similarity will hold over more complex models.

Another issue concerns running times. To run the model with AC flows, we employ formulation (5.8)–(5.14), which must be applied to each island in each of the time-steps of the simulations. In our experiments we have relied on MATPOWER with default settings. The running times are indeed onerous. Whereas the run in the table required CPU time on the order of one tenth of a second using the DC formulation, the model with the AC formulation required approximately 60 seconds. It is not clear to us how to speed up the computations to completely erase this difference. Clearly, switching to a different OPF solver than default MATPOWER might result in an improvement; however even a hundred-fold speedup will still leave the AC-based model at approximately one-fifth the speed of the DC-based model. Another possibility would be to rely on warm-restart capabilities of optimization engines when solving subsequent problems involving a common island. Note that this possibility is not always available because line trips change the structure of an island (even if nominally the set of buses in the island is the same) and, often enough, new islands are created. Moreover, not all optimization engines for OPF can make use of a warm start; this is in particular the case with engines based on interior point methods (sequential linearization methods may fare better).

A final issue regarding this experiment concerns the quality of the approximation of our formulation (5.8)–(5.14) to a true power flow model. Recall that at time-step t the objective function (5.8) attempts to produce generation outputs P_k^t that are close to target values $\bar{P}_k^t$. Table 5.10 reports the following error metric, computed at the end of each time-step t:

$$\mathcal{E}(t) \doteq \max_{k \in \mathcal{G}} \frac{|P_k^t - \bar{P}_k^t|}{S(k,t)}, \quad \text{where} \tag{5.23}$$

$$S(k,t) = \text{total generation in the island holding bus } k \text{ at time } t, \text{ for each } k \text{ and } t,$$

and as before $\mathcal{G}$ is the set of generator buses.

Table 5.10 *suggests* a consistent picture: as the cascade evolves, it becomes more difficult to attain the generation levels $\bar{P}_k^t$—which are just large enough to match the loads assuming zero transmission losses—without significant voltage excursions at the generators. Of course, the table may also reflect an artifact introduced by the particular OPF solver we have relied on, which may not be computing a true global optimum of problem (5.8)–(5.14).

Table 5.11. *Experiment* 5.4.10. *Error metric* (5.23) *under reactive power shortfall.*

t	0	1	2	3	4	5
$\mathcal{E}(t)$	8.0e-4	4.3e-3	1.1e-2	8.1e-2	2.5e-1	2.5e-1

Experiment 5.4.10.

To further explore these issues, we repeated Experiment 5.4.9 with a single alteration, namely, that the cascade-initiating contingency is a compound event:

(a) For every nongenerator bus k, its reactive power demand Q_k^d is increased by a factor of 2.0.

(b) As before, line (9,10) is tripped.

Item (a) is quite significant. First, its impact should be that voltage magnitudes at the loads decrease, making it difficult for the system to deliver power to the loads, in particular given that we have increased the reactive power demands without increasing the reactive power generation limits at the generators. However, a loadflow computation with reactive loads changed as in (a) (but no (b) applied) shows that the system remains feasible, albeit with some low voltages (bus 53 at voltage 0.93).[45]

Running the cascade simulation with parameters as in Experiment 5.4.9 appears to show a similar trajectory, albeit requiring six time-steps, with final yield 46.47%, 88 line trips, and 39 islands (compare with Table 5.9). However, Table 5.11, the counterpart of Table 5.10, presents a different picture.

We notice significant errors in the last three time-steps, suggesting once more that in some of the optimization problems (5.8)–(5.14), the target generation levels $\bar{P}_k^t$ are difficult to attain. A closer look at time-step 4 (or 5) of the simulation reveals an island with selected characteristics described in Table 5.12. In this island, buses 69, 70, and 74 are generators, whereas buses 75 and 118 are loads. However, at time $t = 0$ buses 70 and 75 have zero output (as per the 118-bus data set) and, together with our zero ramp-up constraint ($\rho = 0$), their output remains unchanged throughout the cascade. The island under consideration is created by line trips at the start of time-step 4, resulting in an excess of generation.

In Table 5.12, the rows labeled "$P^{g,3}$" and "$P^{g,4}$" show generator output levels after the islanding model applied in time-steps 3 and 4, respectively, whereas rows labeled "$\bar{P}^{d,3}$" and "$\bar{P}^{d,4}$" shows similar information concerning loads. From these four rows we can see that in time-step 4, the island undergoes a generator ramp-down at generator 69 to match the total load of 179.84 units. The row labeled "Q^g" shows the reactive power generator *limits*, whereas the row labeled "$\bar{Q}^{d,4}$" shows the reactive power demands after islanding in time-step 4. Finally, the rows labeled "P^4" and "Q^4" show the active and reactive power *generation* as per optimization model (5.8)–(5.14), in time-step 4.

Comparing rows $P^{g,4}$ and P^4 of the table we see that the objective function attained by the optimization model is

$$(139.08 - 179.84)^2 + (45.58 - 0)^2 \approx 3738.91.$$

As a fraction of total generation, the mismatch at bus 74 equals 45.58 units, or roughly 25% of the total generation amount (179.84) prescribed by the islanding model; this is the error metric 5.23 for time-steps 4 and 5 reported in Table 5.11. The question is why such a large mismatch takes place. The answer is provided by the next two rows of the

[45] A common practice is to consider a bus tripped if its voltage dips below 0.95, p.u.

Table 5.12. *Experiment* 5.4.10. *Inadequate reactive power capability in time-step* 4.

Bus	69	70	74	75	118
$P^{g,3}$	292.69	0	0	0	0
$\bar{P}^{d,3}$	0	55.46	57.14	39.5	27.73
$P^{g,4}$	179.84	0	0	0	0
$\bar{P}^{d,4}$	0	55.46	57.14	39.50	27.73
Q^g	$-300 \le Q \le 300$	$-10 \le Q \le 32$	$-6 \le Q \le 9$	0	0
$\bar{Q}^{d,4}$	0	16.81	22.69	18.50	25.21
P^4	139.08	0	45.58	0	0
Q^4	62.39	−10	9	0	0
LFP	188.77	0	0	0	0
LFQ	50.11	−13.98	37.84	0	0

table. Here, we computed the loadflow solution for a problem with loads as given by the quantities $\bar{P}^{d,4}$ and generation as per the quantities $P^{g,4}$, using bus 69 as the slack bus. Rows "LFP" and "LFQ" show the resulting active and reactive power injections at generators. First, note that in addition to the fact that this solution does achieve zero output at generators 70 and 74, its objective function on problem (5.8)–(5.14) equals

$$(139.08 - 188.77)^2 \approx 79.75,$$

which is actually smaller than that attained by the outputs in row P^4. It follows that the vector obtained in the loadflow computation should be infeasible for problem (5.8)–(5.14), and, in fact, we can see from the table that the reactive power outputs for buses 70 and 74 violate the constraints stated in row Q^g of the table.

In summary, therefore, we would be justified in declaring this island "infeasible" (though we did not do so in this experiment). The infeasibility is, in this case, *not* due to an inadequacy of our islanding model, which as discussed before assumes zero transmission losses (and thus in this case generator 69 needs to output 188.77 units and not the somewhat smaller amount 179.84 prescribed by the islanding model). It is a feature of the cascade, which was initiated by increasing reactive power loads without a proportional increase in generation capacity. Considering Table 5.11 we can see a large mismatch in time-step 3; tripping the corresponding island results in the cascade attaining yield approximately 33.20%, much smaller than the stated amount 46.47% attained when not applying the mismatch criterion (or under the DC model).

As a final point regarding this cascade, the specific island under discussion above is created in time-step 4 with a large generation excess of 292.69 − 179.84 units, or nearly 40%. Normally this might result in a large frequency increase to generator 69, enough to trip the generator, thus causing a collapse of the island. The elaboration of islanding models in Section 5.5 will include such frequency modeling.

5.5 ▪ Finer Δt cascade models

We now turn to models where the duration parameter Δt is small enough that higher resolution can be attained; however we do not intend to model infinitesimally small durations.

The largest benefit to be gained from using a comparatively small duration parameter Δt arises in the context of race conditions, in particular those that concern the behavior of islands under generation shortfall. One of the more significant shortcomings of the models in Section 5.3 (as well as previous models) is the assumption that generator ramp-up takes place at a much faster time scale than line tripping. This assumption is unwarranted when an island (with a generation shortfall) is created and the island includes a line that was already severely overloaded just prior to the island creation. If the line remains overloaded it will possibly trip (and quickly), but whether or not this happens, and when, may be impacted by the pattern of generator ramp-up in the island—in fact the line may trip before ramp-up is completed. Using a smaller value for Δt provides a means, however imperfect, of modeling this quintessential race condition.

To understand how our model works, consider the dynamics of power injections and frequency in a freshly created island that has generation gap (= generation − load) $S < 0$. The process of frequency recovery is spanned by two "bookends."

Model 5.5.1. Frequency recovery in a newly created island with generation deficit

(a) Immediately after the island has been created, frequency will drop in the island by an amount proportional to the deficit (to first order). Note that load continues to be satisfied, unless some form of load shedding (intended or not) has instantaneously taken place. This implies that a new vector of (active) power injections in the island has been instantiated. Via the simplification (1.50) of the swing equation one obtains that the *actual* power injection at each generator bus k will change, from its pre-islanding value, by an amount proportional to the frequency drop (possibly zero). This is the initial frequency adjustment.

(b) In the second phase of the process (secondary control) a new set of power injections will be instantiated, and frequency will be returned to the (pre-islanding) target value. The process gradually migrates power injections from the values in **(a)** to their final values; a corresponding shift involves power flows. This shift can be significant because of the different attributes of generators involved in each phase. In fact different generators may be involved in (a) and (b).

Example 5.5.2. Power injections in an island with generation deficit

Consider a freshly created island containing four generators which, just prior to the creation of the island, are generating 10, 10, 10, *and* 70 *units of output. Suppose that total load in the island is* 110 *(thus,* $S = -10$ *units). The following scenario balances and restores generation and load:*

(a) *As part of primary frequency control, the generation output mix changes to* 14, 13, 13, *and* 70.

(b) *Following secondary control, the mix becomes* 10, 10, 10, *and* 80.

The description above assumes that the amount of secondary reserves in the island is adequate to make up the generation shortfall. If not, load will be lost. And if the island is further split (due to line trips), the process is interrupted. [8], [25], [106] all contain examples of the process given by Model 5.5.1. We want to provide an *approximate* description of the evolution of power injections and of frequency that takes place, so that, in particular, we can model line and generator tripping that may occur in the midst of the process. The development of a very accurate model of the process would be an interesting task. The model would need to account for the swing equations in all participating generators, that is to say, account for frequency evolution, mechanical power output, and electrical power injections of generators (which are different); inertial properties of the generators; and, finally, the (dampened) control algorithms that are implemented as part of secondary response. We anticipate that in a large-scale setting, the models might prove computationally demanding (in [106] it is stated that numerical difficulties could be experienced). Moreover, there would be a need for extensive modeling, analysis, and calibration, in particular to accurately capture how line tripping is affected by the dynamics of the process. We have already described the difficulties associated with modeling line tripping; these would be multiplied in a dynamic setting. As a simple example, overloaded lines could temporarily trip—not only would the model need to capture this, but in addition it would need to adequately represent the dynamics that ensue when such trip(s) create temporary additional islanding.

At any rate, the numerical procedure that we will rely on will simply construct a *piecewise*-linear interpolation between the values (of power injections and frequencies) at the two "bookends" (a) and (b) in Model 5.5.1. It will not model the dampened control algorithm implemented in practice. As a result, if secondary control is completed, frequency will evolve from the deficit situation in (a) to the nominal setting in (b). To do this, for each generator bus g, we will maintain three quantities:

1. p^I_g, the *true* power injection at g,
2. $\breve{p}_g$, the *pseudo* power injection at g, and
3. p^E_g, the *effective* power injection at g.

We stress that these three quantities do not represent precise physical parameters and instead will be used as approximations to the power injection at g. In the standard situation where bus g is in an island without generation deficit, we will have $\breve{p}_g = p^I_g$, but if there is a deficit all three values will generally differ. However, the following condition will always be maintained, even during the process of secondary control. Let C be an island, and denote by $G(C)$ the set of generators in C and by $L(C)$ the load in C. The condition that will be maintained is

$$\sum_{g \in G(C)} p^E_g = \sum_{g \in G(C)} p^I_g = L(C). \tag{5.24}$$

The values p^I_g and p^E_g will be used to approximate the actual electrical power injection by g into island C at the end of each time-step. Further, the p^I_g will be used as an input to the power flow solver and thus will also impinge on our line tripping model. The values $\breve{p}_g$ will be used to keep track of the output and ramping constraints of the generators and, except at termination of secondary control, will add up to *less* than $L(C)$. All values $\breve{p}_g, p^E_g, p^T_g$, as well as the frequency in island C, will be updated at the start or end of each time-step. Model 5.5.3 presents our precise mechanism; we will discuss it later. Throughout we assume that target system frequency has been normalized to 1.0.

Model 5.5.3. Frequency and power adjustments in one time-step of duration Δt in an island with generation deficit

Notation:

1. The set of available generators in the island is denoted by $\mathscr{A}$.
2. L = total load in the island.
3. For each generator g, denote by $R_g > 0$ its "regulation constant." Let $0 \leq f^{\min} \leq 1$ denote the minimum allowable system frequency.

Step 1.

(a) Write $S = \sum_{g \in \mathscr{A}} \check{p}_g - L < 0$ and $\Delta f = \frac{S}{\sum_{g=1}^{k} 1/R_g}$, the frequency drop versus nominal value.

(b) For each $g \in \mathscr{A}$, set $p_g^E = \check{p}_g - \frac{\Delta f}{R_g}$, and set the frequency of generator g to $1 + \Delta f$. If $p_g^E > P_g^{\max}$, generator g is *tripped* and we reset $\mathscr{A} \leftarrow \mathscr{A} - g$.

(c) If no generators tripped in Step 1(b), continue to Step 2; otherwise repeat Step 1(a) and (b).

Step 2.

(i) In parallel, each generator $g \in \mathscr{A}$ increases its output $\check{p}_g$, at rate v_g per unit time, by at most $P_g^{\max} - \check{p}_g$ units, and for at most Δt time units, or until total output equals L, whichever comes first. (This follows a similar process as Model 5.3.2.) Let τ_g indicate the amount by which the output of generator g is increased, so that Step 2(i) resets $\check{p}_g \leftarrow \check{p}_g + \tau_g$.

If $\sum_{g \in \mathscr{A}} \tau_g = 0$, perform emergency load shedding and **EXIT**.

(ii) Set $\tilde{\Delta} f = \frac{\sum_{g \in \mathscr{A}} \check{p}_g - L}{\sum_{g \in \mathscr{A}} 1/R_g}$ (≤ 0) and write $\phi = \frac{\sum_{g \in \mathscr{A}} \check{p}_g}{L}$. Thus $0 \leq \phi \leq 1$.

(iii) For each generator g, set $p_g^I = \phi\left(\check{p}_g - \frac{\tilde{\Delta} f}{R_g}\right) + (1 - \phi) p_g^E$.

(iv) If $\tilde{\Delta} f = 0$, set all generator frequencies to 1.0.

(v) If $1 + \tilde{\Delta} f < f^{\min}$ *trip*, all generators in the island (remove them from the list of available generators). Island collapses.

In Model 5.5.3, the quantities $\check{p}_g$ approximately model generator output during secondary control; if $v_g = 0$, then generator g does not participate. In general, multiple time-steps will be needed for the generation deficit to be removed. Immediately after an island with generation shortfall is created, the values $\check{p}_g$ in Step 1 are the power injections just prior to the island creation, and so the p_g^E quantities in Step 1(b) model the initial frequency response; thereafter, repeated applications of the model monotonically increase the quantities $\check{p}_g$. We expand on "emergency load shedding" below.

Note that the quantities $\check{p}_g$ are used both to set system frequency (in Step 1(b)) and to keep track of the performance constraints on generator g (in Step 2(a)). The quantities

p_g^E are set in Step 1(b) and can be interpreted (via the swing equation) as an estimate of the electrical power injected by generator g into the grid. Finally, p_g^I is set in Step 2(iv) as an intermediate value between p_g^E and $\breve{p}_g - \frac{\tilde{\Delta f}}{R_g}$. This latter quantity, as will be formally shown below (Lemma 5.5.5), is the value that p_g^E *will* take in the next time-step. In essence, p_g^I is an interpolate (a convex combination) of power injections at bus g. A final point concerns the generator tripping in Step 1(b). As a generator approaches its upper limit the model breaks down; however one way or another the generator should be removed from secondary control. Note that for $\breve{p}_g - \frac{\Delta f}{R_g} > P_g^{\max}$ it should be the case that the generation shortfall is very large, because the quantities R_g are small, and so our rule is consistent with the likely outcome in a real system.

We will discuss further properties of the model following the proof of Lemma 5.5.5; first we consider an example that illustrates the behavior of the model.

Example 5.5.4. *Here we have a freshly created island with 7 generators, load $L = 1000$, and generation deficit $S = -100$. Table 5.13 describes the disposition of the generators in the island.*

Table 5.13. *Generator attributes and output just prior to the creation of the island.*

g	1	2	3	4	5	6	7
$\breve{p}_g = p_g^I$	50	50	50	50	150	200	350
$P_g^{\max}$	80	80	80	80	160	242	400
R_g	0.001	0.001	0.001	0.001	100	100	100
v_g	0	0	0	0	10	2	2
Freq.	1.0	1.0	1.0	1.0	1.0	1.0	1.0

Assuming duration $\Delta t = 10$, the progression of time-steps is as follows, where for simplicity time is measured relative to the contingency.

First time-step postcontingency: time = 0 through Δt. *Since $\sum_g \breve{p}_g = 900$ we have $S = -100$ and Step 1(a) yields $\Delta f = -100/4000.03 \approx -0.025$, so in Step 1(b) the system frequency becomes 0.975 and we set the vector p^E to*

$$(74.99, 74.99, 74.99, 74.99, 150.0, 200.0, 350.0),$$

which adds up to $L = 1000$, Next, in the ramping-up process (Step 2(i)) generator 5 reaches its maximum of 160 units after 1 unit of time; thus $\tau_5 = 10$, $\tau_6 = \tau_7 = 20$. So Step 2(ii) sets $\breve{p}$ to

$$(50, 50, 50, 50, 160.0, 220, 370),$$

which adds up to 950. Thus, in Step 2(ii) we set $\tilde{\Delta f} = -0.0125$ and $\phi = 0.95$. Finally, Step 2(iv) sets the true power injection vector p^I to

$$(63.1249, 63.1249, 63.1249, 63.1249, 159.5, 219.0, 369.0),$$

which also adds up to 1000.

Second time-step: time = Δt through $2\Delta t$. *Now* $\sum_g \check{p}_g = 950$, *so* $S = -50$, $\Delta f = -0.0125$, *system frequency is* 0.9875, *and now (approximately)*

$$p^E = (62.499, 62.499, 62.499, 62.499, 160.0, 220.0, 370.0).$$

In the ramping-up process, generators 6 *and* 7 *increase their output by* 20 *units, and the rest by zero, so in Step* 2(i) *we get*

$$\check{p} = (50, 50, 50, 50, 160, 240, 390)$$

(which adds up to 990*). Now we get* $\tilde{\Delta} f = -0.0025$ *and* $\phi = 0.99$ *and we obtain*

$$p^I = (52.999, 52.999, 52.999, 52.999, 160.00, 239.80, 389.80),$$

which as before adds up to 1000.

Third time-step: time = $2\Delta t$ through $3\Delta t$. *At this point* $\sum_g \check{p}_g = 990$, *so* $S = -10$, $\Delta f = -0.0025$, *and*

$$p^E = (52.499, 52.499, 52.499, 52.499, 160.0, 240.0, 390.0).$$

In Step 2(i), *generator* 6 *ramps up by* 2 *units and reaches its limit, while generator* 7 *ramps up by* 8 *units and at this point total output* = 1000. *So we now obtain*

$$\check{p} = (50, 50, 50, 50, 160, 242, 398),$$

which adds up to 1000. *Thus,* $\tilde{\Delta} f = 0$ *and the true injections* p^I *are identical to* $\check{p}$. *At this point the generation deficit has been closed and system frequency restored to* 1.0. *Figure* 5.19 *shows system frequency as a function of time (notice the sharp decrease near time zero).*

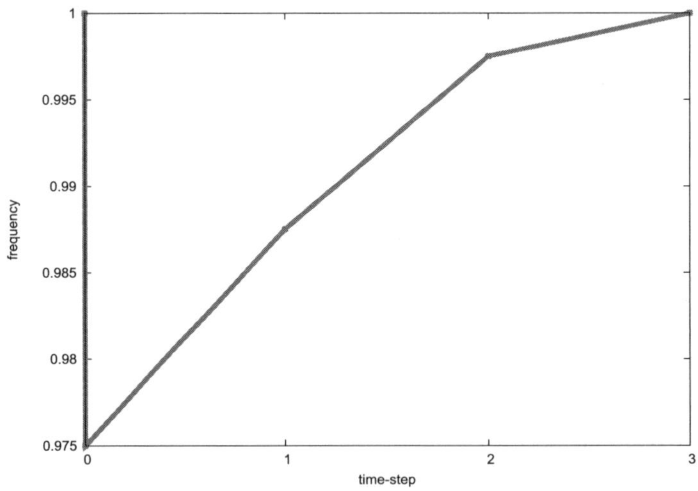

Figure 5.19. *System frequency in Example* 5.5.4.

The following lemma formalizes a number of properties of Model 5.5.3 which we observe in the example.

Lemma 5.5.5. *Consider two consecutive time-steps, say time-steps* R *and* $R+1$, *involving island* C *with generation/load mismatch. Suppose the island remains intact during this period (no line or generator trips). Then we have the following:*

(1) *The quantity* $\tilde{\Delta} f$ *computed in Step* 2(ii) *of time-step* R *equals* Δf *computed in Step* 1(a) *of time-step* $R+1$.

(2) *For any generator* g, *the quantity* $\breve{p}_g - \frac{\tilde{\Delta} f}{R_g}$ *used in Step* 2(iv) *of time-step* R *equals* p_g^E *as computed in Step* 1(a) *of time-step* $R+1$.

(3) *In both time-steps* R *and* $R+1$, *we have* $\sum_{g=1}^k p_g^E = \sum_{g=1}^k (\breve{p}_g - \frac{\tilde{\Delta} f}{R_g}) = \sum_{g=1}^k p_g^I = L$.

Proof. (1) follows because the $\tilde{p}_g$ used to compute Δf in Step 1(a) of time-step $R+1$ are set in Step 2(ii) of time-step R and thus are used to compute $\tilde{\Delta} f$ in Step 2(ii) of time-step R. (2) follows from (1). To prove (3), note that by construction,

$$\sum_{g \in \mathcal{A}} p_g^E = \sum_{g \in \mathcal{A}} \left(\breve{p}_g - \frac{\Delta f}{R_g} \right) = \sum_{g \in \mathcal{A}} \breve{p}_g - \Delta f \sum_{g \in \mathcal{A}} \frac{1}{R_g} = \sum_{g \in \mathcal{A}} \breve{p}_g - \frac{\sum_{g \in \mathcal{A}} \breve{p}_g - L}{\sum_{g \in \mathcal{A}} 1/R_g} \sum_{g \in \mathcal{A}} \frac{1}{R_g}$$

$$= \sum_{g \in \mathcal{A}} \breve{p}_g - \left(\sum_{g \in \mathcal{A}} \breve{p}_g - L \right) = L. \tag{5.25}$$

The fact that $\sum_{g \in \mathcal{A}} (\breve{p}_g - \frac{\tilde{\Delta} f}{R_g}) = L$ is similarly established. Finally, since p_g^I is a convex combination of $\breve{p}_g - \frac{\tilde{\Delta} f}{\rho_g}$ and p_g^E, we also have $\sum_{g \in \mathcal{A}} p_g^I = L$. ■

This lemma formalizes critical properties of the model. In particular, the p_g^I amount to valid power injections (i.e., they meet demand), assuming that transmission losses are neglected. This is important because the p_g^I will be used as the input to the power flow solver (i.e., they will be used as the power injections). Concerning Step 2(iv), note that $\tilde{\Delta} f = 0$ precisely when $\sum_{g \in \mathcal{A}} \breve{p}_g = L$, in which case, additionally, $\phi = 1$. When this happens, Step 2(iii) sets $p_g^I = \breve{p}_g$; this is the condition we guarantee in an island with no generation/load mismatch.

Emergency load shedding

If in an iteration of Model 5.5.3 we have that $\tau_g = 0$ for each generator g, then the generation shortfall cannot be decreased. This situation will arise if each operating generator g with $v_g > 0$ has attained its maximum output $P_g^{\max}$. At this point the system performs emergency load shedding, as in Step 2(a),(b) of Model 5.3.2: if the shortfall is small enough then each load is proportionally reduced to match total generation, and otherwise all loads are lost (island collapses).

Interrupted ramp-up

Since we are using time-steps of limited duration Δt, several steps may be needed to complete the frequency restoration process in an island with generation shortfall. In our complete model (Model 5.5.7; see below) we will intersperse such (partial) ramp-up steps with our line tripping model using power flows computed from the injections p_g^I in Step 2(iii) of Model 5.5.3. Should line(s) trip in the midst of frequency recovery in an island with generation shortfall, several new islands will be created, with different generation shortfall (or excess) characteristics. The solution of systems of swing equations in this case appears especially problematic due to the discrete change; this appears to be particularly difficult with regards to the immediate frequency change. In our implementation we use Model 5.5.3 exactly as stated above to handle a newly created island with generation shortfall, using the quantities $\check{p}_g$ computed in the prior application of Model 5.5.3. For islands with a generation excess we will instead apply Model 5.5.6 given below.

A complexity arises when as a result of line trip(s) in an island with generation shortfall (and in the midst of frequency restoration), a new island is created where there is a perfect balance between generation and demand. The complexity arises because prior to the creation of this island, generators are running at below-normal frequency. For simplicity, we will assume that frequency is instantly restored to normal value in this case.

Island with generation excess

Model 5.5.3 handles an island with generation shortfall. A completely analogous model handles the opposite case, where there is a generation excess and as a result frequencies will exceed 1.0. For the sake of simplicity, we will assume that generators can ramp down much quicker than a time-step duration; nevertheless we incorporate generator tripping through overfrequency.

Model 5.5.6. Generation versus frequency: One time-step of duration Δt in an island with generation excess

Notation:

1. E = generation excess in the island; $f^{\max}$ = maximum allowable system frequency.
2. For notational purposes, assume $P_1^0 - P_1^{\min} \le P_2^0 - P_2^{\min} \le \cdots \le P_k^0 - P_k^{\min}$ (k = number of generators in the island).

(a) If $E > \sum_{g=1}^{k} P_g^0 - P_g^{\min}$, trip all generators in the island.

(b) Otherwise, let $\Delta f = \frac{E}{\sum_{g=1}^{k} 1/R_g}$ (> 0). If $1 + \Delta f > f^{\max}$, trip all generators in the island.

(c) Otherwise, let the index k^* and value $x^* \ge 0$ be such that $\sum_{g \le k^*}(P_g^0 - P_g^{\min}) + (k - k^*)x^* = E$, $P_k^0 - P_k^{\min} \le x^*$ for $k \le k^*$, and $P_k^0 - P_k^{\min} \ge x^*$ for $k > k^*$.

For $1 \le g \le k^*$, the output of generator g is set to $P_g^{\min}$, and for $k^* < g \le k$ it is set to $P_g^0 - x^*$.

In cases (a) and (b) all loads in the island are lost. Case (c) on the other hand assumes that the ramping-down process is faster than the time-step duration Δt and that all generators ramp down at the same speed.[46] Further, when case (c) applies then generator frequencies are left unchanged; in other words the calculation in case (b) is only used to determine if generators will be tripped.

[46]This feature is easily modified to model, for example, some generators that cannot ramp down, by appropriately changing the $P_g^{\min}$ quantities.

5.5.1 ▪ Complete model

We can now update our complete cascade model from the basic version given in Section 5.3.4, which was itself based on Model 4.3.1 and of course on previous work found in the literature. The most significant change is that since we are explicitly addressing the (potentially small) time-step duration Δt the model is now closer to a true "numerical integration" approach.

Model 5.5.7. Cascade model II—Time-steps with limited duration Δt

Step 0. Input: an exogenous contingency that removes equipment (lines, generators) from operation.

Initialize $p_g^I = \check{p}_g$ for each generator g to its current output. Set $T = 0$ and $L = 1$.

Step 1. Comment: Time-step L begins at time T.

Step 2. For each island C with generation deficit, as measured using injections $\check{p}_g$, apply Model 5.5.3.

Step 3. Compute new power flows in each island, using the p_g^I as generator injections.

Step 4. Apply line tripping and generator tripping models.

Step 5. If no lines or generators were tripped in Step 4, and no lines are overloaded, and all islands have generation/load balance, **STOP**.

Step 6. Otherwise, set $T \leftarrow T + \Delta t$, set $L \leftarrow L + 1$, and go to Step 1.

In contrast with the previous models we have analyzed, the use of a limited duration Δt allows this model to intersperse generator ramp-up/ramp-down (in the midst of secondary control) with line tripping. This is interesting since a line trip might take place in an island which is not yet "stable." From this perspective, using very small Δt provides greater resolution. Of course we cannot use extremely small Δt because (among several other factors) the steady-state power flow model becomes at that point inaccurate.[47]

5.6 ▪ Numerical experiments using finer time scale models

Here we first describe some basic experiments to indicate the impact of the finer time scale on the detailed timeline of a cascade. **In all figures below, we will indicate flows in bold and line limits in gray.**

Example 5.6.1. *We will now perform several experiments, all involving the example in* Figure 5.20 *with generator injections and load withdrawals indicated by arrow directions as before. Next to each bus we show its injection (for a generator) or load amount. The operating parameters for the generators are given in Table* 5.14, *with* $P_g^{\min} = 0$ *for each generator* g. *Consider the contingency that removes one of the two lines* $(0,1)$, *as well as line* $(2,3)$. *We will perform two experiments using different values for the duration parameter* Δt *and a third that includes randomness in line tripping.*

Experiment 5.6.2.

First, let the duration $\Delta t = 1$. As we will see, this corresponds to the separated time scale model discussed above. The outcome of this experiment is described next, assuming

[47] However, we stress that Model 5.5.7 *does* remain viable for small Δt provided we substitute an appropriate power flow "model" and we refine Model 5.3.2 to better approximate the swing equation and other dynamics.

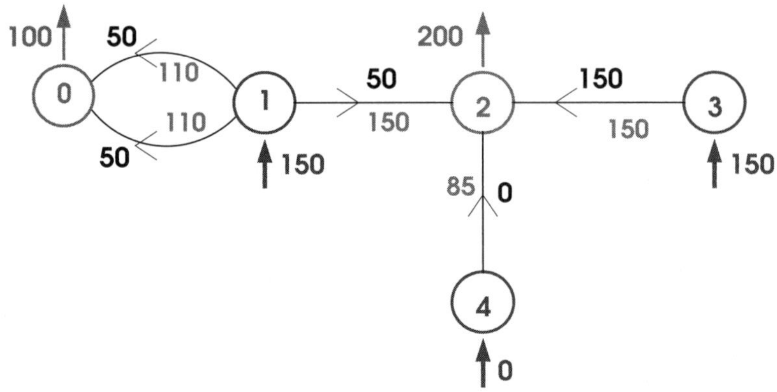

Figure 5.20. *Precontingency state of system in Example* 5.6.1.

Table 5.14. *Generator attributes for Example* 5.6.1.

g	1	3	4
$P_g^{\max}$	300	150	100
R_g	0.001	0.001	0.0005
v_g	200	200	250

the deterministic line tripping model and no hysteresis ($\eta_e = 1$ for each line e, as per (5.1)–(5.2)).

Immediately following the contingency, the island made up of buses 0, 1, 2, and 4 has a generation shortfall of 150 units. In 1/3 time units, generator 1 will increase generation by 200/3 units, whereas generator 4 will increase its generation by 250/3 units, thereby together making up the shortfall in less time than one time-step. The resulting disposition of the system is shown in Figure 5.21; since all flows are within limits, the cascade stops, after one time-step, with yield = 100%. Using $\Delta t = 1$ (in fact, any $\Delta t \geq 1/3$) amounts

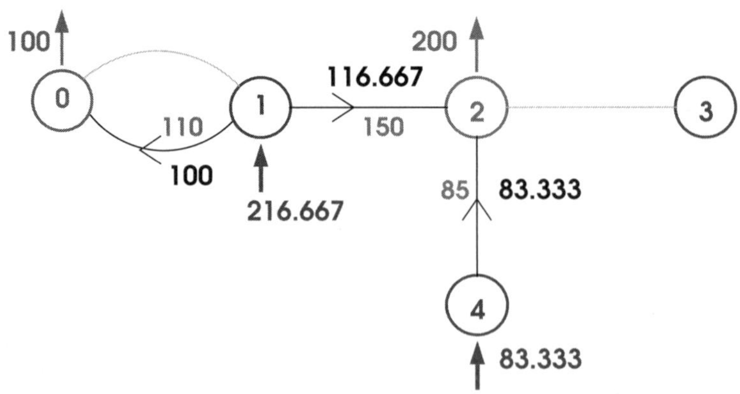

Figure 5.21. *Final condition of system in Example* 5.6.1 *when* $\Delta t = 1$.

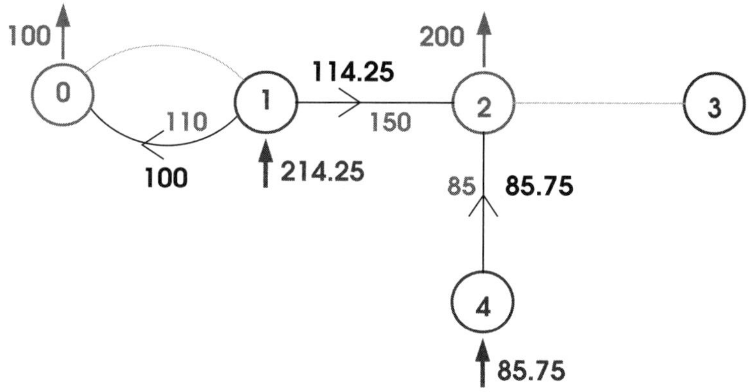

Figure 5.22. *Example* 5.6.1 *when* $\Delta t = 0.3$ *at the end of time-step* 1.

to using the separated time scale model because ramp-up within the island will take place within one time-step and there is no interaction with the line tripping process.

Experiment 5.6.3.

For the second experiment with this system, suppose instead that $\Delta t = 0.3$—this was chosen specifically so that the ramp-up that takes place in the case $\Delta t = 1$ cannot complete. We assume that a generator will trip if its frequency falls outside the range $[0.8, 1.2]$. As discussed prior to Model 5.5.3, just before the contingency we have $\breve{p}_1 = 150, \breve{p}_4 = 0, \breve{p}_5 = 150$. Immediately following the contingency the island consisting of buses $0, 1, 2, 4$ has a generation shortfall of 150 units; applying rule 1(a) of Model 5.5.3, the frequency change Δf in the island therefore equals $-150/(1/1.0e^{-3} + 1/5.0e^{-4}) = -0.05$. Using rule 1(b) we then have $p_1^E = 150 + 0.05/1.0e^{-3} = 200$ and $p_4^E = 100$. Next, in this island the generators ramp up to close the shortfall; in $\Delta t = 0.3$ the outcome is that $\breve{p}_1 = 210, \breve{p}_4 = 75$ (and thus still shy of closing the shortfall). Applying Step 2(ii) of Model 5.5.3 we have $\tilde{\Delta f} = -0.005, \phi = 285/300 = 0.95$, and applying Step 2(ii) we get $p_1^I = 0.95 * (210 + 0.005/1e^{-03}) + 0.05 * 200 = 214.25$ and $p_4^I = 85.75$ (which add up to 300, as per Lemma 5.5.5). Thus, after this partial ramp-up, the power flows are as shown in Figure 5.22 (recall that flows are shown in bold and limits, in gray). Generators 1 and 4 in the island end up with frequency 0.95 and therefore do not trip; however, line $(4, 2)$ *does* trip.

This yields the island with load buses 0 and 2 and a total load of 300 units, and the generator at bus 1 with $\breve{p}_1 = 210$, $p_1^E = 200$, and $p_1^I = 214.25$. The next time-step (time-step 2) of the cascade handles this island, which has a shortfall of $300 - \breve{p}_1 = 90$ units; summarizing the process, in $\Delta t = 0.3$ time units this generator will ramp up by 60 units, and at the end of time-step 2 we will have $\breve{p}_1 = 270$ and $p_1^E = p_1^I = 300$, with power flows as shown in Figure 5.23 and generator 1 at frequency 0.97. Thus, line $(1, 2)$ trips. This leaves the load at bus 2 isolated (demand lost) and creates an island consisting of buses 0 and 1. This island, handled in time-step 3, starts with a generation *excess* of 170 units (recall that as per Model 5.5.3 the quantities $\breve{p}_g$ are used to account for shortfall or excess amounts). Thus generator 1 reaches frequency $1 + 70/1000 = 1.17 < 1.2$ and does not trip; assuming instantaneous ramping down of generators, the time-step will end with the generator at 1 at output = 100 frequency and with line $(0, 1)$ carrying 100 units of flow. Since the system has demand/generation balance and all lines are within limits, the cascade stops with yield = 33.333%. See Figure 5.24. Figures 5.21 and 5.24 highlight the

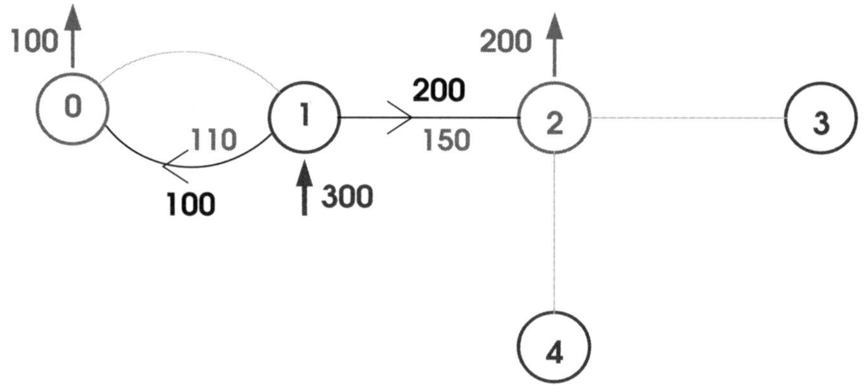

Figure 5.23. *Example* 5.6.1 *when* $\Delta t = 0.3$ *at the end of time-step* 2.

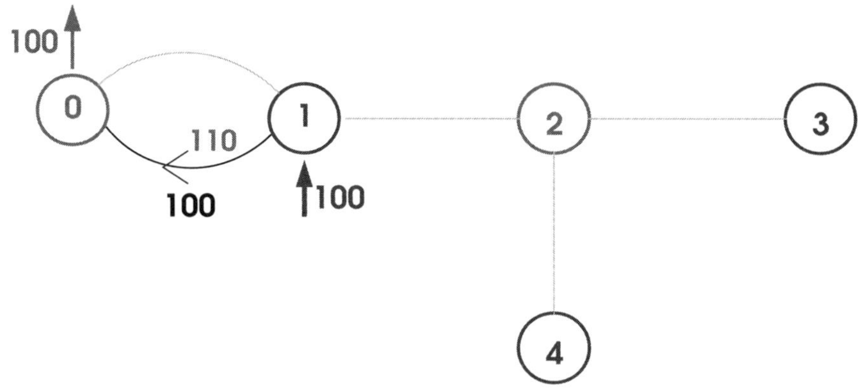

Figure 5.24. *Final condition of system in Example* 5.6.1 *when* $\Delta t = 0.3$.

more nuanced behavior that is obtained as the duration becomes small. Even if we were to introduce randomness in the line tripping process, the outcome of Experiment 5.6.2 would not change (since there is no line tripping), and with some probability the outcome of Experiment 5.6.3 would still be the state shown in Figure 5.24 (we will consider this point below in Experiment 5.6.4). Further, for any $\Delta t < 0.3$ the outcome will be exactly as in Experiment 5.6.3. In general, decreasing Δt in the analysis of a system can be expected to produce "richer" behavior; of course the accuracy of the model will gradually break down (and the computational load will proportionally increase).

Experiment 5.6.4.

Here we add stochastics to the line tripping model; we still assume that $\Delta t = 0.3$. Recall that under Experiment 5.6.3, line $(4,2)$ trips during time-step 1 because its flow (85.75) exceeds the line limit (85). Since the excess over the limit is small, the line may realistically not trip. For this experiment we consider the line tripping model given by (5.5a)–(5.5b) with the following parameter choice: for each line e, $p_e = 0.5$, $m_e = 10$, and $\epsilon_e^U = \epsilon_e^L = 0.01$. An examination of the possible evolutions of the cascade shows the following outcomes, noting that the behavior of the system is fixed through the end of time-step 1 (as per Figure 5.22):

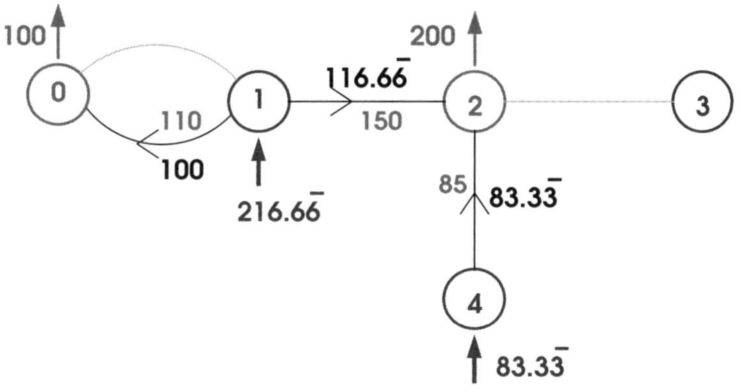

Figure 5.25. *Example* 5.6.1 *when* $\Delta t = 0.3$ *when line* $(4,2)$ *does not trip.*

- Our choice for the line tripping model is such that line $(4,2)$ will trip during time-step 2 with probability 0.5; this yields the system as per Figure 5.23. The overload of line $(1,2)$ is so significant that, with probability rapidly converging to 1 (as time-step count increases or, equivalently, time goes to infinity), that the line will trip and we obtain the same cascade as per Experiment 5.6.3, culminating in the state depicted in Figure 5.24 (and yield $= 1/3$).

- On the other hand, line $(4,2)$ does not trip during time-step 1 also with probability 0.5. Recall from Experiment 5.6.3 that at the end of time-step 0 generators 1 and 4 are ramping up, and that at the end of time-step 0 we have $\check{p}_1 = 210, \check{p}_4 = 75$, so there is still a shortfall of 15 units of demand. Thus, the ramping-up process continues during time-step 1 and closes the shortfall with the disposition of flows shown in Figure 5.25. Here, we can see that line $(4,2)$ has relative load approximately $0.98 (\approx 83.33/85)$ and is therefore relatively exposed to risk—to be precise, under our line tripping model its one time-step tripping probability is approximately 0.39377. Moreover, our cascade termination rule requires $\lceil 1/\Delta t \rceil = 4$ trip-free time-steps. Thus, either line $(4,2)$ will not trip in time-steps 1 through 4 and the cascade terminates as per Figure 5.25, or it will trip, with probability

$$0.5 * (1 - 0.39377)^3 \approx 0.886,$$

  and then we obtain the same outcome as in Experiment 5.6.3. In summary, therefore, the cascade terminates as per Figure 5.24 with probability 0.886 and yield 33.333%, or it terminates with probability 0.114 as per Figure 5.25 with yield 100%.

Example 5.6.5. *Here we will conduct three experiments to exhibit various race conditions between line tripping events and generator ramp-ups, and as such we will highlight the usefulness of using small enough duration* Δt. *See Figure* 5.26.

Table 5.15 *lists the generator attributes (*$P_g^{\min} = 0$ *for each generator* g*). We will ignore generator tripping due to under- or overfrequency and assume no hysteresis in the line tripping process. We consider the cascade that is set off by the failure of line* $(0,1)$*. We assume that no emergency load shedding is allowed.*

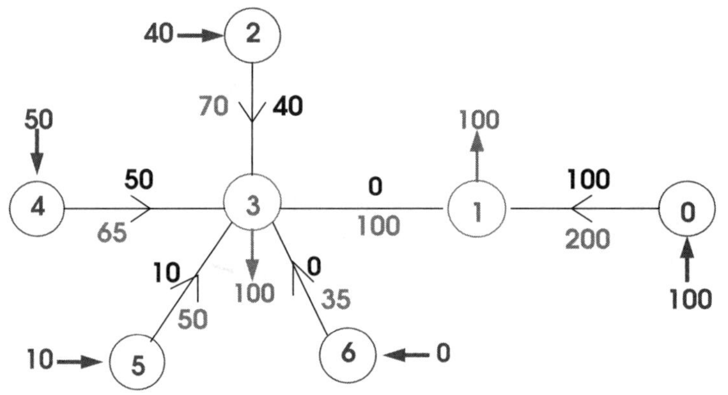

Figure 5.26. *Precontingency state of system in Example* 5.6.5.

Table 5.15. *Generator attributes for Example* 5.6.5.

g	0	2	4	5	6
$P_g^{\max}$	100	200	100	200	100
R_g	0.005	0.0015	0.002	0.001	0.001
v_g	10	80	20	30	100

Experiment 5.6.6.

Here we assume that $\Delta t = 1$, which in this case amounts to the separated time scale model. Further, the line tripping model assumes the parameters $p_e = 0.9$, $m_e = 10$, and $\epsilon_e^U = \epsilon_e^L = 0.01$ for each line e. With the chosen duration, we have that the power flows and injections, postcontingency, are (approximately) as given in Figure 5.27. We see that lines $(2,3)$ and $(6,3)$ are heavily overloaded—in fact their line tripping probabilities are greater than 0.9999. Thus, overwhelmingly, those lines will trip in time-step 1 of the cascade. We also see that line $(1,3)$ is 100% loaded and, according to our model, it will therefore trip with probability 0.9 in time-step 1.

If line $(1,3)$ does trip in time-step 1, then in time-step 2 the operating subset of the grid will simply consist of buses 4, 5, and 3 with total demand 100; thus generators 4 and 5 will need to ramp up, and, using the parameters shown in Table 5.15, this results in generator 4 ramping up to 66 units and generator 5, to 34 units. This will result in line $(4,3)$ tripping, either in this time-step or shortly after, since its relative load is approximately $1.015 > 1.01$. After that, line $(5,3)$ will be severely overloaded and will thus also trip, resulting in zero load being satisfied.

If line $(1,3)$ does not trip in time-step 1, then the situation is even more dire: buses 4 and 5 will need to ramp up to 200 units and again both line $(4,3)$ and line $(5,3)$ will trip.

In summary, with probability very close to 1, the $\Delta t = 1$ case produces yield 0.

Experiment 5.6.7.

Now we will assume $\Delta t = 0.1$ and we will use the same line tripping model as in Experiment 5.6.6, but we will prorate it to the shorter duration (refer to (5.6)). Note that the given contingency creates an island with a generation deficit of 100 units; with the ramp-up parameters as given, in one time-step (of duration 0.1) this shortfall is reduced by 23

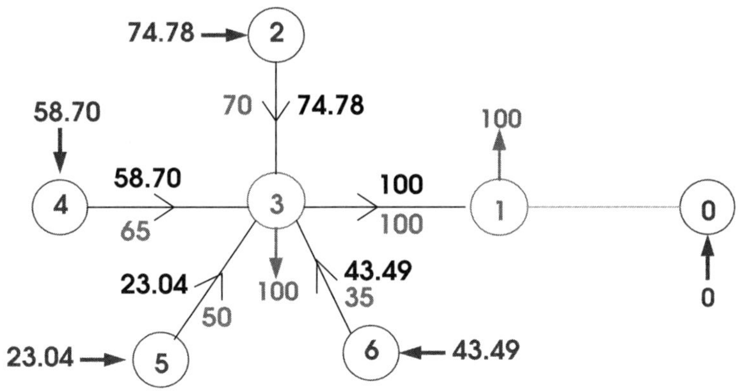

Figure 5.27. *Postcontingency state of system in Example* 5.6.5 *when* $\Delta t = 1$.

($= 0.1(80+20+30+100)$) units and therefore five time-steps will be needed to remove the deficit.

Figure 5.28 shows the power flows and injections at the end of time-step 0. We see from this figure that line $(3,1)$ is again 100% loaded and is therefore highly stressed. Note that so long as a generator remains connected to bus 3, if line $(3,1)$ has not yet tripped, then bus 1 will receive 100 units of demand, since we do not permit load shedding. Thus either line $(3,1)$ will eventually trip, or the system will collapse (i.e., zero yield)—the question is how the generator ramp-up process will affect the likelihood of line $(3,1)$ tripping before other lines, and what value the yield will take should this event happen. To address this issue, consider Table 5.16, which shows the relative line loads and corresponding tripping probabilities at the start of time-steps 1 through 5, *assuming for each time-step that no lines have tripped in prior time-steps.*

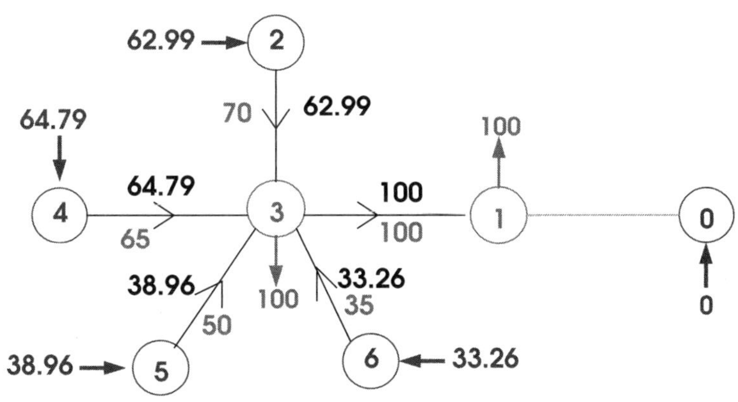

Figure 5.28. *System in Example* 5.6.5 *when* $\Delta t = 0.1$ *at end of time-step* 0 *(= start of time-step* 1*).*

Several points are worth noting with regard to Table 5.16. The tripping probabilities were computed by prorating the line tripping model used for the case $\Delta t = 1$, to $\Delta t = 0.1$. For example, the flow on line $(3,1)$ is 100, so the relative line load for $(3,1)$ is 1.00. Thus the tripping probability for this line, for $\Delta t = 1$, is 0.9. Thus, the tripping probability for $\Delta t = 0.1$ is that number q such that

$$1-(1-q)^{10} = 0.9,$$

Table 5.16. *Relative line loads and tripping probabilities in Experiment* 5.6.7.

	Relative line loads					Tripping probabilities				
Line	(3, 1)	(2, 3)	(4, 3)	(5, 3)	(6, 3)	(3, 1)	(2, 3)	(4, 3)	(5, 3)	(6, 3)
time-step 1	1.0	0.900	0.997	0.779	0.950	0.206	1.2e-5	0.206	1.2e-6	9.1e-5
time-step 2	1.0	0.950	0.969	0.684	1.038	0.206	9.0e-5	9.7e-4	1.8e-7	0.595
time-step 3	1.0	1.005	0.941	0.589	1.125	0.206	0.206	5.0e-5	1.6e-8	0.719
time-step 4	1.0	1.051	0.913	0.494	1.211	0.206	0.645	1.7e-5	5.9e-10	0.752
time-step 5	1.0	1.068	0.903	0.461	1.242	0.206	0.676	1.4e-10	1.4e-10	1.00

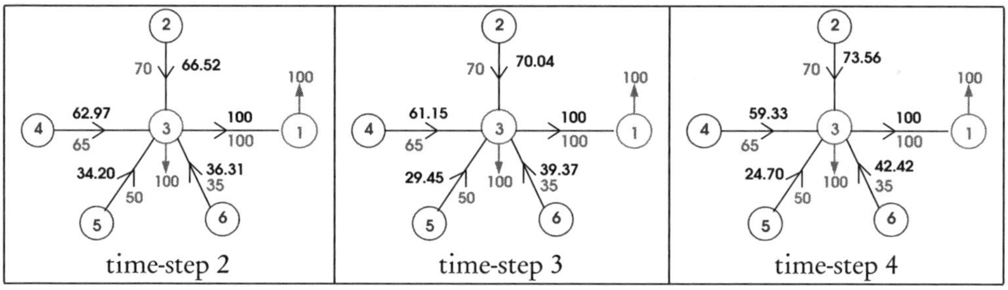

Figure 5.29. *Flows at start of time-steps* 2–4 *under Experiment* 5.6.7 *assuming no line trips.*

that is to say, $q = 1-(1-0.9)^{0.1} \approx 0.206$. The other tripping probabilities were similarly computed. In particular, line (6, 3) will trip at the start of time-step 5 with overwhelming probability.

Also, as noted above, assuming no lines trip in time-steps 1 through 4, the ramp-up amounts in time-steps 0 through 4 will be $23, 23, 23, 23, 8$. At the start of time-step 5 (or end of time-step 4), the flow amounts are exactly as in Figure 5.27; i.e., though if no lines trip during time-steps 1 through 4, the system is in exactly the same state as at the end of time-step 0 for the case $\Delta t = 1$. See Figure 5.29.

Finally, note that the relative line loads shown in Table 5.16 are monotonically increasing for lines $(2,3)$ and $(6,3)$ and monotonically decreasing for the others. In time-steps 0 through 4 the system is ramping up from the flow amounts shown (for all lines other than $(0,1)$) in Figure 5.26 to those shown in Figure 5.27; the monotonic behavior is explained by the fact that generators 2 and 6 have, by far, the highest ramp-up rate (the quantities v_g in Table 5.15).

Table 5.16 crystallizes the difference between cases $\Delta t = 1$ and $\Delta t = 0.1$. When $\Delta t = 1$, the postcontingency flows are as in Figure 5.27, and whether or not line $(3,1)$ trips, lines $(2,3)$ and $(6,3)$ *will* trip, resulting in all demand lost. However, considering Table 5.16 we can see that when $\Delta t = 0.1$ there is significant probability that line $(3,1)$ will trip *before* any of the other lines do. Should that happen, the island comprising buses 2–6 will now have a generation *excess*, and thus, after a ramp-down, the system will achieve yield 0.5.

In fact, the probability that line $(3,1)$ trips in time-steps 1–5 is $1-(1-0.206)^5 \approx 0.68$, and a more detailed analysis shows that there are multiple cases where line $(3,1)$ *and* another line trip in the same time-step and nevertheless the cascade terminates with yield 0.5. Simulation shows that the probability that yield equals 0.5 is approximately 0.421 (and all other positive yield cascades have negligible probability), so that the expected yield is 0.2105. This (significant) difference with the case $\Delta t = 1$ is entirely due to multiple race conditions arising between the tripping of line $(3,1)$ and of other lines, which arises during the ramp-up process following the initial contingency.

Experiment 5.6.8.

Here we will consider **generator overfrequency tripping**. All data is as per Experiment 5.6.7 but in addition we assume that a generator will trip if its (normalized) frequency exceeds 1.05. To better understand how this can affect the course of the cascade, consider Table 5.17, which shows, at the start of each time-step of the cascade, the pseudopower injections at each generator (other than 0), *assuming that no line or generator trips have taken place in prior time-steps.*

Table 5.17. *Pseudopower outputs in Experiments* 5.6.7 *and* 5.6.8.

Generator	2	4	5	6	Total
time-step 1	48.00	52.00	13.00	10.00	**123.00**
time-step 2	56.00	54.00	16.00	20.00	**146.00**
time-step 3	64.00	56.00	19.00	30.00	**169.00**
time-step 4	72.00	58.00	22.00	40.00	**192.00**
time-step 5	74.79	58.70	23.04	43.48	200.00

As a comment on this table, note that at the start of time-step 1, the pseudopower amounts add up to 123—this equals the precontingency combined output of generators 2, 4, 5, and 6 of 100 units, plus the 23 units ramped up postcontingency. And at the start of time-step 5, the pseudopower amounts add up to the total demand of 200 (ramp-up completed).

Now if line $(3,1)$ were to trip before any other line, under Experiment 5.6.7 the system would stabilize with yield $= 0.5$. Under Experiment 5.6.8, however, this line trip may create an island with a generation excess and thus result in a frequency increase which might be large enough to cause generator tripping. For example, suppose line $(3,1)$ trips at the start of time-step 3 and no other line trips at or before time-step 3. From Table 5.17 we see that the excess of pseudopower over demand in the island comprising buses $2,3,4,5,6$ equals $169 - 100 = 69$; the sum of inverse R_g parameters for generators in this island (see Table 5.15) is $1/0.0015 + 1/0.002 + 1/0.001 + 1/0.001 \approx 3166.67$, and, consequently, the new system frequency in the island will be $1 + 69/3166.67 \approx 1.02179$. Since this value exceeds the upper frequency limit of 1.015, all generators in the island trip and the system collapses (zero yield). On the other hand, suppose line $(3,1)$ trips at (or before) time-step 2, and no other line trips then or before. Then Table 5.17 indicates that the resulting system frequency will be, approximately, 1.0145; thus generators do not trip and the cascade will stop with yield 0.5. Finally, suppose that both line $(3,1)$ and line $(5,3)$ trip at the start of time-step 3 and are the first to trip. The system frequency increase will now be 1.02123 and again we have system collapse.

We conclude that whether or not the system collapses is due to a collusion between timing and combination of line trips. Earlier tripping of $(3,1)$ and delayed tripping of other lines favors yield 0.5, and otherwise the yield is zero. The probability of yield 0.5 is approximately 0.353, significantly different than under Experiment 5.6.7.

Experiment 5.6.9.

Now we will add **generator underfrequency tripping** to the model, which otherwise is as in Experiment 5.6.8; we will assume generators trip when system frequency falls below 0.96. An example where this factor comes into play is as follows. Suppose that during time-step 1 of the cascade, line $(3,1)$ does not trip, but line $(4,3)$ does. The resulting island has demand 200, but as per Table 5.17 the total pseudopower injection is $48+13+10=71$, i.e., a shortfall of 129 units, and so frequency response sets system frequency to $1-129/(1/0.0015+2/0.001)=1-0.048375<0.96$, resulting in system collapse. Note that this situation can arise only relatively early in the cascade; e.g., the creation of the same island in time-step 2 would only entail a shortfall of 108 units, and system frequency would be $1-102/(1/0.0015+2/0.001)=0.96175$, which is large enough. In other words, these are relatively rare events; simulation shows that the probability of yield $=$ 0.5 is now approximately 0.350.

The above experiments highlight various race conditions; it should be clear that examples with rapidly mounting complexity can arise. Consider, for example, the cascades involving "ladder" networks introduced in Section 5.4.1. Such models will require large-scale simulation and data analysis as the primary agent for investigation. The numerical integration approach we follow, which relies on breaking up time into small segments of duration Δt, permits the efficient estimation of line flows and line and generation tripping simulation throughout the ramp-up process. The examples highlight the difficulties entailed in implementing continuous-time simulation models, as opposed to the numerical integration techniques we follow. Continuous-time models would require closed-form expressions for power flows, generator injections, and system frequencies. Such models would also require the estimation of line tripping probabilities as a function of time for lines with time-varying flows. While it may be possible to derive such models, we posit that they will encompass significant complexity, enough to make the discrete-time approach attractive.

Chapter 6

Online cascade control

In this chapter we discuss some initial work toward optimization-based algorithms for control or abatement of cascades. The algorithms we will consider focus on minimizing the amount of demand lost, though other possibilities are reasonable, for example minimizing the time (or cost) of restoring the system to its precascade status.

A basic question regarding cascade abatement is the nature of the control actions—entirely precomputed, or derived on the basis of real-time observations? Given a transmission system, there is an exponential number of possible initiating events for a cascade, with very different trajectories followed thereafter, and it is difficult to conceive of completely precomputed actions that prove effective and yet not overly conservative, regardless of the nature of the cascade.

Thus we converge to a view where an optimal control action is computed on the basis of real-time, postcontingency observations. Note that even some seemingly simple (if dramatic) courses of control, such as deliberate islanding (or separation into independent subgrids) to "isolate" the cascade, would also fall into this category to render an effective response. In the case of islanding, in particular, not only would the islands have to prove viable (postcontingency) but care would need to be taken that the "isolated" part of the grid does not suffer such a large loss of load that altogether the grid sheds far more load than might be possible through a more subtle control. Current national-scale power grids, with generation (including, especially, renewables) often placed far from consumption, could make such goals difficult. Instead an adaptive islanding scheme, computed postcontingency, appears more attractive.

Consider, thus, a scheme where some amount of load shedding is optimally computed postcontingency. A question which arises then is *when* to shed the load. Immediately after the contingency has been identified? In fact, should the load shedding even happen all at once, or perhaps should it be done in some gradual fashion? Below we will argue that, for a number of reasons, it appears best to consider control strategies that do not necessarily shed all load at the onset of the cascade, and possibly not all at once in any case. This will place the algorithms we study firmly within the classical mold of online optimal control algorithms, which have been a fixture in many areas of engineering (e.g., in aerospace) for a very long time. From the perspective of optimization we will end up with stochastic or robust multistage combinatorial algorithms, of course in a nonstandard application setting. To motivate a discussion leading to these observations, we first consider a more basic issue: what should be the underlying mathematical nature of a control algorithm?

Under postcontingency conditions a power grid could immediately attain a highly unstable state. Here we say "could" rather than "would" because perhaps most notably the 2003 cascading failure of the U.S. Northeast and Canada (see Section 4.2.1) did *not* experience such conditions, as we have discussed above, instead progressing over several hours to a dramatic phase, but nonetheless operating in relatively stable fashion during that period (and continuously delivering power to tens of millions of customers during the process). Nevertheless, there is a possibility of a very severe contingency which could result in an extremely fast cascade, perhaps lasting just a few seconds from contingency to collapse. Another setting where a very fast collapse is experienced, of course, is the terminal phase of an otherwise slow-moving cascade. It can be argued that a significant portion of the mathematically well-founded research on cascading failures, in the power engineering literature, has been aimed at one of these two of these settings, i.e., conditions where the grid is quickly deteriorating, or generally *transient* conditions.

A different approach is discussed in [127] and [150] (also see the PhD thesis [149]). These works argue (as has been done elsewhere) that the risk of cascading failures has increased as market mechanisms transfer increasing amounts of power over large distances. The approach in [150] incorporates a stress-test of transmission systems, whereby power transfers are increased (in a range of alternative patterns) until some limit is exceeded and a relay trips. Following this event, a simplified cascade model is used to determine the likelihood of a cascade. The authors propose such a stress-test to evaluate the susceptibility of a grid to cascading failures. A fast online scheme for detecting a cascade, based on a sparse set of measurements ("sparse sensing") is proposed in [149], as well as a methodology for enhancing economic dispatch (OPF) to render cascade-resilient states. In [127] the authors propose the online application of sophisticated controls (such as phase angle and voltage regulators) to avert a cascade and describe realistic scenarios where the application of such controls proves successful.

6.0.1 ▪ Dynamical system analysis of power grids

A challenge for analysis of power grids in the immediate aftermath of a contingency is that it is not even clear that the steady-state AC power flow model applies, let alone approximations such as the DC model. The first line of defense for a power grid is generator stability—generators provide power at a given voltage level. Following a contingency, generators would trip due to frequency instability (typically frequency drop), or voltage instability (voltage loss), or related conditions that impede synchronicity, such as large phase angle differences across ends of lines. Generator trips would happen quite quickly (possibly in subsecond fashion), and an appropriate mathematical model is needed to explain this behavior. We have already discussed some of these points in Chapter 1, in particular Section 1.4. An initial, non-steady state model focused on generator stability could be constructed using the so-called swing equation (1.51). One such equation would be needed for each generator. Additional equations would also be needed to model voltage magnitudes. Overall one obtains a set of partial differential equations that are used to explain system behavior, in other words, a *dynamical system*.

One can briefly outline a mathematical treatment for postcontingency stability analysis as follows. The swing equation relates the second derivative of the phase angle at a generator bus to the net power imbalance at that bus. Combining this statement with the AC active power flow equation (1.10) one obtains, at a generator bus k

$$A_k \frac{\partial^2 \theta_k(t)}{\partial^2 t} = p_k - \sum_{(k,m)\in\delta(k)} \Big(|V_k|^2 g_{\{k,m\}} - |V_k||V_m| g_{\{k,m\}} \cos\theta_{km}(t) - |V_k||V_m| b_{\{k,m\}} \sin\theta_{km}(t) \Big). \tag{6.1}$$

In this equation the sum is over all lines incident with bus k; furthermore, $V_i = V_i(t)$ for any bus k and as before $\theta_{km}(t) = \theta_k(t) - \theta_m(t)$. The quantity p_k indicates the net output of generator k (generation minus colocated load), which could also be time dependent, for example as a result of automatic generation or load shedding controls. In this context the authors would add, to the left-hand side of (6.1), a first-order term to model damping behavior, obtaining

$$A_k \frac{\partial^2 \theta_k(t)}{\partial^2 t} + D_k \frac{\partial \theta_k(t)}{\partial t} = p_k - \sum_{(k,m)\in\delta(k)} \Big(|V_k|^2 g_{\{k,m\}} - |V_k||V_m| g_{\{k,m\}} \cos\theta_{km}(t) - |V_k||V_m| b_{\{k,m\}} \sin\theta_{km}(t) \Big). \tag{6.2}$$

D_k is an appropriate constant. Equation (6.1) already incorporates some basic assumptions, namely that the time scale for phase angle changes (as provided by the left-hand side) is slower than the time scale needed for the steady-state AC power flow model to be at least approximately valid (on the order of a second). Additionally, and ideally, one would also account for reactive power flows. When bus k is not a generator bus we can still write an equation like (6.1), with appropriate A_k and D_k. Note that as we did in equation (1.52), by introducing the auxiliary variable $w_k = w_k(t) = \frac{\partial \theta_k}{\partial t}$, (6.2) becomes a first-order differential equation.

The system of all equations (6.2) constitutes a *dynamical system* (more properly, a *differential* dynamical system), which broadly speaking can be defined as a set of differential equations that describe the behavior of a system over time. Because of the very broad applicability of such systems, a vast literature exists on this topic; see [161], [209]. In a general setting we have a vector $x = x(t) \in \mathbb{R}^n$ and a system of first-order differential equations of the form

$$\frac{\partial x}{\partial t} = f(t, x), \tag{6.3a}$$

$$x(0) = x_0, \tag{6.3b}$$

where $f : \mathbb{R} \times \mathbb{R}^n \rightarrow \mathbb{R}^n$ describes a set of laws governing the evolution of a system over time (for example, laws of physics), with the vector x used to describe the "state" of the system. The vector $x_0 \in \mathbb{R}^n$ indicates the initial conditions. Solving (6.3) would then indicate the evolution of the underlying system over time. Of particular interest are *steady-state* solutions or, more properly, equilibrium points (also known as fixed points, among other terms), that is to say vectors $x^* \in \mathbb{R}^n$ such that

$$0 = f(t, x^*)$$

for any $t > 0$. This means that a system in state x^* will remain there, and if we can argue that a solution $x(t)$ to (6.3) satisfies

$$x(t) \rightarrow x^* \quad \text{as } t \rightarrow +\infty, \tag{6.4}$$

then we are guaranteed that the system converges to the equilibrium point x^*. Equilibria are also classified as to their *stability*, that is to say for which points x_0 in the neighborhood of x^* condition (6.4) is satisfied.

The concepts of stability and equilibria are key in the context of power systems. Power grids are designed to operate so that phase angles are constant, meaning that the left-hand side of (6.2) is zero at each bus k. From a power systems perspective (through the use of the swing equation) this means that active power generation exactly matches active power loads plus losses. Clearly, from a dynamical systems perspective, when the left-hand side of (6.2) equals zero we have an equilibrium point. An especially interesting case regarding equilibria arises when we consider a power grid in postcontingency state: what, if any, is the new steady state going to be? To model a contingency, we would assume that it takes place at time $t = 0$ with the system at $t = 0^-$ assumed to be in a (known and stable) steady state and (thus) with the right-hand side of each equation (6.2) evaluating to zero. The contingency would be represented by indicating an altered system at time $t = 0^+$; for example, in the case of a contingency where some subset of lines are removed from operation, this would be indicated by missing terms in some of the sums in (6.2), thereby causing the left- and right-hand sides of (6.2) to evaluate to nonzero values in the case of some set of buses. The dynamical system behavior would then take over with the system perhaps converging to a new steady state. But which one?

This is a question of critical importance because, postcontingency, an equilibrium point may not exist, i.e., there is no steady state and the power system will fail. Even more interesting, one could have a case where postcontingency multiple equilibria exist—which one will be attained, and why? From a power systems perspective some (or all) of the equilibria may correspond to solutions that are not practicable, i.e., voltage or frequency values outside acceptable ranges. From a mathematical perspective, a contingency can frequently be modeled or approximated as the change in system behavior as some parameter continuously changes (for example, we might model a line trip by having its reactance grow large). In the dynamical systems setting we thus have, rather than (6.3a), a more general equation of the form

$$\frac{\partial x}{\partial t} = f(t, x, \mu), \tag{6.5}$$

where μ is a parameter and we are interested in the equilibria of the dynamical system as μ varies around a given value μ_0. This leads to the analysis of *bifurcations*, a widely studied topic in the dynamical systems literature. A bifurcation is said to exist at $\mu = \mu_0$ if for any $\epsilon > 0$ there exist values μ_1, μ_2 such that $|\mu_i - \mu_0| \leq \epsilon$ for $i = 1,2$ and yet the number of equilibria for the dynamical system at $\mu = \mu_1$ and at $\mu = \mu_2$ are different (see [161, Chapter 8] for a thorough coverage). A classical example is provided by the *logistic model*

$$\frac{\partial x}{\partial t} = 4x(1-x) - \mu,$$

where $x \in \mathbb{R}$ and μ is a parameter. For $\mu < 1$ this system has the two solutions $x = \frac{1}{2}(1 \pm \sqrt{1-\mu})$, for $\mu = 1$ it has only one solution, and for $\mu > 1$, it has none. What will happen to the system if the parameter μ suddenly increases from less than 1 to greater than 1?

In a more general setting, say μ_0 is a bifurcation point of (6.5) such that for each $\mu < \mu_0$ there are two equilibria, which we indicate by x^1_μ and x^2_μ. Assuming the function f is well behaved, then for any two values $\mu < \mu' < \mu_0$ with $\mu' - \mu$ small enough, we will expect that x^1_μ and $x^1_{\mu'}$ are close and that x^2_μ and $x^2_{\mu'}$ are also close. In other words, when $\mu < \mu_0^-$ we have two well-behaved (and qualitatively different) "branches" of system behavior. When the system is in branch x^1_μ or x^2_μ, it will (should?) stay in that branch if μ experiences a small change (and stays smaller than μ_0). This behavior is aligned with standard ideas on hysteresis in physical systems. In a power systems setting, for example, in general we would expect[48] that a very small change in the reactance of one line will

[48] Perhaps incorrectly.

only cause a small change in the resulting power flow, but not a large change in the broad system behavior (e.g., if the output of a given generator is large before the change in μ, it should also be large after the change). From a mathematical standpoint, we would expect these ideas to be supported by the inverse function theorem, so long as this function is well defined, and this will be the case if the Jacobian of (6.5) (with respect to x) is invertible at x_0, a property that does not hold at bifurcation points (see the above citations).

Returning to our generic example, suppose that μ_0 is a bifurcation point and that for $\mu > \mu_0$ there are three equilibria. Suppose that initially $\mu < \mu_0$ and the system starts on branch x^1_μ. Further suppose μ smoothly increases past μ_0—where will the system end up and, most important, why? How can the change be explained by physical laws, given that (6.5) already incorporates the underlying physics? As we have noted, questions regarding stability and bifurcations take a prominent role in the dynamical systems literature; perhaps unsurprisingly, the term "chaos" is sometimes employed in connection with complex bifurcations.

Many authors in the power systems literature have addressed these questions; see, e.g., [66], [207], [65], [89], [64], [81], [80]. There are many contexts in which a power grid is viewed as a parameterized dynamical system, for example, in order to study system behavior as a function of a load, or a line parameter. When (6.2) has no equilibria we would expect some sort of systemic breakdown, for example, a "voltage collapse" condition or lack of synchrony—in practice, conditions that will lead the grid to, essentially, shut down. A point of focus for research has been solution procedures for (6.2); *continuation methods*, which solve the dynamical system for a choice of the parameter, and then seek to adapt the solution as the parameter is changed by a small amount, have received some attention. A general goal is to derive a precise description of system behavior following some disturbance. A particularly challenging issue concerns behavior past a bifurcation point; one intellectually compelling methodology for analyzing this situation borrows ideas from the physical sciences to posit that a new equilibrium will be established according to a "minimum-energy" rearrangement of the system. How accurately this paradigm applies to a large-scale grid, which would be governed by multiple automated control mechanisms, is not clear.

In the context of this chapter, we are interested in methodologies that explain the behavior of a grid following a major contingency, including in particular the impact of control mechanisms. Such methodologies could be used in simulation mode to study proposed controls; they could also be used in real time to evaluate the risk posed by an event that has just transpired. It is of interest to consider cases where stability is not necessarily achieved, and rather than simply predicting a generic systemic breakdown the methodology should evince *how* the breakdown would take place. Likewise, the methodology should provide an appropriate algorithmic framework to predict discrete events, such as generator and line trips. In the case of islanding, further modeling would be needed. In summary, we are considering an algorithmic procedure that goes beyond the standard dynamical system analysis and attempts to incorporate discrete events, and would furthermore need to be accurate and *fast* (i.e., near real time) in a large-scale setting. These goals give rise to significant challenges, some of which we already introduced in Chapter 5, and which can be summarized as follows, ranked from least to most significant:

(1) First, from a purely methodological standpoint, the solution of a basic system of the type (6.2) is highly nontrivial—this task should be as hard as solving an AC power flow system, and because of the context, the term "solution" carries with it the need for some accuracy and performance guarantees. These difficulties are echoed in the literature.

(2) Even assuming that there is an efficient solution scheme for the system, one would need a vast amount of data to set up the system and its initial conditions. These data would include all phase angles, initial frequency amounts, loads, and line information. In terms of real-time usage of a solution methodology, it is not clear that all such information would be available in today's power systems. Assuming that the solution process is updated as the cascade proceeds (e.g., some recomputation is performed), then fresh data would have to be continuously delivered even as the grid degrades.

(3) A related issue concerning data is accuracy. While conceivably today's SCADA systems could be upgraded to, e.g., deliver highly complete and accurate phase angle and frequency data in real time (such a capability does not currently exist in actual practice), it is less clear that all other data would be equally available and accurate. This would include, for example, data concerning exposure of power lines to ambient conditions, which would make possible accurate prediction of the time when an overloaded line will trip. As in (2), when a transmission system degrades during a cascade it is quite likely that, for lack of a better term, more exposure to "noise" will take place.

(4) Assuming that issues (1)–(3) have been resolved, there is one last issue that appears most complex. This concerns the notion of *race conditions*, which was examined in greater detail in Section 5.1. Race conditions arise when multiple discrete events, in particular equipment tripping events, may be near-simultaneous in actual time, even if they are governed by phenomena in different time scales (slow for thermal line trips; much faster for generation trips). As amply shown by simulations (as will be seen later) and observed by several authors, the particular order in which discrete events take place can have material impact on system behavior. An explanation (or prediction) for the ordering of events seems to fall beyond the scope of physics and engineering and may be due to poorly explained exogenous factors such as weather, erroneous control actions, and response of control mechanisms to faulty data or missing data.

The difficulties we have outlined, in our view, argue for methodologies for the analysis and control of cascades that sacrifice the putative accuracy gained by a mechanistic understanding of systems of the form (6.2) and instead focus on broad risk measures. In particular we are interested in methodologies that can take into account data paucity and information errors, as well as the race conditions described in (4). Of course, it is important to incorporate as much realistic modeling as possible, subject to computational practicability. These two somewhat competing goals will underlie the work below.

The power grid community has addressed some of these concerns. A discrete-time procedure that alternates between computing (time-dependent) solutions to (6.2) as well as other grid power flow parameters, and computations to determine discrete changes in the grid (e.g., trips) is given in [106, Section 11.5]. All modeling is deterministic, and nonbifurcation seems to be assumed. Essentially, the procedure in [106] can be viewed as a numerical integration algorithm. Also see [45]. These procedures blend approximate methods for solving the dynamical system (6.2) with practical engineering considerations.

6.1 ▪ Optimal affine control

In this section we introduce a family of online control algorithms used to limit the impact of a cascading failure. These algorithms, based on work in [30], incorporate some of the viewpoints described in the previous section, namely a focus on broad features of a cascade using a simplified model of the cascade, at the cost of losing some accuracy. In Section 6.2 we will consider extensions of the algorithms given in this chapter that incorporate some realistic modeling elements, however still falling short of extremely detailed power system modeling.

The principles underlying the algorithms are as follows:

1. We are interested in control schemes that shed load on the basis of real-time data, with the goal of stopping a cascade with a minimum of demand lost. One can conceive of algorithms in this mold that are available a priori, i.e., algorithms independent of the nature of the cascade. Since there is an exponential number of cases for the initial event that causes the cascade, and since simulations show that the nature of the cascade can depend strongly on the initial event,[49] instead we focus on control algorithms that are computed soon after the onset of the cascade. We assume that initially there is sufficient time to *compute* an appropriate control algorithm; once computed, the control will be *applied* as the cascade unfolds, using real-time observations. Such a method should thus be viewed as a data-driven control algorithm.

2. The DC approximation for power flows is used and a purely deterministic model will be used. In Section 6.2 these assumptions will be bypassed. As discussed in Section 1.2.4, the DC model on a directed graph G is described by a system of equations of the form

$$Nf \;=\; \beta, \quad YN^T\theta - f \;=\; 0, \tag{6.6}$$

where N is the node-arc incidence matrix of G, f is the vector of active power flows, β_i is the net supply (generation minus load) at bus i, and $Y = \operatorname{diag} y_{\{i,j\}}$ is the diagonal matrix of line susceptances. This system is feasible if and only if for any component K, $\sum_{i \in K} \beta_i = 0$, in which case the vector f is unique, as discussed before. For future reference we introduce some notation.

Definition 6.1.1. *Let G be a grid. A vector β such that $\sum_{i\in K} \beta_i = 0$ for any component K of G and such that $\beta_i \geq 0$ ($\beta_i \leq 0$) for each generator i (resp., for each load i) is called a net-supply vector over G. We also denote by $\hat{f}(\beta, G)$ the unique flow solving* (6.6).

An important property of the DC approximation is as follows: we clearly have that for any real λ,

$$\hat{f}(\lambda\beta, G) = \lambda \hat{f}(\beta, G). \tag{6.7}$$

Without loss of generality we will assume that each bus v is either a generator bus (the set of such buses is denoted by $\mathcal{G}$), or a load bus ($\mathcal{L}$, resp.), or neither a load nor a generator. Thus, for a net-supply vector β and bus v, if $v \in \mathcal{G}$ we have $0 \leq \beta_v =$ generation at v, if $v \in \mathcal{L}$ then $0 \geq \beta_v = -$demand at v, and $\beta_v = 0$ otherwise.

3. We rely on a discrete-time model for the cascade, based on Model 4.3.1. Thus, a simplified cascade template is given in Template 6.1.2.

[49] Easily verified. Also see, e.g., [182] and [92].

Template 6.1.2. Basic cascade template

Input: A power grid with graph G in postcontingency state. Set $G^1 = G$.
For $t = 1, 2, \ldots$ **Do**
(comment: time-step t of the cascade)
Step 0. Adjust loads and generation so that each island of G^t is balanced, obtaining net-supply vector β^t.
Step 1. Set $\hat{f}^t = \hat{f}(\beta^t, G^t) =$ vector of power flows in G^t.
Step 2. Set $\mathcal{T}^t =$ set of lines of G^r that become tripped in time-step t. If no lines are overloaded, cascade terminates.
Step 3. Set $G^{t+1} = G^t - \mathcal{T}^t$.

We stress that G^1 is the grid *postcontingency* and that G^1 will be assumed to be known.

4. We have previously discussed two central issues in the description of cascades: the modeling of line tripping and of load balancing in freshly created islands. Here, we will use the deterministic rule that

$$\text{line } j \text{ becomes tripped at time } t \text{ if } |\hat{f}_j^t| > u_j, \tag{6.8}$$

where u_j is the limit of line j. To handle islanding (Step 0 of the above template), let K be any island in G^t and define

$$G(K) \doteq \sum_{v \in \mathcal{G} \cap K} \beta_v^{t-1}, \tag{6.9a}$$

$$L(K) \doteq - \sum_{v \in \mathcal{L} \cap K} \beta_v^{t-1}. \tag{6.9b}$$

(Here, when $t = 1$ the vector β^0 is the net-supply vector precontingency.) We apply the following rule, a significant simplification of Models 5.3.2 and 5.3.4. Let $v \in K$. If $G(K) < L(K)$ we set

$$\beta_v^t = \begin{cases} \frac{G(K)}{L(K)} \beta_v^{t-1} & \text{if } v \in \mathcal{L}, \\ \beta_v^{t-1} & \text{otherwise.} \end{cases} \tag{6.10}$$

And if $G(K) > L(K)$ instead we set

$$\beta_v^t = \begin{cases} \frac{L(K)}{G(K)} \beta_v^{t-1} & \text{if } v \in \mathcal{G}, \\ \beta_v^{t-1} & \text{otherwise.} \end{cases} \tag{6.11}$$

Finally, if $L(K) = G(K)$, then $\beta_v^t = \beta_v^{t-1}$. In all three cases island K will be balanced, i.e., $\sum_{v \in K} \beta_v^t = 0$, and the restriction of β^t to K is a net-supply vector over K. We also write β^0 to indicate the precontingency net-supply vector.

5. Our notion of online control is summarized by Template 6.1.3, where T is a fixed (and known) time limit.

Template 6.1.3. Cascade control

Input: a power grid with graph G in postcontingency state. Set $G^1 = G$.

Step C. Compute control algorithm.

For $t = 1,2,\ldots,H-1$, **Do** (comment: controlled time-step t of the cascade)

Step 0. Adjust loads and generation so that each island of G^t is balanced, obtaining net-supply vector β^t.
Step 1. Set $\hat{f}^t = \hat{f}(\beta^t, G^t) =$ vector of power flows in G^t.
Step 2. Apply control (if allowed).
Step 3. Set $g^t =$ vector of resulting power flows in G^t.
If no lines are overloaded, cascade terminates.
Step 4. Set $\mathscr{T}^t =$ set of lines of G^t that become tripped in time-step t.
Step 5. Set $G^{t+1} = G^t - \mathscr{T}^t$.

Termination (time-step H). For each island K of G^H, perform steps 0 and 1 above on K. If any line in K is overloaded, proportionally shed demand in K until all line overloads are eliminated. Formally, with $\Psi_K^H \doteq \max\{1, \max_{j \in K}\{|\hat{f}_j^H|/u_j\}\}$, each bus v of K resets its net-supply to β_v^H/Ψ_K^H.

An implementation of Step C in the template will be detailed below; the control is applied in online fashion in Step 2 assuming all relevant grid data is known. Likewise, Step C assumes that the postcontingency state of the grid are known. The termination step is needed because otherwise one can have examples of cascades that continue for a very long time, an undesirable feature. Our particular termination step can be viewed as an example of emergency load shedding. Note that one could have a cascade that terminates before time-step H because no line overloads exist; in this case no action takes place at termination (such early cascade terminations could also happen under the basic Template 6.1.2) (if a stochastic line tripping rule were used one could have an infinitely long cascade).

A final technical point concerns the "if allowed" clause in Step 2. In cases where the duration parameter for the simulation, Δt, is very small, the nature of the control under consideration may not always permit application of the control. In the experiments we carry out below we will assume that control can always be applied, though the modification of the procedures to handle the general case will be straightforward.

6.1.1 ▪ The optimal scaling problem

We consider an affine control scheme that sheds load as a function of maximum line overloads. To describe the control we use the following notation: for any given node v and time t, let x_v^t denote the maximum relative line overload of a line in the island containing v at time t, that is to say, the maximum quantity $|\hat{f}_j^t|/u_j$ over all lines j of the component that contains v at time t.[50]

[50]In this section we will adhere to the standard graph-theoretic terminology of "component" rather than "island" and "node" rather than "bus."

The general control law we consider is described, for each time step t and node v, by a pair of values $(b_v^{(t)}, s_v^{(t)})$ to be computed in Step C of the template. This control is applied in Step 2 of the above template as follows:

- First, for any load node v we reset its demand from $-\beta_v^t$ to

$$\min\{1, [b_v^{(t)} + s_v^{(t)}(1 - x_v^t)]^+\}\,(-\beta_v^t) \quad \text{if } x_v^t > 1, \tag{6.12a}$$
$$-\beta_v^t \quad \text{otherwise.} \tag{6.12b}$$

  In (6.12a), $[x]^+$ denotes $\max\{x, 0\}$. To understand this rule, consider the case where, for all v, $b^{(t)} = 1$ and $s_v^{(t)} > 0$. Then if $x_v^t > 1$ we shed demand at v by a multiplicative factor proportional to the excess relative overload $x_v^t - 1$, while if $x_v^t \leq 1$, we do not shed demand at v. In fact, when $b^{(t)} = 1$ and $s_v^{(t)} \geq 0$, (6.12b) is redundant; however, the general form of the affine rule allows for $s_v^{(t)} < 0$; i.e., the rate of load shedding decreases with increasing overload. In all cases, the "min" in (6.12a) implies that demands are never allowed to increase.

- Rule (6.12) specifies how to shed load. As the second step of the control mechanism, generator output in each island is proportionally decreased (per generator) so that total generation equals total load (this is similar to the scheme in (6.11)).

- In view of (6.10) and (6.11) the resulting scheme does not permit generators to increase output. This, and additional shortcomings, will be discussed later, following Algorithm 6.1.20.

The control scheme (6.12) is quite general. In what follows we will assume a more restricted version of the control which we term *island-wide* control, where for each time t and island K of G^t, there is a pair $(\tilde{b}_K^{(t)}, \tilde{s}_K^{(t)})$ such that $(b_v^{(t)}, s_v^{(t)}) = (\tilde{b}_K^{(t)}, \tilde{s}_K^{(t)})$ for each node v of K. This is a simpler control (and possibly easier to implement); however, the more general form of the control allows for *distributed* implementation where each node (under the assumption of island-wide control) performs its own control computation (in Step C) using its own observations and possibly arriving at different estimates of the island-wide pairs $(\tilde{b}_K^{(t)}, \tilde{s}_K^{(t)})$.

A subtle detail that arises at this point concerns the specification of an island-wide control. As stated, we need to compute a pair $(\tilde{b}_K^{(t)}, \tilde{s}_K^{(t)})$ for each time t and island K, but, in principle, there is an exponential number of such pairs for any given t (since there is an exponential number of potential islands). However, we are assuming a *deterministic* cascading process, and as we will argue next, this implies that only a polynomial number of pairs $(\tilde{b}_K^{(t)}, \tilde{s}_K^{(t)})$ need to be specified. We can prove this as follows. First, at time $t = 1$ we only need to specify $(\tilde{b}_K^1, \tilde{s}_K^1)$ for each actual island K that was produced as a result of the contingency. Having made such a specification, Template 6.1.3 implies that a unique set of islands will materialize at $t = 2$. Continuing inductively we see that at most Hn pairs $(\tilde{b}_K^{(t)}, \tilde{s}_K^{(t)})$ need to be specified to completely describe a control (though of course the set of such pairs needs to be consistent). Without the determinism assumption, this analysis breaks down, and, as we will see later, one needs to further simplify the description of the control algorithm.

Assuming an island-wide control, let K be an island of G^t, and let x_K^t be the maximum relative line overload over all lines of K (at time t). Then the multiplier in (6.12a) equals

$$\lambda_K^{(t)} \doteq \min\{1, [b_K^{(t)} + s_K^{(t)}(1 - x_K^t)]^+\} \tag{6.13}$$

for all nodes $v \in K$, and the control rule (6.12) can be restated by saying that

$$\text{for each node } v \in K \cap \mathscr{L}, \begin{cases} \text{reset its demand from } -\beta_v^t \text{ to } \lambda_K^{(t)}(-\beta_v^t) & \text{if } x_K^t > 1, \\ \text{leave its demand unchanged} & \text{otherwise.} \end{cases} \tag{6.14}$$

Conversely, given a value $0 \leq \lambda_K^{(t)} \leq 1$ it is simple to find a pair $(b_K^{(t)}, s_K^{(t)})$ satisfying (6.13). Hence we can equivalently refer to the control by rule (6.14). As discussed above, to specify the control we need only state, for each t, a value $\lambda_K^{(t)}$ for each island K resulting from an application of the control at time $t-1$ (or from the contingency at $t = 1$). We will term such a control *consistent*. From a formal perspective we can state the following property.

Lemma 6.1.4. *Suppose that for each $1 \leq t \leq T$ we are given values $0 \leq \lambda_K^{(t)} \leq 1$ (for $K \in I^t$) with the following properties:*

(1) *For each t and $K \in I^t$, the restriction of the graph G to K is connected.*

(2) *For each t and each $K \in I^t$ denote the set of nodes in K by $V(K)$. Then for each t, $\cup_{K \in I^t} V(K) = V$, the set of all nodes.*

Then we can verify in polynomial time that the family $\{\lambda_K^{(t)} : K \in I^t\}$ is consistent.

Proof. To verify consistency we simply run the control Template 6.1.3 using rule (6.14) in Step 2. ■

Remark. *Another way to put this result is that consistency belongs in the computational class NP.*

We can now state the optimal control problem of interest.

Problem 6.1.5. Optimal scaling problem

Compute a consistent island-wide control $\{\lambda_K^{(t)} : K \in I^t\}$ that maximizes the sum of demands at termination.

Example 6.1.6. *Consider the network in Figure 6.1. Here nodes 1, 3, 5, and 7 are generators, with $\beta_1^0 = 15, \beta_3^0 = 5, \beta_5^0 = 14, \beta_7^0 = 2$, and 2, 4, 6, and 8 are loads, with $\beta_2^0 = 10, \beta_4^0 = 10, \beta_6^0 = 8, \beta_8^0 = 8$. Line (2,3) has limit 4, and line (6,7) has limit 3. All other lines have large limits. Consider the contingency made up of the two lines (1,7) and (5,3), shown as dashed lines in Figure 6.1. This contingency splits the network into two islands (labeled A and B in the figure), both of which are balanced (island A has total supply and demand equal to 20, while for island B the corresponding value is 12). Thus, postcontingency, line (2,3) carries 5 units of flow and line (6,7) carries 6 units. We assume $H \geq 2$.*

Suppose we apply a control with time-step 1 parameters $\lambda_A^{(1)} = \lambda_B^{(1)}$. Following Step 2, the disposition of G^1 will be as shown in Figure 6.2. Thus if $\lambda_A^{(1)} = \lambda_B^{(1)} = 3/4$ we obtain the situation shown in Figure 6.3.

With this choice of the control parameters, line (6,7) will trip in Step 4 of time-step 1. Hence, following Step 0 of time-step 2 the grid will be in the state shown in Figure 6.4.

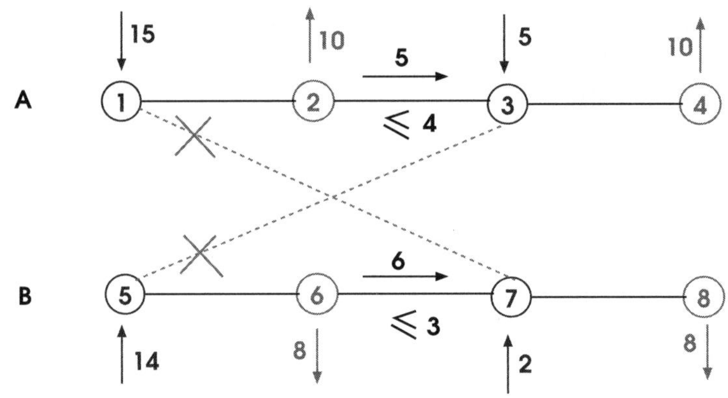

Figure 6.1. *Example* 6.1.6. *Flow pattern, loads, and generation, postcontingency.*

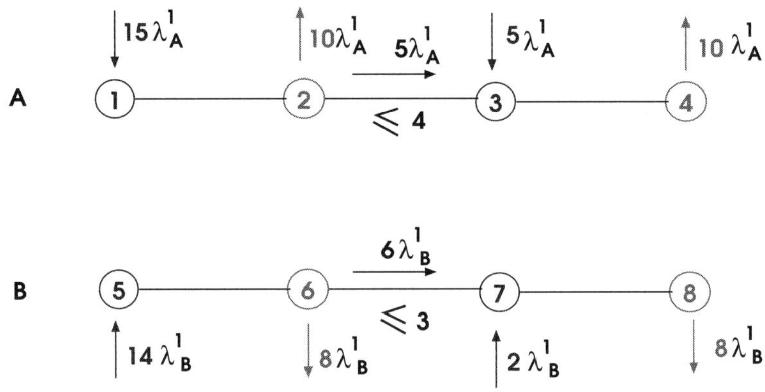

Figure 6.2. *Example* 6.1.6. *Step* 2, *time-step* 1.

We can see that no lines are overloaded. Hence the cascade will terminate exactly in this condition, with total demand equal to $15+6+3/2 = 45/2$ *and, hence, yield* $= 22.5/36 = 0.625$. *The loss of* 27.5% *of demand is explained by the load shedding of* 25% *taking place in the scaling step of time-step* 1, *followed by the load balancing in the two islands obtained from island B (also in time-step* 1*).*

Suppose now that we use $\lambda_A^{(1)} = \lambda_B^{(1)} = 0.95$. *Considering Figure* 6.2 *we can see that postcontrol, both line* $(2,3)$ *and line* $(6,7)$ *will be overloaded and will trip in Step* 4 *of time-step* 1. *Following Step* 0 *of time-step* 2 *the grid will therefore be in the state shown in Figure* 6.5. *Again no line overloads are present and the cascade will terminate with total demand* $=$ 23.75, *i.e., yield* 65.972%.

If on the other hand we were to choose $\lambda_A^{(1)} = \lambda_B^{(1)} = 0.5$, *then again by considering Figure* 6.2 *we have that postcontrol, in time-step* 1, *no lines will be overloaded and the cascade will terminate, obviously with yield* 50%. *In all these experiments we have assumed that* $\lambda_A^{(1)} = \lambda_B^{(1)}$. *Of course this may not be optimal, and in the current setup it is not necessary, and we can optimize over islands A and B separately. Considering island A, we can see that if* $\lambda_A^{(1)} > 4/5$, *then line* $(2,3)$ *will trip in time-step* 1, *and that following load balancing in time-step* 2 *there will be no further line overloads (in any subisland of island A) with total demand (in A) equal*

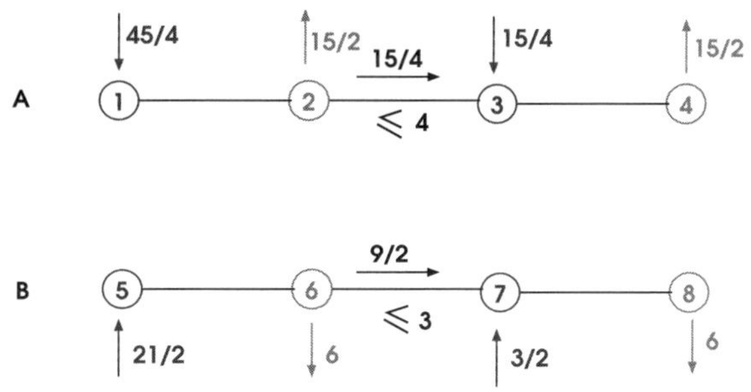

Figure 6.3. *Example* 6.1.6. *Step* 2, *time-step* 1, *with* $\lambda_A^{(1)} = \lambda_B^{(1)} = 0.75$.

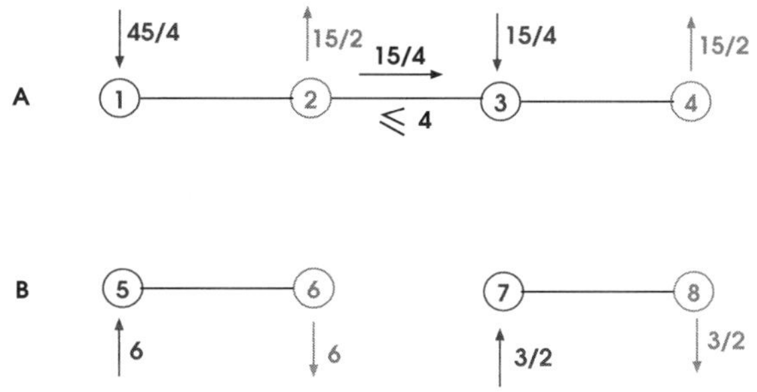

Figure 6.4. *Example* 6.1.6. *Post-step* 0, *time-step* 2, *with* $\lambda_A^{(1)} = \lambda_B^{(1)} = 0.75$.

to

$$10\lambda_A^{(1)} + 5\lambda_A^{(1)} = 15\lambda_A^{(1)}.$$

Clearly this quantity is maximized at $\lambda_A^{(1)} = 1$. *If on the other hand we choose* $\lambda_A^{(1)} \leq 4/5$ *then* $(2,3)$ *will not trip; no overloads will exist in island A; and, as far as this island is concerned, the cascade terminates with total demand (in A)*

$$20\lambda_A^{(1)},$$

which is maximized at $\lambda_A^{(1)} = 4/5$ *for a total of* 16 *units of demand. Turning now to island B, we see that if* $\lambda_B^{(1)} > 1/2$ *then line* $(6,7)$ *will trip in time-step* 1 *and as with island A the cascade will terminate in island B with total demand (in B) of*

$$8\lambda_B^{(1)} + 2\lambda_B^{(1)} = 10\lambda_B^{(1)},$$

again maximized at $\lambda_B^{(1)} = 1$. *With* $\lambda_B^{(1)} \leq 1/2$ *all overloads are eliminated and the cascade terminates with total demand (in B)*

$$16\lambda_B^{(1)},$$

maximized at $\lambda_B^{(1)} = 1/2$. *In summary, therefore, the optimal control is given by*

$$\lambda_A^{(1)} = \frac{4}{5}, \quad \lambda_B^{(1)} = 1,$$

with total demand, at termination, equal to $16 + 10 = 26$ *units, or* 72.22%. *Note that in this case the behavior is independent of* $H \geq 2$.

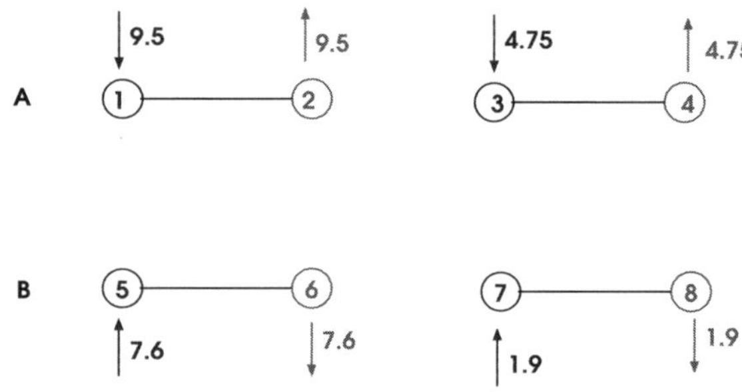

Figure 6.5. *Example* 6.1.6. *Post-step* 0, *time-step* 2, *with* $\lambda_A^{(1)} = \lambda_B^{(1)} = 0.9$.

6.1.2 ▪ Polynomial solvability of the optimal scaling problem under fixed time horizon

A salient feature in Example 6.1.6 is that in each island (A or B) there are two behavior regimes according to whether line $(2,3)$ ($(6,7)$, resp.) trips in time-step 1. And in each regime the demand at termination is an increasing linear function of the $\lambda^{(1)}$. We will show that a more general version of these properties always holds. The main result is Theorem 6.1.7, stated next. This theorem, as well as Lemmas 6.1.9, 6.1.15, 6.1.17, and 6.1.18, is drawn from [30], though we will primarily provide proof sketches.

Theorem 6.1.7. *For each fixed value $H \geq 1$ there is a polynomial-time algorithm to solve the optimal scaling problem.*

We will obtain this result in a number of steps. For each line j we will assume that its limit u_j is positive. We denote the set of lines by $\mathcal{A}$. First, it will prove useful to embed the optimal scaling problem in a broader family of problems.

Definition 6.1.8. *For $0 \leq w \leq 1$ real, let $\Theta_G^{(H)}(w|\beta)$ denote the demand, at termination, resulting from applying an optimal scaling control in an H time-step cascade on the graph G, assuming that the initial net-supply vector is $w\beta$.*

Setting $w = 1$ we recover the problem of interest. As the central step in the proof of Theorem 6.1.7, we will show that $\Theta_G^{(H)}(w|\beta)$ is a nondecreasing, piecewise-linear function of w with a polynomial number of pieces. This will be done by induction on H with the base case handled next, and with Lemma 6.1.17 (given later) handling the general case.

Lemma 6.1.9. *Let β be a net-supply vector and $\tilde{D}$ the sum of demands implied by the vector β. Suppose G is connected. Then,*

- *if $w \leq 1/x$, then $\Theta_G^{(1)}(w|\beta) = w\tilde{D}$, and*
- *if $w > 1/x$, then $\Theta_G^{(1)}(w|\beta) = \frac{w}{wx}\tilde{D} = \frac{1}{x}\tilde{D}$,*

i.e., $\Theta_G^{(1)}(w|\beta)$ is a nondecreasing piecewise-linear function of w with two pieces. ■

For the remainder of the analysis we will have $H > 1$. We can assume that G^1 is connected since each postcontingency component ($t = 1$) is handled separately by our control Template 6.1.3. We write $\hat{f} = \hat{f}^1$ $(= \hat{f}(G^1, \beta))$. A useful parameter introduced in [30] is as follows.

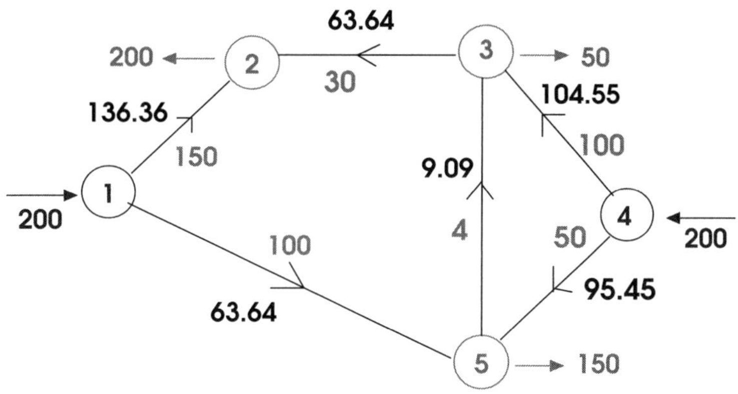

Figure 6.6. *Example 6.1.11. Flow pattern, loads, and generation, postcontingency.*

Definition 6.1.10. *A critical point is a real* $0 < \gamma < 1$ *such that for some line* j, $\gamma|\hat{f}_j| = u_j$.

We assume $u_j > 0$ for all j; thus let $0 < \gamma_1 < \gamma_2 < \cdots < \gamma_p$ be the set of all distinct critical points. Here $0 \le p \le m$. Write $\gamma_0 = 0$ and $\gamma_{p+1} = 1$.

To informally understand the significance of these parameters, assume for the moment that $w = 1$. Then, postcontrol, the flow on any line j will equal $\lambda^{(1)}\hat{f}_j$. As a result, if $\lambda^{(1)} \in (\gamma_{i-1}, \gamma_i]$ (for an appropriate index i), then in Step 4 of Template 6.1.3 the set of lines that trip is the same, regardless of the particular choice of $\lambda^{(1)}$. In fact, if line j is such that $\gamma_k|\hat{f}_j| = u_j$ for some $k < i$, then

$$\lambda^{(1)}|\hat{f}_j| > \gamma_{i-1}|\hat{f}_j| \ge \gamma_k||\hat{f}_j|/\gamma_k = u_j,$$

and line j trips. On the other hand, if $\gamma_k|\hat{f}_j| = u_j$ for some $k \ge i$ then a similar calculation shows that line j does not trip. This invariant will make it easy to optimize over $\lambda^{(1)}$.

Example 6.1.11. *Figure 6.6 shows a postcontingency system. As previously, generator arrows pointing into (or out of) buses indicate generator injections (load withdrawals). All reactances have unit value. Line limits are shown in gray and flows indicated in black, rounded to two digits (arrows show direction).*

There are four critical values, arising from lines $(4,3)$, $(4,5)$, $(3,2)$, *and* $(5,3)$, *respectively; i.e.,* $\gamma_1 = 100/104.55 \approx 0.96$, $\gamma_2 = 50/95.45 \approx 0.52$, $\gamma_3 = 30/63.44 \approx 0.47$, *and* $\gamma_4 = 4/9.09 \approx 0.44$. *Using* $w = 1$, *i.e., with the generation and demands as given, suppose we choose* $\lambda^{(1)} = 0.6$. *Then immediately postcontrol, the system will be in the condition shown in Figure 6.7.*

In the general inductive framework of the proof we assume that the demands and supplies have already (prior to time-step 1) been scaled by w; thus the invariant will arise whenever $\lambda^{(1)}w \in (\gamma_{i-1}, \gamma_i]$. This observation motivates the following definition.

Definition 6.1.12. *For* $1 \le h \le p$ *let* $F^h = \{j \in \mathscr{A} : \gamma_h|\hat{f}_j| = u_j\}$.

As a summary of the above observations, the set of lines that trip in time-step 1 is $\cup_{h \le i^*-1} F^h$; in other words, we obtain the same network $G^2 = G^1 - \cup_{h=1}^{i^*-1} F^h$ for every pair $\lambda^{(1)}, w$ with $\lambda^{(1)}w \in (\gamma_{i^*-1}, \gamma_{i^*}]$.

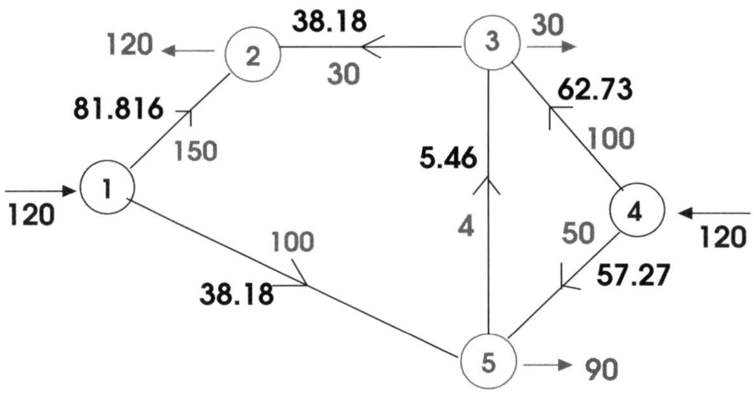

Figure 6.7. *Example* 6.1.11. *System postcontrol in time-step* 1 *assuming* $\lambda^{(1)} = 0.6$.

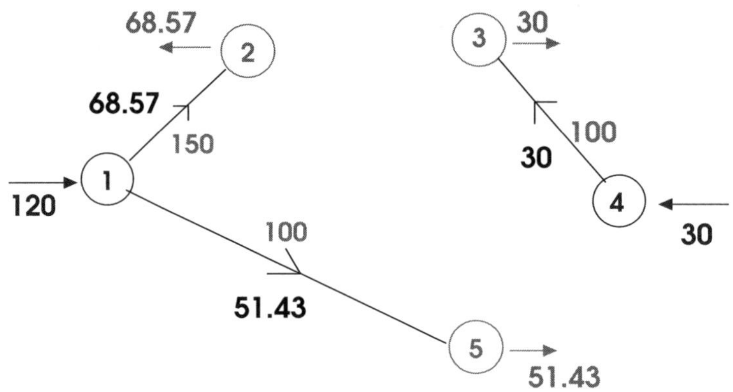

Figure 6.8. *Example* 6.1.14. *System at end of time-step* 1 *assuming* $\lambda^{(1)} = 0.6$.

Notation 6.1.13. *For an index* $r \leq p+1$, *write* $\mathcal{K}(r) =$ *set of components of* $G^1 - \cup_{h=1}^{r} F^h$.

Using this notation we have that after Step 4 of time-step 1 the grid is split into the set of components $\mathcal{K}(i^* - 1)$.

Example 6.1.14 (Example 6.1.11, continued). *With* $w = 1$, *assume we choose* $\lambda^{(1)} = 0.6$ *so that immediately postcontrol we obtain the system as shown in Figure* 6.7. *We see that lines* $(3,2)$, $(5,3)$, *and* $(4,5)$ *will trip. The tripping will cause the system to separate into two components: one with nodes* 1, 2, *and* 5 *with total supply* 120 *and total demand* 210, *and one with buses* 3 *and* 4, *with total supply* 120 *and total demand* 30. *Thus, in the first component demands will be proportionally decreased by a factor of* $120/210 = 4/7$ *and in the second component the supply at node* 4 *will be decreased by a factor of* $30/120 = 1/4$ *to* 30 *units. See Figure* 6.8.

The situation in this example is formalized by the following result.

Lemma 6.1.15. *Consider any component* $K \in \mathcal{K}(i^* - 1)$. *There exists a net-supply vector* $\breve{\beta}^{(K)}$ *over* K, *independent of* $\lambda^{(1)}$ *and* w, *with the following property: after application of Step* 4 *of time-step* 1, *for any node* $v \in K$ *the net supply at* v *equals* $\lambda^{(1)} w \breve{\beta}_v^{(K)}$.

Table 6.1. *Lemma* 6.1.15 *applied to Example* 6.1.11. *Net-supply values* β *and* $\breve{\beta}^{(K)}$.

	$K=\{1,2,5\}$			$K=\{3,4\}$	
v	1	2	5	3	4
β_v	200	150	200	50	200
$\breve{\beta}_v^K$	200	$\frac{4}{7}$ 200 = 114.29	$\frac{4}{7}$ 150 = 85.71	50	$\frac{1}{4}$200 = 50

Proof sketch. Postcontrol at time-step 1, the net supply of any node $v \in K$ will become $\lambda^{(1)} w \beta_v$. Thus in Step 4 we will apply the rules specified by (6.10) and (6.11). Suppose, in particular, that component K has an excess of demand. Then all supplies in K will be unchanged, and all demands will be scaled by the factor $r < 1$, where

$$r = \frac{\sum_{s \in \mathcal{G} \cap K} (\lambda^{(1)} w \beta_s)}{\sum_{s \in \mathcal{L} \cap K} (-\lambda^{(1)} w \beta_s)} = -\frac{\sum_{s \in \mathcal{G} \cap K} \beta_s}{\sum_{s \in \mathcal{L} \cap K} \beta_s}.$$

A similar analysis applies to the case where K has an excess of demand. ■

Example 6.1.16. *Table* 6.1 *shows the application of Lemma* 6.1.15 *to Example* 6.1.11. *Note that if we multiply the* $\breve{\beta}_v^K$ *values shown in this table by* 0.6, *we obtain the net-supply vector shown in Figure* 6.8.

As a result of this observation we have the following lemma.

Lemma 6.1.17. (a) *Let* H *be a fixed integer. Then, as a function of* w, $\Theta_G^{(H)}(w|\beta)$ *is nondecreasing, and piecewise-linear and has a number of break points that grows polynomially in the number of nodes.* (b) *For every time-step t the optimal scaling value,* $\lambda^{(t)}$, *is either equal to* 1 *or equal to a critical point, in which case, postcontrol, some line will be at capacity.*

Proof sketch. The proof is by induction on H. Consider the optimal choice of $\lambda^{(1)} = \lambda^{(1)}(w)$. As discussed above, the set of lines that will trip is $\cup_{h \le i^*-1} F^h$; using Lemma 6.1.15, the supply-demand vector for each component in the resulting graph is a multiple of $\lambda^{(1)}$. By induction, therefore, we will set $\lambda^{(1)} = \gamma_{i^*}$. Optimizing over the choice of i^* now yields the desired result. For a formal proof, see [30]. ■

As a result of the above, we can now prove the following lemma.

Lemma 6.1.18. *The optimal scaling problem can be solved in time* $O(m^H)$, *where* m = *number of lines.*

Remark. *In* [30] *the following stronger result is proved.*

Theorem 6.1.19. *Given a net-supply vector* β *over* G, $\Theta_G^{(H)}(w|\beta)$ *has at most*

$$\frac{m^{H-1}}{(H-1)!} + O\left(m^{\max\{1,H-2\}}\right)$$

breakpoints, which can be computed in $O(\log m)$ *time per breakpoint.*

Clearly Theorem 6.1.19 implies Lemma 6.1.18, but we will focus on the lemma here for brevity.

Proof of Lemma 6.1.18. Let $\hat{f} = \hat{f}(G^1, \beta))$. Given $w \geq 0$, let $\lambda^{(1)} = \lambda^{(1)}(w)$ be the optimal time-step 1 scale; we know by Lemma 6.1.17(b) that either $\lambda^{(1)} = 1$ or there exists a line j such that

$$w\lambda^{(1)}\hat{f}_j = u_j,$$

or, in other words,

$$w\lambda^{(1)} = \gamma_i$$

for some critical point γ_i (see Definition 6.1.10). In other words we only need to consider at most $m+1$ choices for $\lambda^{(1)}$. For each such choice a unique set of components will be in existence at the start of time-step 2. The result now follows by induction on H. ■

The proof of Lemma 6.1.18 immediately suggests a simple algorithm (Algorithm 6.1.20) for the optimal scaling problem. The algorithm would be separately applied to each component, postcontingency, assuming load-generation balance.

Algorithm 6.1.20. Algorithm for optimal scaling problem

Input. A connected grid G, a net-supply vector β over G, and an integer $H \geq 1$.

Output. $\Theta_G^{(H)} \doteq \Theta_G^{(H)}(1|\beta)$, i.e., the maximum H time-step demand that can be attained by an optimal scaling control with input G and net-supply vector β.

Step 0. Let $\hat{f} = \hat{f}(G, \beta)$.

Step 1. If $H = 1$, apply Lemma 6.1.9 to each component of G and **EXIT**.

Step 2. Let Γ be the set of critical points, i.e., the set of reals γ such that there is a line j with $\gamma\hat{f}_j = u_j$. Write $\hat{\Gamma} \doteq \{\gamma \in \Gamma : 0 < \gamma \leq 1\} \cup \{1\}$.

Step 3. For each $\lambda \in \hat{\Gamma}$:

(3.a) Apply scaling control using parameter λ.

(3.b) Let $\mathcal{K}$ be the set of components created as a result of this control.

(3.c) For each component $K \in \mathcal{K}$, perform the following steps.

(3.c.1) Let $\hat{\beta}^{(K)}$ be the net-supply vector over K obtained after step **(3.a)** and net-supply balancing in K. Using notation as in Lemma 6.1.15, $\hat{\beta}^{(K)} = \lambda\breve{\beta}^{(K)}$.

(3.c.2) *Recursively* compute $\Theta_K^{(H-1)}(1|\hat{\beta}^{(K)})$, denoted $\Theta_K^{(H-1)}$ for simplicity.

(3.d) Set $\mathcal{D}(\lambda) \doteq \sum_{K\in\mathcal{K}} \Theta_K^{(H-1)}$.

Step 4. Set $\lambda^{(1)} = \text{argmax}_{\lambda\in\hat{\Gamma}}\, \mathcal{D}(\lambda)$, and $\Theta_G^{(H)} = \mathcal{D}(\lambda^{(1)})$.

Remarks

1. The complexity of the algorithm is proportional to the workload required to solve a DC loadflow system on the input grid G, times the number of breakpoints in the function $\Theta_G^{(H)}(w|\beta)$, which by Theorem 6.1.19 is polynomial in the size of G for each fixed H. We conjecture that the optimal scaling problem is NP-hard when H can be part of the input to the problem, i.e., arbitrarily large.

2. The structure of the optimal control as per Lemma 6.1.17(b) reveals it as a quintessentially nonrobust control. Any error in the input data parameters is likely to cause a misapplication of control, by having lines that are binding (in the sense of Lemma 6.1.17) trip when they should not. Some thought about this issue shows that an easy fix for this difficulty is not obvious when the function $\Theta_G^{(H)}(w|\beta)$ has multiple breakpoints that are tightly clustered. We will later address *robust* versions of the optimal scaling problem, as well as versions of the control that permit more complete models of grid behavior (i.e., generator response). Note that the identification of initial faults is unto itself an error-prone process which is being ignored in the above analysis. See [142] for a discussion and algorithms.

3. Algorithm 6.1.20 is stated in explicit enumerative form. However, one can devise a close variant that works in a fashion similar to branch-and-bound.

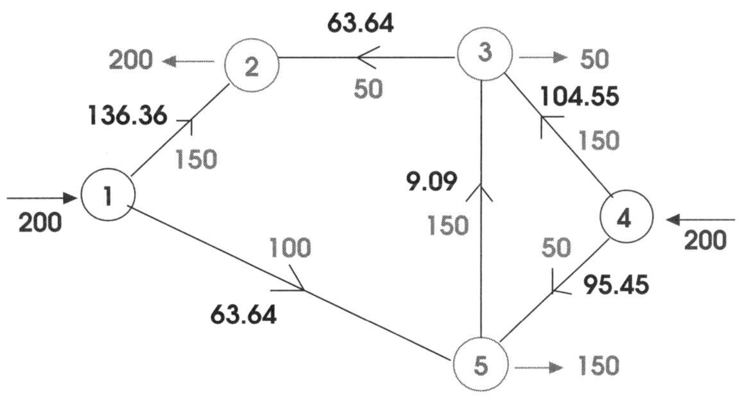

Figure 6.9. *Example* 6.1.21. *Flow pattern, loads, and generation, postcontingency.*

Example 6.1.21. *Here we consider a variation on Example* 6.1.11. *Refer to Figure* 6.9. *We will apply Algorithm* 6.1.20 *to this system for the case* $H = 3$. *The set* $\hat{\Gamma}$ *is given in Table* 6.2.

Table 6.2. *Example* 6.1.21. *Set* $\hat{\Gamma}$.

i	1	2	3
$\lambda_{1,i}$	0.524	0.786	1.0

Proceeding first with $\lambda_{1,3}$, *we see that in Step* (3.a) *of the algorithm, lines* (3, 2) *and* (4, 5) *trip, resulting in a single component, and thus all loads and generations are unchanged in Step* (3.c.1). *We denote this component by* $K(3)$. *The disposition of power flows in this component is depicted in Figure* 6.10. *Note the reversal of flow direction on line* $(3,4)$ *relative to the root case.*

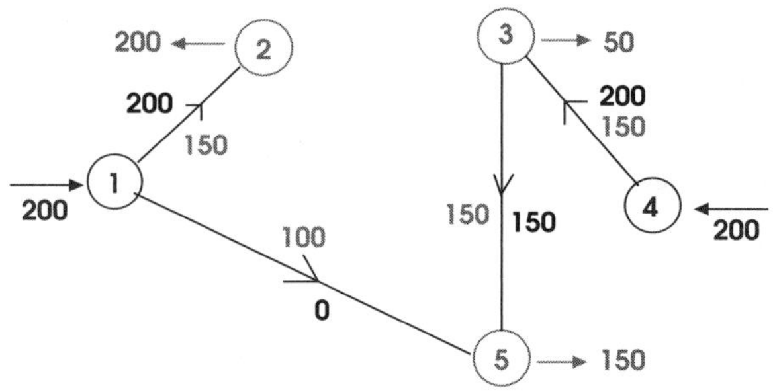

Figure 6.10. *Example* 6.1.21. *Flow pattern, loads, and generation,* $K(3)$.

Considering Figure 6.6, *we can see that in Step* 3 *of the algorithm the set* $\hat{\Gamma}$ *will be obtained by enumerating lines* $(4,5)$ *and* $(3,2)$, *yielding* $\gamma_1 = 50/95.45 \approx 0.524$ *and* $\gamma_2 = 50/63.64 \approx 0.786$ *(resp.). Hence* $|\hat{\Gamma}| = 3$, *and the different values of* $\lambda \in \hat{\Gamma}$ *to be enumerated in Step* 3 *are listed in Table* 6.2, *sorted in increasing order.*

We can compute $\Theta^{(2)}_{K(3)}$ *by inspection, as follows.* **(a)** *If we set* $\lambda = 1$ *then lines* $(1,2)$ *and* $(3,4)$ *will trip, isolating load node* 2 *and generator node* 4, *and yielding the component with nodes* 1,3, *and* 5 *with total supply and demand of* 200 *units. Thus at the start of the third time-step the flow on line* $(1,5)$ *will equal* 200 *units, and the flow on* $(3,5)$ *will be* 50 *units. Since* $200 > 100$, *the termination criterion will scale all supplies and demands by a factor of* $100/200 = 0.5$. *Hence the total demand satisfied at termination will equal* $0.5 \times 200 = 100$ *units.* **(b)** *If on the other hand we set* $\lambda = 150/200 = 3/4$ *all lines are within limits and the algorithm terminates with total demand* $= 3/4 \times 400 = 300$. *We conclude that case* (b) *attains the optimum, and*

$$\Theta^{(2)}_{K(3)} = 300, \tag{6.15}$$

in other words, 75% *of the initial demand.*

Returning to Table 6.2, *we can therefore conclude that the choice of scaling parameter* $\lambda_{1,1} = 0.524$ *is suboptimal, since applying this control will immediately scale all demands by* $0.524 < 0.75$, *with possibly more load shed at later time-steps.*

Hence, from the three values in Table 6.2 *there remains to consider the case where we choose* $\lambda_{1,2} = 0.786$ *and any recursively defined subcases. Scaling demands and supplies by* $\lambda_{1,2}$ *will result in a flow pattern where line* $(4,5)$ *is still overloaded and will trip, producing the case in Figure* 6.11. *We will refer to this system as* $K(2)$.

Table 6.3. *Example* 6.1.21. *System* $K(2)$. *Set* $\hat{\Gamma}$.

i	1	2	3
$\lambda_{2,i}$	0.7273	0.9546	1.0

Note that lines $(3,4)$ *and* $(3,2)$ *are overloaded, leading to the* $\hat{\Gamma}$ *set in Table* 6.3. *Proceeding recursively, we have to examine system* $K(2)$ *over a two time-step cascade using each of the values in* Table 6.3 *as the first-round scale. Each of these choices will scale demands and supplies and (possibly) cause some line tripping. The state of the network as a result of each choice is shown in Table* 6.4.

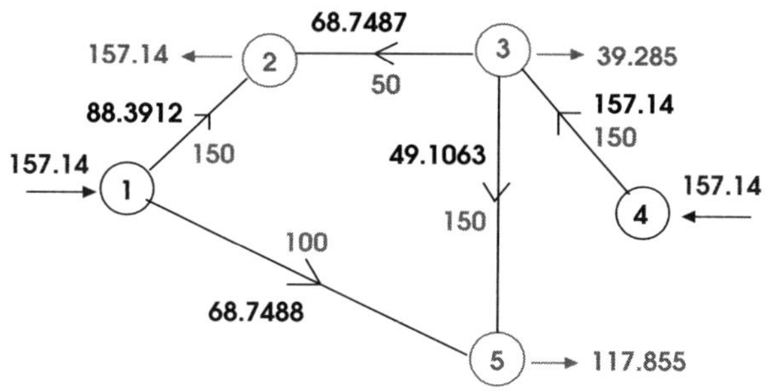

Figure 6.11. *Example* 6.1.21. *Flow pattern, loads and generation, system* $K(2)$.

Table 6.4. *Example* 6.1.11. *One time-step behavior starting from* $K(2)$ *using scale as in Table* 6.3.

Case $i=1$, $\lambda_{2,1}=0.7273$	Case $i=2$, $\lambda_{2,2}=0.9546$	Case $i=3$, $\lambda_{2,3}=1.0$
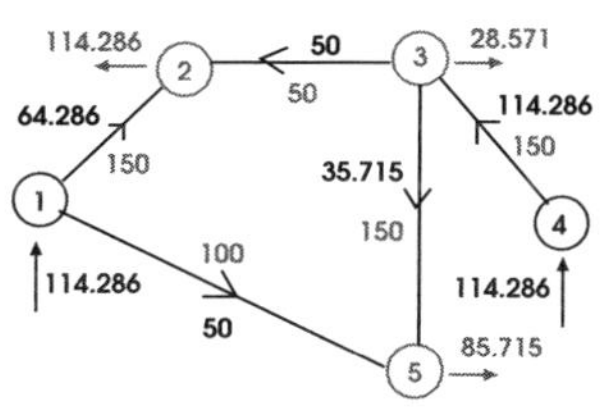 (1) $\lambda=0.7273$	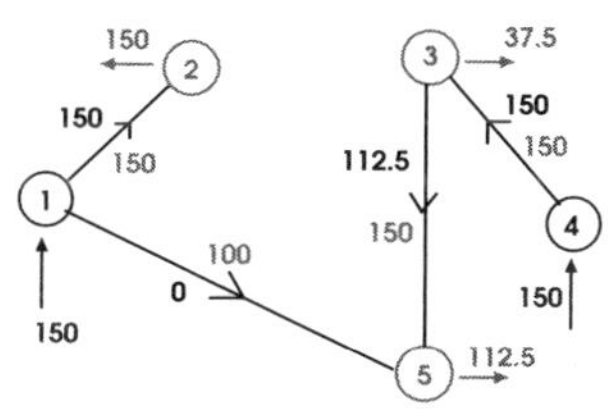 (2) $\lambda=0.9546$	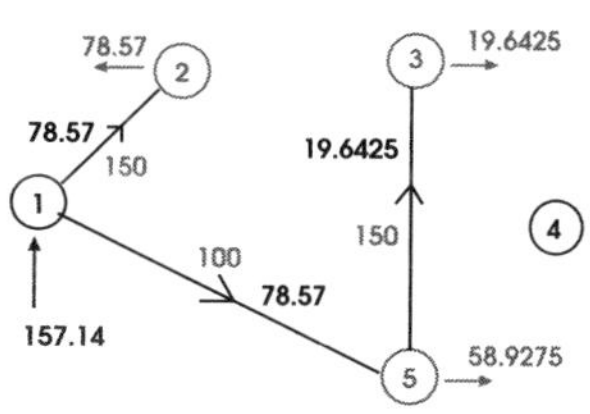 (3) $\lambda=1.0$
$\Theta^{(2)}=228.572$	$\Theta^{(2)}=300$	$\Theta^{(2)}=157.14$

If we choose $\lambda_{2,3}=1.0$, *then lines* $(3,2)$ *and* $(4,3)$ *trip, causing node* 4 *to become isolated, and the net-supply balancing action in Step* (3.c.1) *will cause the loads in the large component to be scaled down, by a factor of* 2, *to match the output of the generator at node* 1, *thus delivering a total of* 157.14 *units of demand. If we instead choose* $\lambda_{2,1}$ *or* $\lambda_{2,2}$, *then as shown in Table* 6.4 *all the flows are within line limits and the cascade terminates immediately with* Θ *values as shown in the table. Thus, scaling by* $\lambda_{2,2}$ *proves to be the best of the three choices, and, in summary,*

$$\Theta^{(2)}_{K(2)} = 300; \tag{6.16}$$

in other words, as in (6.15), *the cascade is stopped with* 75% *of the load saved.*

This completes the analysis of the problem. We conclude that

$$\Theta^{(3)} = 300,$$

attained with two different controls:

$$(\lambda_1, \lambda_2) = (1, 0.75) \quad \textit{and} \quad (0.786, 0.9546).$$

Computational complexity issues

In Theorem 6.1.7 we showed that, if the number of time-steps H is bounded, the optimal scaling problem can be solved in polynomial time. We would theorize, as a result, that the optimal scaling problem can be equivalently formulated as a compact linear program, thus keeping in line with many other (polynomially solvable) optimization problems.

Conjecture 1. For each fixed H there is a polynomial-size linear programming formulation for the optimal scaling problem.

An important question concerns the case where H is part of the input, as in the following conjecture.

Conjecture 2. For general H the optimal scaling problem is NP-hard.

6.2 ▪ Robust controls

Example 6.1.11 echoes the remarks made after the statement of Template 6.1.3. In particular, it is clear that the control computed by the template may well be nonrobust. Consider, for example, the control $(\lambda_1, \lambda_2) = (0.786, 0.9546)$. Case $i = 2$ in Table 6.4 shows the state of the network immediately following the application of control at time-step 2. Note that lines $(1,2)$ and $(3,4)$ attain their respective limits. Hence, noise or any other form of error, even in very small quantities, may well cause either of these lines to trip (in time-step 2, according to our model); each of these trips would reduce the amount of served demand by 150 units. Similar considerations also apply to the first optimal control.

These observations motivate a relevant question: can we devise a robust form of the scaling control? Here, by "robust" we mean *insensitive to all forms of data errors, so long as such errors take place in small quantities*. Ideally, we should produce simple, common substitutes for all forms of error (again assumed to take place in small magnitudes), or, in other words, we should develop a model for uncertainty in the cascading context that yields robust controls no matter what precise form the data errors take. Aiming toward this goal, we now revisit the point that not only is the control in Template 6.1.3 nonrobust, but it intrinsically requires a deterministic setting—in a stochastic setting the control does not even make sense, since it assumes a particular islanding structure at every time-step in the planning horizon. Under a stochastic data regime the decisions made at time t will contribute to the uncertainty in island structure at later time-steps; that is to say, we will have a case of *endogenous uncertainty* with possibly an exponential number of scenarios to be examined.

Stochastic programming under endogenous uncertainty

In the typical stochastic programming setting considered in the literature, uncertainty is *exogenous*, that is to say, independent of the decisions under consideration. In the *endogenous* case stochastics can be altered as a result of decisions. This explicit classification was initially given by Jonsbraten [132], who also proposed an algorithm; an early reference is Pflug [184]. More recent work includes Ahmed [2] and Goel and Grossman [112]. Not surprisingly, endogenous uncertainty arises in stochastic network interdiction problems; see Held and Woodruff [121].

In the case of endogenous uncertainty it is possible to obtain an explicit decision tree (with stochastic paths), thus effectively rendering the problem into a standard stochastic program. The downside of this approach is the possibly large size of the resulting formulations.

We will next discuss an approach that will yield a robust version of the control methodology described in the previous section. With this objective in mind, we will first modify the control scheme so that while still deterministic, it does not require a particular realization of islanding for its application. This is accomplished by modifying the scheme (6.12) (while still applying Template 6.1.3) as follows. The new control will be embodied by constants $s^{(t)} \geq 0$ for $t = 1, \ldots, H-1$; at time t the demand of any load node v is reset from $-\beta_v^t$ to

$$\min\{1, [1 + s^{(t)}(1 - x_v^t)]^+\}\,(-\beta_v^t) \quad \text{if } x_v^t > 1, \tag{6.17a}$$

$$-\beta_v^t \quad \text{otherwise.} \tag{6.17b}$$

We stress that the constant $s^{(t)}$ used in the scaling is now independent of the node v; however the actual scaling constant in (6.17a) *does* depend on v. We will term a control of the form (6.17) a *time-driven* control.

Problem 6.2.1. Optimal time-driven control problem

Compute a time-driven control $\{s^{(t)} : 1 \leq t \leq H-1\}$ that maximizes the sum of demands at termination.

As in the case of the optimal scaling problem, to solve the time-driven variant we consider a broader family of problems.

Definition 6.2.2. *For $w \geq 0$, denote by $\Delta_G^{(H)}(w|\beta)$ the final total demand resulting from applying an optimal time-driven control in an H time-step cascade on graph G, where the initial net-supply vector is $w\beta$.*

The problem of interest is to compute $\Delta(H)_G(1|\beta)$. Note that $\Delta_G^{(H)}(w|\beta) \leq \Theta_G^{(H)}(w|\beta)$, since the standard scaling control is less constrained than its time-driven version. This inequality can be strict, as shown by the following example.

Example 6.2.3. *Consider Example* 6.1.6 *(which uses $H = 2$), shown again postcontingency in Figure* 6.12 *for convenience. In the analysis of Example* 6.1.6 *we saw that an optimal solution to the scaling problem under island-wide control attained a yield of* 72.22% *or* 26 *units and was obtained by setting $\lambda_A^{(1)} = \frac{4}{5}, \lambda_B^{(1)} = 1$. Since $x_A^1 = \frac{5}{4}$ and $x_B^1 = 2$, an optimal solution to the island-wide scaling problem is thus obtained by setting $s_A^{(1)} = \frac{4}{5}$ and $s_B^{(1)} = 0$ (see* (6.13)*).*

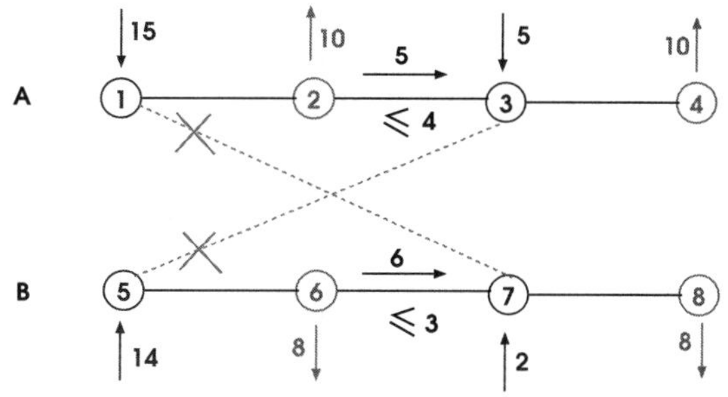

Figure 6.12. *Example* 6.2.3. *Flow pattern, loads, and generation, postcontingency.*

Suppose we consider a time-driven control. Note that in time-step 1*, such a control will scale all loads, generation, and flows in components A and B by*

$$\lambda_A^{(1)} = 1 - \frac{s^{(1)}}{4} \quad \textit{and} \quad \lambda_B^{(1)} = 1 - s^{(1)},$$

respectively. Suppose first that $s^{(1)} \geq \frac{4}{5}$. *Since in that case* $\lambda_A^1 \leq 4/5 = 1/x_A^1$ *and* $\lambda_B^1 \leq 1/5 < 1/x_B^1$*, it follows that postcontrol all line flows are within their limits and so the cascade terminates with a yield of*

$$20(1 - s^{(1)}/4) + 16(1 - s^{(1)}).$$

By choosing $s^{(1)} = \frac{4}{5}$ *we can thus obtain a yield of* $16 + 16/5 = 19.2$ *units.*

Suppose next that $1/2 \leq s^{(1)} < \frac{4}{5}$. *Then line* $(3,2)$ *will trip but* $(6,7)$ *will not. Hence component A splits into two subcomponents, one containing nodes* 1 *and* 2*, where the generation at node* 1 *is therefore reduced to* $10(1 - s^{(1)}/4)$ *units to attain balance, and one consisting of nodes* 3 *and* 4*, where load at* 4 *is reduced to* $5(1 - s^{(1)}/4)$ *to attain balance. The yield in this case will thus equal*

$$15(1 - s^{(1)}/4) + 16(1 - s^{(1)}).$$

By choosing $s^{(1)} = 1/2$ *we therefore get a yield of* 21.125 *units.*

Finally suppose $s^{(1)} < \frac{1}{2}$. *Postcontrol, both line* $(3,2)$ *and line* $(6,7)$ *will trip, and we attain a yield of*

$$15(1 - s^{(1)}/4) + 10(1 - s^{(1)})$$

units. With $s^{(1)} = 0$ *this quantity equals* 25 *units, which is thus the overall optimum.*

In summary, in this example we have $\Theta_G^{(2)}(1|\beta) = 26$ *but* $\Delta_G^{(2)}(1|\beta) = 25$.

Nevertheless, we have the following analogue of Lemma 6.1.17, with (nearly) identical wording.

Lemma 6.2.4. *Let G be a grid and* β *a net-supply vector on G.* (a) *For each fixed H,* $\Delta_G^{(H)}(w|\beta)$ *is a monotonically nondecreasing piecewise-linear function of* w *with a polynomial number of breakpoints.* (b) *The optimal choice for* $s^{(1)}$ *is either* 0 *or such that post-control some line will be at capacity (and recursively for every time-step).*

The proof of this result amounts to a step-by-step repetition of Lemma 6.1.9 through Lemma 6.1.17. Also note that with $s^{(1)} = 0$ the scaling constant in (6.17) equals 1. Lemma 6.2.4 can be used to obtain an algorithm for the time-driven optimal control problem that runs in polynomial time for each fixed H, along the lines of Algorithm 6.1.20.

6.2.1 ▪ Applying the sample average approximation framework

Even though the control scheme (6.17) can in fact be applied even in the case of data errors, it is not robust, as highlighted by Lemma 6.2.4(b). With this particular weakness in mind, let us consider a model where the line capacities u_j are subject to (small) errors of a stochastic nature. To be specific, we assume that for each line j, the random variable u_j lies in some interval $[\tilde{u}_j - \epsilon_j, \tilde{u}_j + \epsilon_j]$ with some known distribution.

Definition 6.2.5. *Given a grid G with net-supply vector* β *and given a vector with nonnegative values* $(s^{(1)}, \ldots, s^{(H-1)})$*, let* $\Upsilon_G^{(H)}(\beta, s^{(1)}, \ldots, s^{(H-1)})$ *denote the yield resulting from applying the time-driven control* $(s^{(1)}, \ldots, s^{(H-1)})$ *to the pair G,* β.

Using this notation, we have that

$$\Delta_G^{(H)}(1|\beta) = \max_{s^{(1)},\ldots,s^{(H-1)}} \Upsilon_G^{(H)}(\beta, s^{(1)},\ldots,s^{(H-1)}).$$

The robust problem of interest we consider can now be written as

$$\max_{s^{(1)},\ldots,s^{(H-1)}} \mathbb{E}\Upsilon_G^{(H)}(\beta, s^{(1)},\ldots,s^{(H-1)}), \tag{6.18}$$

where the expectation is over the distribution of data errors.

We stress that the goal underlying this computation is robustness with respect to small data errors. The ϵ_j should be assumed small, and one may even argue that any "reasonable" distribution of data errors will suffice to attain our goal, for example, the uniform distribution, and we can further assume that the random variables u_j are pairwise independent. One could approach problem (6.18) through a variety of methods; here we will focus on the SAA approach previously discussed in Chapter 3.
Formally, we have the following:

- A *sample* of size $N \geq 1$ will consist of a family of N vectors $(e^1,\ldots,e^N)$, where each vector e^i includes a value e^i_j for each line j, drawn from the distribution for u_j.

- Given a sample $(e^1,\ldots,e^N)$, the sample approximation to problem (6.18) is

$$\max_{s^{(1)},\ldots,s^{(H-1)}} \frac{\sum_{i=1}^N \Upsilon_{G(i)}^{(H)}(\beta, s^{(1)},\ldots,S^{H-1})}{N}, \tag{6.19}$$

  where for $1 \leq i \leq N$, $G(i)$ is the grid G modified so that each line j has limit e^i_j.

We conclude that to apply the SAA framework we need to (efficiently) maximize the quantity

$$\sum_{i=1}^N \Upsilon_{G(i)}^{(H)}(\beta, s^{(1)},\ldots,S^{H-1}).$$

To address this problem, suppose $s^{(1)}$ has been fixed. Then, using an analogue of Lemma 6.1.15, for *any* choice of the parameters $s^{(2)},\ldots,S^{H-1}$ and any i,

$$\Upsilon_{\hat{G}(i)}^{(H)}(\beta, s^{(1)},\ldots,S^{H-1}) = \Upsilon_{\hat{G}(i)}^{(H-1)}(w(s^{(1)})\hat{\beta}, s^{(2)},\ldots,S^{H-1}),$$

where $\hat{G}(i)$ is the subgraph of G obtained by deleting those lines in $G(i)$ that trip when using $s^{(1)}$ in (6.17a) at $t = 1$, $\hat{\beta}$ is a net-supply vector on each component of $\hat{G}(i)$, and $w(s^{(1)}) \geq 0$ is appropriately defined.

Moreover, clearly,

$$\max_{s^{(1)},\ldots,s^{(H-1)}} \sum_{i=1}^N \Upsilon_{G(i)}^{(H)}(\beta, s^{(1)},\ldots,S^{H-1}) = \max_{s^{(1)}} \max_{s^{(2)},\ldots,s^{(H-1)}} \sum_{i=1}^N \Upsilon_{\hat{G}(i)}^{(H-1)}(w(s^{(1)})\hat{\beta}, s^{(2)},\ldots,S^{H-1}).$$

Thus, as in the last section, one can prove the following result.

Theorem 6.2.6. *For each fixed H the sample approximation problem* (6.19) *can be solved in polynomial time. The number of DC-flow systems solved by the algorithm grows linearly in N, the number of samples.*

6.2.2 ▪ Broader robust models

In Chapter 5 we discussed several forms of uncertainty that pervade cascade behavior, in particular the line tripping model and race conditions. Here, we are interested in building control algorithms that can be robust under various forms of noise and uncertainty, in particular under more general cascade models than that in Template 6.1.2. In particular refer to the discussion in Section 5.5.

A second goal should be that of *speed*. Even though Theorem 6.2.6 (or Lemma 6.1.18 and Theorem 6.1.19 for the nonrobust case) formally states that there is a version of Template 6.1.3 that is theoretically efficient when the number of time-steps is small, in the case of large networks and $H \geq 3$ the worst-case bound in Theorem 6.1.19 is simply too large. And, of course, we may actually want H to take large values to be able to model fairly brief time slices (as time-steps).

Toward attaining both goals, in this section and the next we will consider heuristics to handle *simpler* controls than those studied above. We will instead use a modified form of the control law (6.12), where at time t we reset the demand of a load bus v from $-\beta^t_v$ to

$$\min\{1, [1+s^{(t)}(1-x^t_v)]^+\}\,(-\beta^t_v) \quad \text{if } x^t_v > 1, \tag{6.20a}$$

$$-\beta^t_v \quad \text{otherwise,} \tag{6.20b}$$

where as before x^t_v is the maximum relative line load of any line in the component holding bus v at time t. Thus, the control is parameterized by a single value $0 \leq s^{(t)}$ for each time-step t. We will *further* constrain the control law by restricting the search for an optimal control to specific cases:

(a) $s^{(t)}$ independent of t. This corresponds to a simple (but perhaps compelling) control law that is repeatedly applied until the cascade terminates (one way or another). Note that even though $s^{(t)}$ is constant, the multiplier applied to $(-\beta^t_v)$ in (6.20a) does depend on x^t_v, i.e., it does depend on the component structure. Under this scheme, components with larger line overloads will see greater load shedding.

(b) For some constants $s \geq 0$ and ρ, $s^{(t)} = s\rho^t$. When $\rho > 1$ the control becomes more stringent as a function of time in the sense that when line overloads are greater than 1 more load is shed as t increases. When $\rho < 1$ the opposite is true. Thus an optimal control of this form will become either more or less conservative as the cascade evolves, and which of the two should be the case is an important fact to determine. When $\rho = 1$ we recover case (a).

(c) The function $s^{(t)}$ has a specific structure designed to capture a given structure in the control strategy. An example is a unimodal function $s^{(t)}$, where for some value t^*, $s^{(t)}$ is nondecreasing for $t \leq t^*$ and nonincreasing for $t \geq t^*$ (or vice versa). Case (b) is a special parameterization of this structure.

In all three cases we have controls that are parameterized by a small number of values. This permits the enumeration of large numbers of candidates to compute near-optimal controls.

Experiment 6.2.7.

First we revisit Experiment 5.4.2 (see Section 5.4.2), which involved the familiar 118-bus example in the cascade initiated by disabling line (9, 10). The experiment used what was termed the *separated time scale* model (Section 5.3), in which component dynamics are assumed to take place on a much faster time scale than line tripping; furthermore the line tripping model was deterministic, with no memory. Moreover, we assumed that $\rho_g = 0$ for every generator g; i.e., no generation ramp-up can take place. In summary, this amounts to the simplest possible cascade model. As was shown in Table 5.1, the cascade terminates in five time-steps with a yield of approximately 37%.

Suppose we further restrict the control scheme (6.20a) so that $s^{(t)}$ is a constant value s independent of t, so that the scaling control becomes

$$\min\{1, [1+s(1-x_v^t)]^+\}, \text{ if } x_v^t > 1 \tag{6.21}$$

and 1 otherwise. Note that even under this restriction, the actual control (6.21) is nevertheless still time dependent; in fact it is time and component dependent. How large s is reflects how aggressive our control policy is. Moreover, such a policy should be relatively easy to implement, with the quantities x_v^t as the only measurements required in real time, and these being common values for all nodes in a common component.

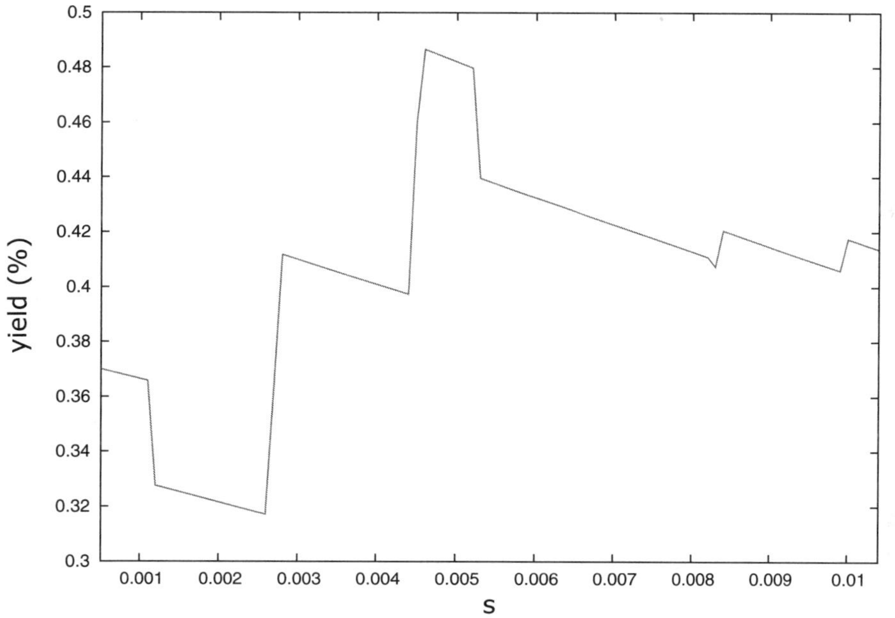

Figure 6.13. *Experiment* 6.2.7. *Yield as a function of the parameter s in control scheme* (6.21)) *on the* 118-*bus case.*

Figure 6.13 shows the yield at termination of the cascade for 100 values of s, starting with $s = 0.0005$ and with increments of 1e-4. The maximum yield is attained at $s = 0.0046$, where yield equals 48.65%. Not surprisingly, the plot is piecewise linear. What we want to point out here is that the optimum control $s = 0.0046$ is quite nonrobust in the sense that an error in the data or the model may make the system attain the behavior shown in Figure 6.13 for values smaller than 0.0046, i.e., attain a yield close to 0.4. In contrast, if we choose a slightly larger value for s, say $s = 0.51$, we will still attain near-optimal behavior when no data errors take place, but we will also be able to weather some data errors without major yield degradation.

Experiment 6.2.8.

This experiment is similar to Experiment 6.2.7; however, it is carried out on the model of the Eastern Interconnection, with 15023 nodes, 23769 lines, 2255 generators, and 7696 loads, that we discussed under Experiment 5.4.8. We consider a cascade initiated by disabling 100 highly loaded lines (this scenario is available with the book materials). As in Experiment 6.2.7, we consider deterministic, memory-free line tripping. We assume that $\rho_g = 1$ for all g; in other words generators can ramp up their output to the maximum allowed, and we further assume that the ramp-up rate is infinitely large. Figure 6.14 shows yield at termination as a function of the parameter s; the figure was obtained by enumerating values of s starting at 0 and with increments of 0.005. The maximum yield is attained at $s = 0.22$, where it approximately equals 67.5%. In contrast, using $s = 0$ (the "do-nothing" control), the yield equals approximately 53.09%.

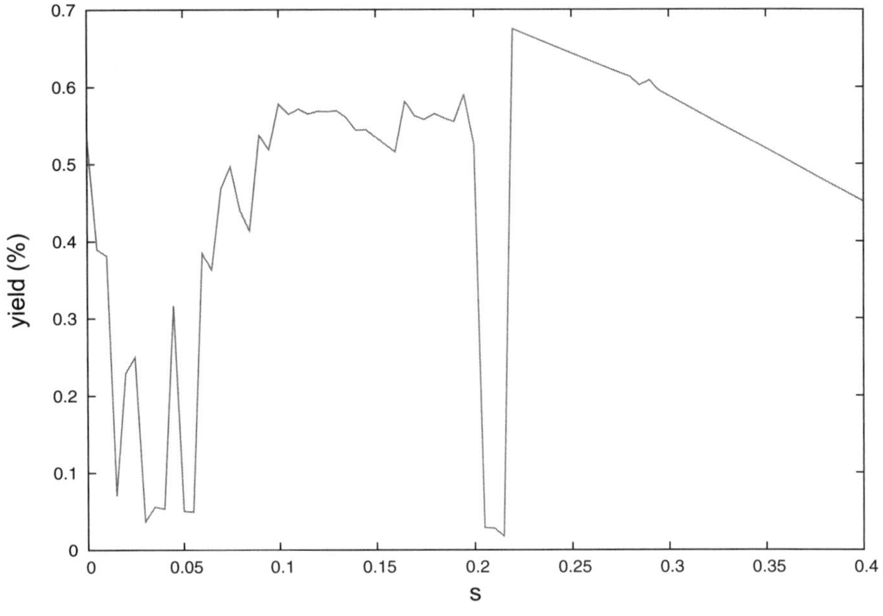

Figure 6.14. *Experiment 6.2.8. Yield as a function of the parameter s in control scheme (6.21) on Eastern Interconnection (15023 buses, 23769 lines, 2255 generators, 7696 loads).*

A glance at Figure 6.14 clearly reveals that the "optimal" control with $s = 0.22$ is, very likely, highly nonrobust. Using $s = 0.215$, the yield at termination is roughly 1.79%. While these yield values are functions of the control parameter s, and not of the underlying data, the control optimization implicit in Figure 6.14 obviously leverages the data, and thus it is clear that even a small variation of the data is likely to cause the control with $s = 0.22$ to significantly deteriorate. (For a related phenomenon, see the remarks following Algorithm 6.1.20.) In view of Figure 6.14 one would guess that using, e.g., $s = 0.25$ would be more prudent, while still attaining high yield (64.35%) in the nominal case studied by the experiment. A similar set of remarks applies to the do-nothing control with $s = 0$. While not optimal in the context of our deterministic experiment, one might still opt for this choice because it avoids the complexity of the control. However, the do-nothing control is also nonrobust, as shown by the figure—using $s = 0.005$, the yield at termination equals 38.96% and using $s = 0.15$ it equals 7.06%, and therefore (if proved robust) the control with $s = 0.25$ appears far more preferable to the do-nothing approach.

There is a feature in Figure 6.14 that has a straightforward explanation. This is the relative smoothness of the yield at termination when $s \geq 0.22$, especially considered with the noisy and seemingly random behavior when $s < 0.22$. The explanation for this behavior is that as s grows the control scheme (6.21) sheds increasing amounts of load. When s is "large enough," this proves sufficient to relieve many overloaded lines, leading to smoothed-out behavior.

To investigate the relative robustness of the different choices for the parameter s, we repeat the tests shown in Figure 6.14 in a stochastic setting. Namely, we employ the stochastic line tripping scheme (5.5). The specific goal of the testing is to evaluate the behavior of each control as a result of small (but random) changes in the data, a proxy for which is plausibly attained by employing stochastic line tripping. Thus, our choice of parameters m_e and p_e in (5.5) for each line e is $p_e = \frac{1}{2}$ and $m_e = 5$ for all e. See Figure 5.1. To refresh the reader's memory, when the relative load of a line exceeds 1.2, the tripping probability for a line is nearly 1.0; when the relative load is less than 0.8, the probability is effectively 0; for relative loads between 0.95 and 1.05, the probability equals 0.5; and otherwise the probability is a increasing smooth function. For each value of s shown in Figure 6.14 we ran 1000 simulations of the cascade under control; Figure 6.15, for each value of s, the average yield, and the average yield plus or minus one standard deviation.

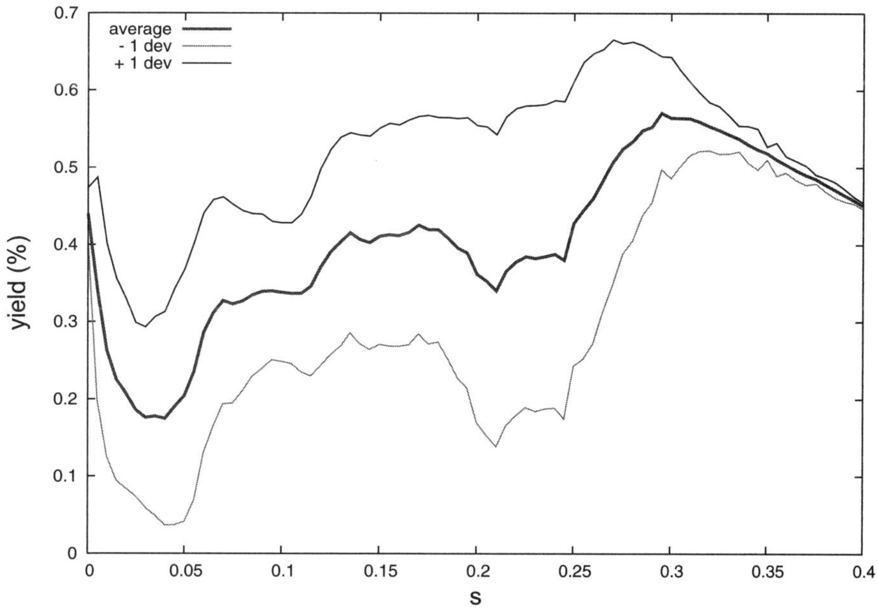

Figure 6.15. *Experiment* 6.2.8. *Yield as a function of the parameter* s *in control scheme* (6.21) *under stochastic line tripping.*

Several salient points emerge from this figure. First, as expected, the deterministic optimal control with $s = 0.22$ significantly decays in the stochastic setting: its average yield decreases to 0.378 from 0.675 in the deterministic case. But, more significantly, the control with $s = 0.22$ now appears quite suboptimal compared to other values. Based on this figure, we would be more likely to pick a value $0.3 \leq s \leq 0.35$. Another interesting point is the comparison of a control in this regime with the do-nothing control ($s = 0$). For example, with $s = 0.33$ the average yield equals 0.543 with a standard deviation of 0.025, whereas with $s = 0$ the respective values are 0.44 and 0.034. Thus the average yield for $s = 0.33$ exceeds that for $s = 0$ by more than three standard deviations.

These comparisons give rise to a significant question: what is an "optimal" control? Borrowing ideas from portfolio theory, let $w \geq 0$ be the "risk-aversion" parameter, and suppose that for a control s with average yield $\bar{Y}(s)$ and standard deviation of yield $\delta(s)$ we assign a risk-weighted (average) yield

$$\bar{Y}(s) - w\delta(s). \tag{6.22}$$

For $w = 0$ we obtain average yield while for w large the risk-weighted yield becomes negative. Notice that, for example, with $w = 1$ the risk-weighted yield (already shown in Figure 6.15 in bold) is maximized at $s = 0.32$.[51] In general, we are interested in optimizing risk-weighted yield as a function of w, with this parameter used to control risk aversion.

Figure 6.16 shows the risk-weighted yield for each of the choices of s considered above and for a range of values of w. As expected, for each choice of w the optimal control is attained for some value $s \geq 0.3$. As w grows, the optimal s value also grows, due to the decreasing variance of yield, as highlighted by Figure 6.15.

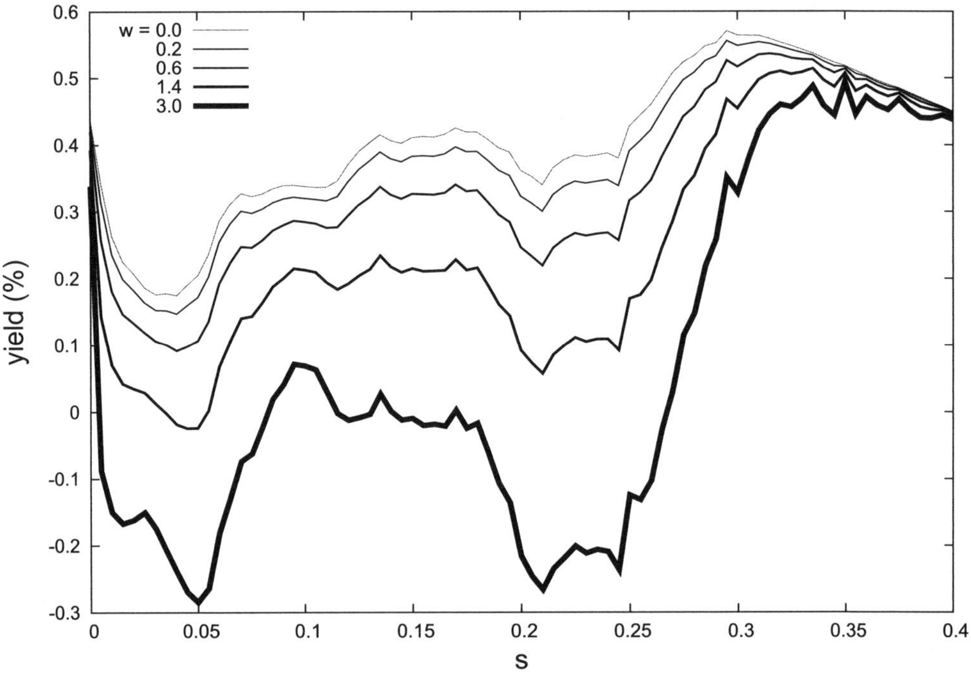

Figure 6.16. *Experiment* 6.2.8. *Risk-weighted yield for different choices of the risk-aversion parameter w, as a function of the parameter s in control scheme* (6.21) *under stochastic line tripping.*

In the remainder of this chapter we will formalize the ideas discussed in Experiment 6.2.8. Rather than focusing on one-parameter controls of the form (6.21), we will consider broader controls that allow for more flexible action as a function of time. We will discuss procedures that use either a deterministic algorithm or an algorithm once again based on concepts derived from the SAA method and, relying on local search, compute an approximately optimal control, where optimality is defined using a confidence interval criterion.

[51] Technical point: in (6.22) we are using the *sample* standard deviation, rather than its true value.

6.2.3 ▪ A second application of the sample average approximation framework

As discussed above, our method for attaining robustness consists of optimizing controls subject to noise, appropriately defined. We consider stochastic noise, of an appropriate type, and we approximately optimize over the stochastics by applying the SAA framework. This approach was used in Section 6.2.1.

Here we will also take the opportunity to introduce a realistic feature that has been absent in our models so far. Consider the control family of the form (6.20). At time t, and for each bus v, this control scheme relies on the parameter x^t_v—the maximum relative line load over all lines in the same component as v. We will introduce noise into the control by assuming that at time t the quantity x^t_v *is observed subject to error*. While many possible models for such errors appear reasonable, we can introduce another element of realism: we will model delays in the (packet) network whose data can be used to estimate x^t_v.

More precisely, we will assume that at time t the communications channel either works correctly and delivers the actual value x^t_v or does not, in which case it delivers the value that was observed the last time the channel worked correctly. Proceeding formally, let $0 \le p_F \le 1$, and for each time $t \ge 0$ we define the random variable

$$\Gamma_t \doteq \begin{cases} 0 & \text{with probability } 1 - p_F, \\ 1 & \text{otherwise.} \end{cases} \tag{6.23}$$

The interpretation of this variable is that the channel malfunctions when $\Gamma_t = 1$. Then the (random) data observation at bus v at time t is defined as

$$\hat{x}^t_v \doteq \begin{cases} x^t_v & \text{if } \Gamma_t = 0, \\ x^{t^*} & \text{if } \Gamma_t = 1, \end{cases} \tag{6.24}$$

where for $t \ge 1$ we define

$$t^* \doteq \max\{1 \le j < t : \Gamma_j = 0\}.$$

We write $\Gamma_0 = 0$. In other words, when the communication channel fails, the measurement that is delivered is that which was obtained the last time the channel worked correctly, i.e., the last correct measurement. This models the behavior of intermittently failing packet networks. We will use the term "skip" to refer to the values Γ_t.

In our simulations of controls of type (6.20) we will use the noisy measurements $\hat{x}^t_v$ rather than the true values x^t_v. Note that a deterministically optimal control (6.20) that applies just the minimum amount of load shedding at a critical time step will likely be degraded under this noise scheme. On the other hand, a control scheme that performs well even under this noise regime will possibly prove robust under other forms of data error (at least, that is our conjecture). By fine-tuning the parameter p_F we can attain various levels of conservatism.

To apply the SAA framework, let H denote as before the number of time-steps in the cascade model. Proceeding formally, we have the following definition.

Definition 6.2.9.

- *Let a skip sequence be a vector*

$$\gamma = (\Gamma_0, \Gamma_1, \ldots, \Gamma_{H-1}) \in \{0,1\}^H.$$

- *Given a control u and a skip sequence γ, let $\Theta(u,\gamma)$ be the yield if the control u is applied using the framework* (6.24) *under the sequence γ.*
- *Let U denote a family of controls. Write*

$$T^*(U) \doteq \max_{u \in U} \mathbb{E}_\gamma \Theta(u,\gamma),$$

where the expectation is taken over the distribution (6.23).

- *A sample of size $N \geq 1$ will be a family of N skip sequences γ^i, $1 \leq i \leq N$. We denote by $\Sigma(N)$ the set of all samples of size N.*
- *Given a sample $\sigma = (\gamma^1, \dots, \gamma^N)$ of size N we write*

$$T(U,\sigma) \doteq \max_{u \in U} \frac{\sum_{i=1}^N \Theta(u,\gamma^i)}{N} \tag{6.25}$$

and

$$T_N^*(U) \doteq \mathbb{E}_{\sigma \in \Sigma(N)} T^*(U).$$

We have (see Section 3.4.1) that

$$T_N^*(U) \geq T^*(U) \quad \text{and} \quad T_N^*(U) \to T^*(U) \text{ with probability 1, as } N \to +\infty.$$

In the SAA framework we exploit these observations by first constructing a number of random samples $\sigma_1, \dots, \sigma_K$ of size N and using the central limit theorem to obtain a confidence interval for an *upper bound* for $T^*(U)$. Second, if u is any control, we construct a confidence interval for $\mathbb{E}_\gamma \Theta(u,\gamma)$, which is a *lower bound* for $T^*(U)$.

Experiment 6.2.10.
As an initial example of this methodology, consider Figure 6.17. This figure repeats the experiment depicted in Figure 6.14, with the exception that here we simulate frequency response in the context of islanding; see Section 1.4 and Model 5.5.3. For each generator g, its "regulatory constant" R_g is set to 0.05. This parameter concerns frequency loss in an island with generation shortage; it can be shown that this choice (standard in power engineering practice) implies that each 0.01 Hz of frequency decrease corresponds to a power shortage of 3.33×10^{-4} per unit of generation. Second, for each generator g we set its frequency limits to 0.99 and 1.01 (p.u.), and we further assume that precontingency, each generator is operating at 90% of its maximum output.

Figure 6.17 shows the yield for controls of the form (6.21), assuming *perfect information*. Figure 6.17 is superficially similar to Figure 6.14. Notice, however, that the do-nothing control with $s = 0$ now proves very poor—the technical reason for this is that postcontingency, there are several islands with large generation deficits, and in the simulation all such generators trip. On the other hand, the controls with $s \approx 0.25$ appear promising (under these controls the resulting load shedding greatly reduces the load imbalances in islands at time zero). Also note the very noisy part of the figure for values of s between 0 and 0.1. In this range we see a mix of line and generator tripping events that are overlapping, creating complex race conditions (as termed in Section 5.1), which are being deterministically settled in the experiment shown in this figure. The total computational (wall-clock) time required to obtain this figure, using 48 parallel processes (running on a variety of Xeon cores), was 28 seconds.

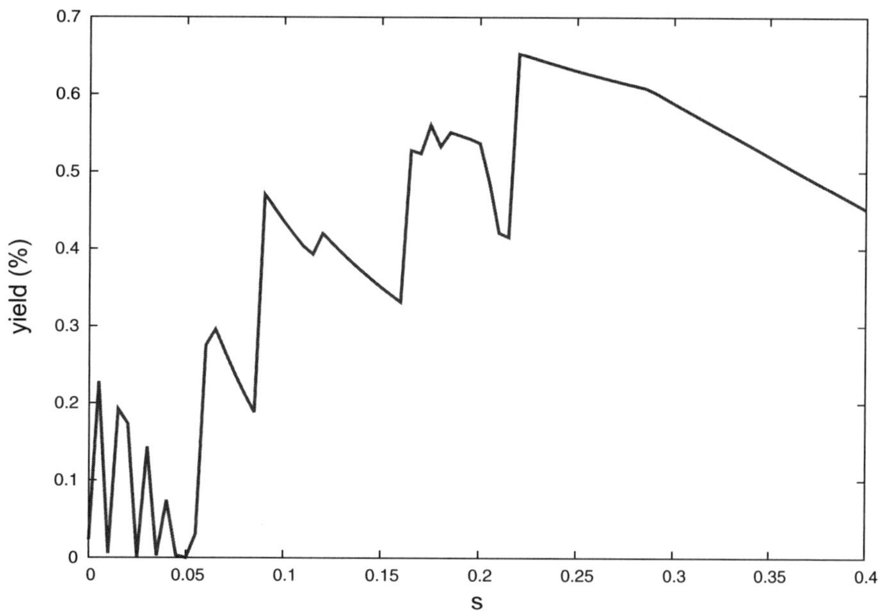

Figure 6.17. *Experiment* 6.2.10. *Yield as a function of the parameter s in control scheme* (6.21) *under frequency response.*

The figure suggests that controls with $s \approx 0.25$ should prove "robust." To investigate this possibility, we now apply the SAA framework introduced above. Given $s \geq 0$, let $u(s)$ denote the control (6.21), and let

$$U = \{u(0), u(0.19), u(0.22), u(0.24), u(0.26)\}.$$

We set $N = K = 100$; that is to say, we construct a total of 10^4 random skip sequences organized into 100 samples of size 100 each, constructed using $p_F = 0.1$. Denoting by σ_k the kth sample, for $1 \leq k \leq K$, we observe that

- the average value of the $T(U, \sigma_k) = 0.6006$;
- the standard deviation of the $T(U, \sigma_k) = 0.01679$;
- with 99% confidence, the interval $[0.5967, 0.6046]$ contains $T^*_{100}(U)$, an upper bound on $T^*(U)$; and
- direct simulation of the control $u(0.22)$ yields that a 99% confidence interval for a lower bound on $T^*(U)$ is given by $[0.5810, 0.6154]$.

The total (wall-clock) computational time expended, on 48 parallel processes, was approximately 6000 seconds, dominated by the simulations. The parallelization of the simulations will clearly scale, linearly, with the number of parallel processes.

We will now carry out a similar experiment but using a more sophisticated family of controls than those in (6.21).

Experiment 6.2.11.

In this experiment we again consider the Eastern Interconnection model used above, though here the initial consistency consists of 10 highly loaded lines. Additionally, at time zero each generator is assumed to be at 80% of capacity, and the generator frequency limits are 0.85 and 1.15. As before, every regulatory constant R_g is set to 0.05. Thus the setting is overall more lenient than that of Experiment 6.2.10. However, we now use the skip model (6.23) with $p_F = 0.5$, a much more conservative choice than the value 0.1 used for Experiment 6.2.10. We again use 10^4 random skip sequences and set $N = K = 100$.

First, we use a heuristic to determine a set of promising controls of the form (6.20a). We consider several families of cases of the structure

$$s^{(t)} = s\rho^t \quad \text{for } 1 \le t \le D, \quad s^{(t)} = 0 \quad \text{for } t > D$$

by enumerating a large number of combinations of parameters $s > 0$ and ρ (possibly negative) and values D. To guide the search, we first consider the state of the grid immediately postcontingency. The maximum relative line load is approximately $x = 1.5 \times 10^3$. Setting $s^{(1)} = x^{-1}$ yields $\lambda^{(1)} = 1 + x^{-1}(1-x) = x^{-1}$, immediately eliminating line overloads at the cost of very small yield. We thus consider a number of controls, both near and far from the choice $s^{(1)} = x^{-1}$:

$$s = 10^{-4} \times 1.5^k,\ 0 \le k \le 19, \quad \rho = 10^{-j},\ 0 \le j \le 5, \quad D = 10,$$
$$s = 10^{-3} \times 1.5^k,\ 0 \le k \le 19, \quad \rho = 2 \times 1.01^{-j},\ 0 \le j \le 5, \quad D = 10.$$

In this form we obtain several hundred controls including both ascending and descending values for the values $s^{(t)}$. We then proceed as follows:

(a) evaluate these controls deterministically (i.e., assuming perfect information). Choose the six controls that attain the highest yield.

(b) For each such control perform a local search that iteratively perturbs each value of $s^{(t)}$ and evaluates the perturbation by computing the average yield when simulated using a sample of 100 randomly generated skip sequences. This step was carried out 100 times.

At termination we have a set $U = \{u1, u2, u3, u4, u5, u6\}$ of controls. We then evaluate U using the SAA framework as in Experiment 6.2.10, using the same notation as there. We have

- average value of the $T(U, \sigma_k) = 0.4289$;
- standard deviation of the $T(U, \sigma_k) = 0.02243$;
- 99% confidence interval on $T^*_{100}(U)$, an upper bound on $T^*(U)$, of $[0.4236, 0.4344]$;
- by direct simulation of the control $u5$, a 99% confidence interval for a lower bound on $T^*(U)$ of $[0.4025, 0.43088]$; and
- the control $u5$ approximately of the form $s^{(t)} = 6.25 \times 10^{-7} \times 4^t$ for $1 \le t \le 7$ and $s^{(t)} = 0$ for $t > 7$.

We see that the (locally optimal) control $u5$ has an *ascending* structure. It does apply control at $t = 1$ (by shedding approximately 0.2% of the load, since $\lambda^1 = 1 + s^{(1)}(1 - x^1) \approx 0.998$) but uses stricter controls on later time periods. As indicated above, its expected yield is approximately 41%; in contrast the do-nothing control with $s^{(t)} = 0$ for all t always attains yield $\approx 6\%$, while the control that immediately stops the cascade at $t = 1$ attains yield $\approx 0.07\%$.

Discussion In this section we have presented a *heuristic* for selecting a finite family of controls on which the SAA method can be applied, by direct enumeration. A technique based on enumeration, local search, and a stochastically limited form of the SAA method (that is to say, one relying on smaller values of K and N) appears promising as a future area of investigation. On a computational platform with 48 cores, the entire computational procedure takes approximately two hours. On a supercomputer the time will be proportionally reduced, and if the values of K and N are reduced the reduction will clearly be greater and likely have little impact.

6.2.4 ▪ Summary

The above discussions focus on online, affine, load shedding algorithms, first presented in [30]. In a purely deterministic setting, an optimal control can be computed in polynomial time under appropriate assumptions. A robust version of the control scheme can be used in the nondeterministic setting, and assuming adequate computing resources a (locally) optimal robust control can be computed by relying on the SAA scheme.

6.3 ▪ The optimal switching problem

In the event of a contingency that could initiate a cascade, a viable alternative to load shedding is that of transmission line *switching*, i.e., the deliberate (controlled) tripping of a line or lines to obtain a stable solution. Transmission line switching has been considered in the literature in the context of generation cost optimization during normal, rather than emergency, situations. See [98], [119], [120], [118], [116]. An interesting extension is considered in [175]. In this context the problem, in its DC approximation, can be formulated as the following mixed-integer program:

$$\min \sum_{k \in \mathcal{G}} f_k(P_k) \tag{6.26a}$$

$$\text{s.t.} \sum_{\{k,m\} \in \delta(k)} p_{km} = P_k - D_k \quad \forall k, \tag{6.26b}$$

$$p_{km} = y_{km}(\theta_k - \theta_m) \quad \forall\, km, \tag{6.26c}$$

$$-p^{\max}_{\{k,m\}}(1 - z_{km}) \leq p_{km} \leq p^{\max}_{\{k,m\}}(1 - z_{km}) \quad \forall\, km, \tag{6.26d}$$

$$-M_{\{k,m\}} z_{\{k,m\}} \leq p_{km} - y_{\{k,m\}}(\theta_k - \theta_m) \leq M_{\{k,m\}} z_{\{k,m\}} \quad \forall\, km, \tag{6.26e}$$

$$z_{\{k,m\}} \in \{0,1\} \quad \forall\, km. \tag{6.26f}$$

In this formulation we use the notation $\delta(k)$ to indicate the set of lines incident with a bus k. The variables are p_{km} for each (oriented) line km; θ_k for each bus k; and for each generator k, P_k (the set of generators is denoted $\mathcal{G}$). Finally for each line km we have the binary variable $z_{\{k,m\}}$ for simplicity. The function f_k is a cost function (typically convex quadratic for standard OPF problems) and D_k is the load at k. Further, $p^{\max}_{\{k,m\}}$ is the limit of line $\{k,m\}$; as usual $y_{\{k,m\}}$ is the susceptance of $\{k,m\}$; and M_{km} is a big M constant, chosen large enough so that (6.26d)–(6.26f) correctly model line switching. This formulation is reminiscent of some we considered in the context of the N-K problem in Chapter 3; see, e.g., (3.28a)–(3.28b). In the form given above the significance of line switching is that it may reduce operating cost, thus giving rise to manifestations of Braess's paradox. When f_k is convex quadratic the above formulation can (nominally) be tackled by many solver systems.

Previous literature [98], [119] has considered the case with f_k linear, with instances of (6.26) (and variations thereof) tackled using commercial solvers. Often, instances of (6.26) are further parameterized by adding a cardinality constraint,

$$\sum_{km} z_{\{k,m\}} \leq \Gamma,$$

where Γ is an integer, and studying the impact of the choice of Γ on the optimal objective of (6.26). A limit on the number of switched lines is reasonable—from an operational perspective, line switching is a complex step that needs to be carefully implemented. In this context, an observation arising from the experiments has been that cost reductions in (6.26) are primarily attained with comparatively small values of Γ and with rapidly declining reductions as Γ is allowed to increase.

Another observation is that the running time entailed in solving (6.26) can be nontrivial. For example, [98] reports running times of several hours on the 118-bus case for values of $\Gamma \geq 9$ and several minutes when $\Gamma = 6$. In the context of OPF-like computations we would ideally obtain a near-optimal solution in subminute time, on much larger grids, with f_k quadratic (rather than linear) and potentially using larger values for Γ. We note, in passing, that the formulation above does rely on a big M construction. In Chapter 3 we discussed alternative ways of modeling line switching by relying on a (combinatorial) dual formulation that does not require using big M constants. However the literature for line switching has, so far, relied on variations of formulation (6.26).

From the perspective of this chapter, line switching would be a step used to attain stability. Thus we would alter formulation (6.26) by removing the objective and simply asking for a feasible solution. A possible variation would allow the operator to shed some load, in which case the problem involves simultaneously load shedding and line switching to attain stability. In any case, the computation would need to be very swift, even in the case of large grids. A basic question, as a result, concerns the basic complexity of problems like (6.26) or pure feasibility versions. In the rest of this chapter we will show that the problem is strongly NP-hard. This fact was stated, without proof, in [37]. Weak NP-hardness was recently proved in [141], and strong NP-hardness (and hardness of approximation) in [147]. Since the switching problem is now of growing interest, we provide a formal proof of this fact.

Formally, a *grid* is a quadruple $(G, x, p^{\max}, b)$, where $G = (N, E)$ is an undirected graph, and

- for each edge $\{k,m\} \in E$, we are given the reactance $x_{\{k,m\}} > 0$ and the limit $p^{\max}_{\{k,m\}} > 0$, and
- for each node $k \in N$, b_k is the *net generation* of k, we assume $\sum_k b_k = 0$.

Using this notation, the DC power flow problem is given by the following system of equations:

$$p_{km} = \frac{\theta_k - \theta_m}{x_{\{k,m\}}} = -p_{mk} \quad \forall \{k,m\} \in E, \tag{6.27}$$

$$|p_{km}| \leq p^{\max}_{\{k,m\}} \quad \forall \{k,m\} \in E, \tag{6.28}$$

$$\sum_{\{k,m\} \in \delta(k)} p_{km} = b_k \quad \forall k \in N, \tag{6.29}$$

which we denote by $DC(G, x, p^{\max}, b)$. As argued before, if this system is feasible, the p component of a solution is unique. If $R \subseteq E$, we denote by $G - R$ the graph with node-set

N and edge-set $E-R$, and by $DC(G-R, x, p^{\max}, b)$ the restriction of the above linear system to $G-R$. The feasibility line switching problem can now be stated as follows:

SWITCH: Is there a subset of edges R such that $DC(G-R, x, p^{\max}, b)$ is feasible?

Note that we assume that the b_k are given parameters. In terms of the above notation, $b_k = P_k - D_k$ and so we assume that both P_k and D_k are fixed. We will comment on this point later.

Theorem 6.3.1. *SWITCH is NP-complete.*

Proof. We will produce a polynomial transformation from Exact 3-Cover (X3C) [104]:

X3C: Given a set V and a set $T \subseteq V^3$, does there exist a subset $T' \subseteq T$ which partitions V?

Next we will describe our transformation from an instance of X3C given by sets V and T. We write $n = |V|$ and we assume $n \geq 3$, $|T| \geq 1$.

The graph $G = (N, E)$ is constructed as follows. For each $v \in V$ there is a node indicated by $\hat{v}$, for each $t \in T$ a node indicated by $\hat{t}$, and there are two additional nodes s and r. We have

- $b_s = n+1$, $b_r = -(n+1)$, and $b_k = 0$ for all other $k \in N$.

The edge-set E of G and the vectors x and $p^{\max}$ are set as follows. First, we have the following:

- For each $t \in T$ and each $u \in T$, E contains the edge $\{\hat{t}, \hat{u}\}$ with $x_{\{\hat{t},\hat{u}\}} = p^{\max}_{\{\hat{t},\hat{u}\}} = 1$.
- Let $t \in T$. Then E contains the edge $\{s, \hat{t}\}$ with $x_{\{s,\hat{t}\}} = 1/3$ and $p^{\max}_{\{s,\hat{t}\}} = 3$.
- Let $v \in V$. Then E contains the edge $\{\hat{v}, r\}$ with $x_{\{\hat{v},r\}} = 1$ and $p^{\max}_{\{\hat{v},r\}} = 1$.
- Finally, E contains the edge $\{s, r\}$, with $x_{\{s,r\}} = 3$ and $p^{\max}_{\{s,r\}} = 1$.

This concludes our construction. We will next prove the following claim.

Claim: There exists a set $R \subseteq E$ such that $DC(G-R, x, p^{\max}, b)$ is feasible if and only if the instance of X3C on (V, T) is feasible.

Clearly, establishing this claim will prove Theorem 6.3. As the first part of the proof, in Lemmas 6.3.2–6.3.6 we assume that we are given a set $R \subseteq E$ and a pair (p, θ) which is a feasible solution to $DC(G-R, x, p^{\max}, b)$ and construct an exact 3-cover $T' \subseteq T$. Without loss of generality we further assume $\theta_s = 0$.

Lemma 6.3.2. *We have that* $\{s, r\} \notin R$; $p_{sr} = 1$; *and for every* $v \in V$, $\{\hat{v}, r\} \notin R$ *and* $p_{\hat{v},r} = 1$.

Proof. This follows because $p^{\max}_e = 1$ for all $e \in \delta(r)$ and so

$$\sum_{e \in \delta(r)} p^{\max}_e = -b_r. \quad \blacksquare$$

Corollary 6.3.3. (i) $\theta_r = -3$ *and* (ii) *for every* $v \in V$, $\theta_{\hat{v}} = -2$.

Lemma 6.3.4. *Let $t \in T$ be such that $\{s,\hat{t}\} \notin R$, and suppose that for some $v \in T$, $\{\hat{t},\hat{v}\} \notin R$. Then* (i) $\theta_{\hat{t}} = -1$ *and* $p_{s\hat{t}} = 3$, *and* (ii) *for every $u \in T$, $\{\hat{t},\hat{u}\} \notin R$ and $p_{\hat{t}\hat{u}} = 1$.*

Proof. To prove (i) notice that if $\theta_{\hat{t}} < -1$, then since $x_{\{s,\hat{t}\}} = 1/3$ we will have $p_{s\hat{t}} > 3 = p^{\max}_{\{s,\hat{t}\}}$. Likewise if $\theta_{\hat{t}} > -1$, we will have by Corollary 6.3.3(ii) that $p_{\hat{t}v} > 1 = p^{\max}_{\{\hat{t},v\}}$. (ii) Follows from (i) and Corollary 6.3.3(ii). ■

Lemma 6.3.5. *Let $t \in T$ be such that $\{s,\hat{t}\} \in R$. Suppose that for some $v \in T$, $\{\hat{t},\hat{v}\} \notin R$. Then* (i) $\theta_{\hat{t}} = -2$ *and* (ii) $p_{\hat{t}\hat{u}} = 0$ *for each $u \in T$ such that $\{\hat{t},\hat{u}\} \notin R$.*

Proof. Write $X = \{u \in T : \{\hat{t},\hat{u}\} \notin R\}$ so that $v \in X$. If $\theta_{\hat{t}} \neq -2$, then $\theta_{\hat{t}} - \theta_{\hat{u}}$ is either positive for every $u \in X$ or negative for every $u \in X$, contradicting

$$0 = b_{\hat{t}} = \sum_{u \in X} p_{\hat{t}\hat{u}}.$$

Statement (ii) follows from Corollary 6.3.3(ii). ■

Lemma 6.3.6. *Let $v \in V$. Then there is a unique $t \in T$ with $v \in T$ and $\{s,\hat{t}\} \notin R$. In other words, the set $\{t \in T : \{s,\hat{t}\} \notin R\}$ is an exact 3-cover.*

Proof. Write $Y(v) = \{t \in T : v \in T, \{\hat{t},\hat{v}\} \notin R\}$. The result follows by noting that

$$1 = p_{\hat{v}r} = b_{\hat{v}} - \sum_{t \in Y(v)} p_{\hat{v}\hat{t}} = -\sum_{t \in Y(v)} p_{\hat{v}\hat{t}} = \sum_{t \in Y(v)} p_{\hat{t}\hat{v}}$$

and applying Lemma 6.3.4(ii) and Lemma 6.3.5(ii). ■

Now we turn to the converse. That is to say, we assume that there is an exact 3-cover $T' \subseteq T$ and construct from it a set $R \subseteq E$ such that $DC(G-R, x, p^{\max}, b)$ is feasible. To do so we simply set $R = \{\{s,\hat{t}\} : t \in T'\}$, and

- $\theta_s = 0$, $\theta_r = -3$;
- $\theta_v = -2$ for each $v \in V$; and
- for each $t \in T$ we set $\theta_t = -1$ if $t \in T'$ and $\theta_t = -2$ otherwise.

With these choices we will have that (p,θ) is feasible for $DC(G-R, x, p^{\max}, b)$, where the following hold:

- For each $t \in T$, if $t \in T'$, then $p_{s\hat{t}} = 3$ and $p_{\hat{t}\hat{u}} = 1$ for each $u \in T$. If $t \notin T'$, then $p_{s\hat{t}} = 0 = p_{\hat{t}\hat{u}}$ for each $u \in T$.
- $p_{sr} = 1$, and for each $v \in V$, $p_{\hat{v}r} = 1$.

These conditions are feasible since T' is an X3C. This concludes the NP-completeness proof. ■

Remarks

All reactances and line limits are simple numbers. There is only one generator and only one load. The cardinality of the set of edges to remove to make the system feasible is unique, i.e., $|T| - 2|V|/3$. The proof can easily be modified to accommodate cases where some limited load shedding is allowed, or where the output of generators can be adjusted.

On the basis of this result, and those in [98], [119], as well as the results we discussed in Chapter 3, it would appear that the solution of a line switching problem as a means to thwart a cascade may not yet be viewed as a viable idea. A point worth noting is that in the event of a contingency, line switching alone may be unable to attain a feasible solution, unless some load shedding is allowed. What may prove computationally feasible is the determination of a small number of lines to switch off, so that in the resulting grid, line overloads are significantly reduced, making load shedding controls such as those described in this chapter.

6.4 ▪ Toward optimal controls using derivative-free optimization

Here we will consider the case of island-wide scaling schemes discussed in Section 6.1.1, where at time t, if K is an island in existence at that point, then all loads in that island are scaled by the value

$$\lambda_K^{(t)} \doteq \min\{1, [1+s_K^{(t)}(1-x_K^t)]^+\}, \tag{6.30}$$

where $s^{(t)}$ is the control. This is a special case of the control scheme (6.13).

However, here we are interested in computing optimal controls of this type under the more general cascade models discussed above. This task is now a significantly more complex problem. In particular it is not clear that the assumptions used to obtain the results in Section 6.1.2, in particular Theorem 6.1.7, are valid.

Instead here we will highlight a different optimization methodology, namely derivative-free optimization. In particular we will carry out experiments using *mesh adaptive direct search* [10], as implemented in NOMAD [84]. This methodology addresses constrained optimization problems of the form

$$\min \ \mathcal{F}(x) \tag{6.31a}$$

$$\text{s.t.} \quad x \in \mathcal{B}, \tag{6.31b}$$

where $\mathcal{B} \subseteq \mathbb{R}^n$. In our use of NOMAD, $\mathcal{B}$ is simply a box-constrained set. As the name indicates, the methodology carries out a search based on enumerating regularly spaced mesh points in $\mathbb{R}^n$ (in our case, in $\mathcal{B}$). In the typical step a finite set of mesh points x^k is evaluated, that is to say the values $\mathcal{F}(x^k)$ are computed through a "black-box" interface. Additionally, the algorithm searches for new points (trial points) chosen from a progressively refined mesh. We refer the reader to [10] and [84] for background and citations. In our use of NOMAD, the black-box calls consist of calls to our cascade simulator.

An initial set of tests that we perform is as follows.

Experiment 6.4.1.

We use exactly the same setup as in Experiment 5.4.8, that is to say we use a model of the Eastern Interconnection, using an initial contingency made up of 55 outaged lines, and using deterministic line tripping. As detailed above the cascade terminates after 22 time-steps, with a yield of 58.30%. Two tests were performed:

- In one case, generator tripping is not allowed.
- In another case, every generator g has "regulation constant" $R_g = 0.01$ per unit. This constant was introduced in Model 5.5.3 to approximate frequency response.

Recall that this means that if an, e.g., freshly created island has load imbalance S (in gigawatts), and a total of N generators, then the change in frequency in that island will equal $\frac{S}{100N}$. Moreover, the minimum (resp., maximum) allowed frequency of a generator is 57Hz (resp., 63Hz) (p.u).

In both cases, we consider a cascade with an unlimited number of time-steps. However, the control (6.30) is restricted to the first five time-steps. In other words, the quantities $s^{(t)}$ for $1 \leq t \leq 5$ are the variables of the optimization problem. We assumed that $s^{(t)} \in [0,1]$ and that the initial vector for the search was given by $s^{(t)} = 0.5$, in both cases for $1 \leq t \leq 5$. In our runs NOMAD was allowed the (default) maximum number of 300 iterations.

The outcome of these studies was as follows. In the first case (no generator tripping allowed) the yield of the control computed by NOMAD was 83.35%, with the cascade terminating in 22 time-steps. The control itself is shown in Table 6.5.

Table 6.5. *Experiment* 6.4.1. *Control computed by NOMAD assuming no generator tripping.*

t	1	2	3	4	5
s^t	0.144	0.089	0.042	0.000	9.063e-08

Note that in this particular cascade, the maximum relative line load postcontingency is approximately 1.40. Using (6.30) and Table 6.5 we see that at $t = 1$ all loads are scaled by a factor of $1 + 0.144(1 - 1.4) \approx 0.94$.

In the second study (generator tripping allowed), the computed control attains a yield of 78.76% (with the cascade stopping in four time-steps) and is shown in Table 6.6.

Table 6.6. *Experiment* 6.4.1. *Control computed by NOMAD allowing generator tripping.*

t	1	2	3	4	5
s^t	0.443	0.134	0.364	0.259	0.850

This control, in contrast with that shown on Table 6.5, scales all loads at $t = 1$ by the factor $1 + 0.443(1 - 1.4) \approx 0.82$; in other words it is more aggressive and it maintains that trend across all time periods.

A final, interesting test is to evaluate the "naive" control in Table 6.5 under the setup for the second experiment. It is confirmed that the control is in fact inadequate: the resulting yield is zero, with large numbers of generators tripping at various points of the cascade. The reverse experiment is also of interest: we evaluate the "frequency aware" control in Table 6.6 under the first setup, i.e., no generator tripping. The result is that yield is unchanged at 78.76%, a natural consequence of the control aggressively reducing line overloads. Table 6.7 illustrates the comparison between the two controls under generator frequency tripping.

In Table 6.7, "starting max. rel. line load" is the maximum relative line load at the start of each time-step, "ending yield" is the yield at the end of a time-step, and "shed load" is the load shed as per control (6.30). The table shows that the frequency-aware control sheds much more load in the first time-step but much less thereafter. Perhaps surprisingly, the naive control maintains high yield through time-step 5; however, there is a collapse at time-step 6. This fact illustrates the complexity of the control scheme we are considering: it is computed at time 0, yet it must be aware of conditional events that may happen considerably into the future.

Table 6.7. *Experiment* 6.4.1. *Comparison between naive and frequency-aware controls under frequency tripping.*

t	1	2	3	4	5	6
starting max. rel. line load, **naive control**	1.41	1.53	2.04	4.66	2.09	9.62
starting max. rel. line load, **freq.-aware control**	1.41	1.17	1.04	0.99	–	–
naive control: line trips,	18	19	32	87	28	45
generator trips	0	89	4	2	33	4
freq.-aware control: line trips	3	3	1	0	–	–
generator trips	0	0	0	0	–	–
ending yield (%), **naive control**	94.08	89.60	85.57	83.14	83.14	0.00
ending yield (%), **freq.-aware control**	81.73	79.88	78.77	78.77	–	–
shed load (%), **naive control**	6.00	4.48	3.39	0.00	0.00	0.00
shed load (%), **freq.-aware control**	18.27	1.85	1.11	0.00	–	–

6.5 ▪ Postscript. Fast computation of optimal, robust controls

An interesting remark, in view of the results of Experiment 6.4.1, concerns the *running time* of NOMAD. The above runs were performed using one current core (Xeon E5-2667, 3.30GHz), and in both cases the running time was on the order of five minutes. Given the size of the network under consideration, i.e., roughly 23,000 lines, the results can therefore be viewed as highly encouraging.

However, the reader should keep in mind the restricted scope of the experiment, primarily the small number of time-steps under consideration. To properly consider all features described above, in particular how the small time-scale generator ramp-up model interacts with the line tripping model, one would need to use the limited duration cascade Model 5.5.7 with small Δt, which would lead to a much larger number of time-steps. We hope to take up these issues in future work.

Appendix A

Glossary

$\delta(v)$, $\delta(v,G)$ **:** Where v is a node of graph G, these expressions indicate the set of edges incident with v. Section 1.2

η_e **:** Hysteresis parameter for line e used to introduce memory in line tripping models. Section 5.3.2.

x_v^t **:** In the context of a cascade, the maximum relative line overload, at time t, in the island (component) containing bus v.

$\succeq$ **relation:** Given two symmetric matrices A and B we say $A \succeq B$ if $A-B \succeq 0$, i.e., every eigenvalue of $A-B$ is nonnegative. Section 2.2.1.

Π-model: Simplified model of a power line, suitable for power flow computations. Section 1.2.

• operator: Also known as the Frobenius inner product, or "matrix dot product," operator. Given two real-valued $m \times n$ matrices A and B, we define $A \bullet B = \sum_{i,j} a_{ij} b_{ji} =$ trace(AB^T). Section 2.2.1.

∘ operator: Also known as the Hadamard product. Given two real-valued $m \times n$ matrices A and B, $A \circ B$ is the $m \times n$ matrix whose (i,j) entry equals $a_{ij} b_{ij}$. Section 3.1.1.

Active power: The real part of complex power. Corresponds to intuitive notions of "power" (energy produced or consumed per unit time). For a line $\{k,m\}$, the active power injected into the line at k is denoted p_{km}. Section 1.2.

ACE or area control error: A measure of load/generation imbalance in a control area. Section 1.4.

Admittance: For a line $\{k,m\}$, its (series) admittance is the complex number $y_{\{k,m\}} = z_{\{k,m\}}^{-1} = g_{\{k,m\}} + j\, b_{\{k,m\}}$, where $z_{\{k,m\}}$ is the series impedance. The quantity $g_{\{k,m\}}$ is the conductance and $b_{\{k,m\}}$ is the susceptance. Section 1.2.

Benders decomposition: Decomposition methodology for optimization problems that seeks to leverage problem structure. Section 3.1.1.

Bilinear expression: A bilinear expression in an optimization problem is of the form $x_1 x_2$, where each x_i is a variable. Section 3.1.1.

Conductance: Physical attribute of conductors; the real component of admittance. The conductance of a line $\{k,m\}$ is denoted as $g_{\{k,m\}}$. See definition of admittance. Section 1.2.

cav$_P[f]$: Convex overestimator of function f over the polytope P. Section 3.1.1.

DC approximation: Linearization of power flow problems. Section 1.2.4.

Derivative-free methods: Methodologies for optimization problems where derivative information is unavailable; often involving the minimization of functions which are only available through computationally expensive procedures, such as simulations. See [72]. Section 3.3.2.
Effective line load, $\tilde{f}_e^t$ **:** For a line e at time t of a simulated cascade, an estimation of line load that takes into account the history of flows on the line. Section 5.3.2
Effective line relative load: For a line e, with limit u_e, at time t of a simulated cascade, the quantity $\tilde{f}_e^t / u_e$, where $\tilde{f}_e^t$ is the effective line load. Section 5.3.2
Filter method: Method for constrained optimization that effectively constructs a bijective optimization problem which seeks to simultaneously minimize objective value and constraint violations, while relying on trust-region methodology. See [100], [170]. Section 2.1.1.
Frequency: In a transmission system in steady state all physical quantities (voltages, current levels, power flows) are sinusoidally time-varying quantities with (nearly) the same frequency.
Homotopy method: Method for computing a zero of a function $F(x)$ by computing zeros of the family of functions $F(x,t) \doteq tF(x) + (1-t)F_0(x)$ where $0 \le t \le 1$ and $F_0(x)$ is an auxiliary function for which a zero is known. Given a zero for $F(x, \hat{t})$ with $\hat{t} < 1$, a *continuation* method may be used to compute a zero for $F(x, \hat{t} + \delta)$ with small $\delta > 0$. See [77] for background related to polynomial equations. Section 2.1.
Impedance: For a line $\{k,m\}$, its (series) impedance is the complex number $z_{\{k,m\}} = r_{\{k,m\}} + j x_{\{k,m\}}$, where $r_{\{k,m\}}$ is the resistance and $n_{\{k,m\}}$ is the reactance. Section 1.2.
Islanding: The separation of a transmission system into two or more disconnected components. Section 1.4.
Line switching problem: Mixed-integer optimization problem which seeks to compute a set of transmission lines to switch off to improve grid operation. Section 6.3.
Maximum severity attack problem: Optimization problem in which a hypothetical adversary seeks to maximize the degradation of a transmission system through a constrained attack. Section 3.1.
McCormick relaxation: Method for convexifying a bilinear expression. Section 3.1.1.
Minimum effort attack problem: Optimization problem in which a hypothetical adversary seeks to degrade a transmission system down to a desired level, at minimum effort. Section 3.1.
MIP: Mixed-integer program. Optimization problem involving continuous and integer-valued variables. Chapter 3.
Moment relaxation: Generalization of SDP relaxation of use in polynomial optimization problems. Section 2.2.3.
N-1 criterion: Requirement stipulating that a transmission system must survive the loss of any one element (line, generator, transformer, etc.). Section 1.3.1.
OPF problem: Optimization version of power flow problem, typically consisting of computing a solution that minimizes cost of generation subject to delivering desired levels of power (active and reactive) to loads. Section 1.2.3.
Per-unit system (p.u.): Numerical scaling system for physical quantities such as voltages. Section 1.2.1.
Power flow problem: The computation of a solution to a system of nonlinear constraints describing power flows on a grid. Section 1.2.2.
Primary control: Part of the three tiers of primary, secondary, and tertiary control that are used to maintain frequency stability. Section 1.4.1.

QCQP: Quadratically constrained quadratic objective optimization problem, possibly nonconvex. Section 2.2.1.
Reactance: Physical attribute of conductors; the imaginary component of the impedance of a line. The reactance of a line $\{k,m\}$ is denoted as $x_{\{k,m\}}$. Section 1.2.
Reactive power: The imaginary part of complex power. Local shortages of reactive power are associated with decaying voltage levels. For a line $\{k,m\}$, the reactive power injected into the line at k is denoted q_{km}. Section 1.1.
Relative line load: Given a line with limit u and (active) power flow f, the relative line load is $|f|/u$. Section 4.3.1.
Resistance: Physical attribute of conductors; the real component of the impedance of a line. The resistance of a line $\{k,m\}$ is denoted as $r_{\{k,m\}}$. Section 1.1.
SAA or sample average approximation method: Body of techniques for optimization over uncertainty that draw large samples of representative data and optimize the average behavior over the samples. Section 3.4.1.
SDP, or semidefinite program: Convex optimization problem with variables arranged into matrices that are required, as a constraint, to be positive semidefinite. Section 2.2.1.
SDP relaxation: Technique that obtains, given a QCQP, a (convex) semidefinite programming relaxation. Section 2.2.1.
Secondary control: Part of the three tiers of primary, secondary, and tertiary control that are used to maintain frequency stability. Section 1.4.1.
SOCP: Second-order cone program. Convex optimization problem of the form

$$\begin{aligned} \min_{x \in \mathbb{R}^n} \quad & f^T x \\ \text{s.t. } & \|A_i x + b_i\| \le c_i^T x + d_i, \quad 1 \le i \le m. \end{aligned}$$

Section 2.2.1.
Stochastic programming: Broad family of mathematical techniques for optimization under uncertainty. Section 3.4.1.
Susceptance: Physical attribute of conductors; imaginary component of admittance. The susceptance of a line $\{k,m\}$ is denoted as $b_{\{k,m\}}$. See definition of admittance. Section 1.2.
Swing equation: Differential equation describing the relationship between the angular speed of a rotor (e.g., a motor) and the difference between generated and consumed power. Section 1.3.1.
Tertiary control: Part of the three tiers of primary, secondary, and tertiary control that are used to maintain frequency stability. Section 1.4.1.
Thermal loss: A transmission line with positive resistance will incur thermal losses of power. Section 1.3.
Tripping probability function: Given a transmission line e, with relative load x, probability that the line will trip during a period of time length Δt, denoted $\Pi_e^{(\Delta t)}(x)$. Section 5.3.2.
Trust-region method: Method for optimization that approximates an objective function with a convex function within a ball and sequentially adjusts the ball and the approximation to converge to a (possibly local) optimum. Section 2.2.1.
Trust-region subproblem: Minimization of a quadratic function over an ellipsoid, often arising as a step in trust-region methods for general optimization. Section 2.2.1.
vex$_P[f]$: Convex underestimator of function f over the polytope P. Section 3.1.1.
Yield: In a simulated cascade or contingency, proportion of demand that is being delivered at cascade termination or postcontingency. Section 3.1, Chapter 6.

Appendix B

Complete power flow derivations

Here we provide complete derivations for power flow equations to account for transformers in a common setting, extending the presentation in Section 1.2. We will also allow nonzero shunt susceptances. The resulting equations are rather more complex, and full derivations are not often included in power engineering textbooks. The material here is adapted from [41].

We consider a line (k,m), but to simplify notation we will drop the indices k and m. Thus, the series impedance is $z = r + jx$ and the series admittance is $y = z^{-1} = g + jb$, where $g = r/(r^2+x^2)$ and $b = -x/(r^2+x^2)$. Likewise, there is a shunt admittance $y^{sh} = g^{sh} + jb^{sh}$ associated with the line. Finally, there is a *transformer* with *ratio*

$$N \doteq \tau e^{j\sigma},$$

where $\tau > 0$ is the *magnitude* and σ is the *phase shift angle*. To understand how transformers work, consider Figure B.1, which is a more general version of Figure 1.1. In Figure B.1 the transformer is placed next to bus k (other conventions/models are possible); the transformer scales voltages by a factor of $1/N$ and currents by a factor of N^* (the complex conjugate of N). Thus, the voltage at point k_1 equals V_k/N, and the current injected into the line at k_1 equals $N^* I_{km}$. Further, between buses k_1 and m we have a standard Π-model. Using this insight, we obtain

$$I_{km} = \frac{1}{\tau} y \left[\frac{1}{\tau} V_k - e^{j\sigma} V_m \right] + \frac{1}{2\tau^2} y^{sh} V_k \tag{B.1}$$

$$= \frac{1}{\tau} y \left[\frac{1}{\tau} V_k - (\cos\sigma + j\sin\sigma) V_m \right] + \frac{1}{2\tau^2} (g^{sh} + j b^{sh}) V_k. \tag{B.2}$$

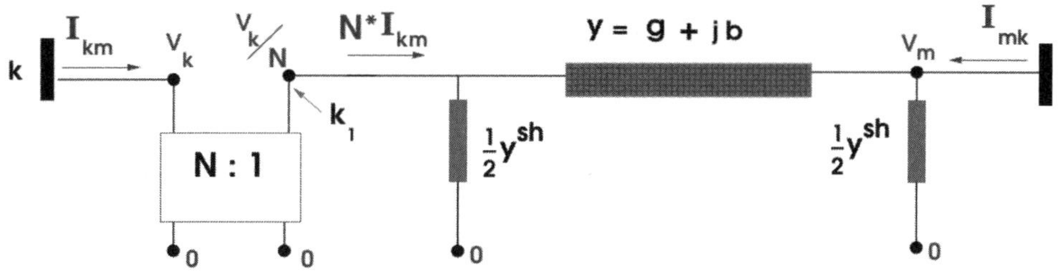

Figure B.1. *π-model, including transformer and shunt admittance.*

In rectangular coordinates (where each bus i has voltage $V_i = e_i + jf_i$) this can be further expanded as

$$\begin{aligned}
&I_{km} \\
&= \frac{1}{\tau}(g+jb)\left[\frac{1}{\tau}(e_k+jf_k)-(\cos\sigma+j\sin\sigma)(e_m+jf_m)\right]+\frac{1}{2\tau^2}(g^{sh}+jb^{sh})(e_k+jf_k) \\
&= \frac{1}{\tau}(g+jb)\left[\frac{1}{\tau}(e_k+jf_k)-e_m\cos\sigma+f_m\sin\sigma-j(e_m\sin\sigma+f_m\cos\sigma)\right] \\
&+\frac{1}{2\tau^2}\left[g^{sh}e_k-b^{sh}f_k+j(b^{sh}e_k+g^{sh}f_k)\right] \\
&= \frac{1}{\tau}(g+jb)\left[\frac{1}{\tau}e_k-e_m\cos\sigma+f_m\sin\sigma+j\left(\frac{1}{\tau}f_k-e_m\sin\sigma-f_m\cos\sigma\right)\right] \\
&+\frac{1}{2\tau^2}\left[g^{sh}e_k-b^{sh}f_k+j(b^{sh}e_k+g^{sh}f_k)\right].
\end{aligned} \tag{B.3}$$

From this expression we obtain

$$\begin{aligned}
\operatorname{Re} I_{km} = {} & \frac{g}{\tau}\left[\frac{e_k}{\tau}-e_m\cos\sigma+f_m\sin\sigma\right] \\
& -\frac{b}{\tau}\left[\frac{1}{\tau}f_k-e_m\sin\sigma-f_m\cos\sigma\right] \\
& +\frac{1}{2\tau^2}\left[g^{sh}e_k-b^{sh}f_k\right],
\end{aligned} \tag{B.4}$$

$$\begin{aligned}
\operatorname{Im} I_{km} = {} & \frac{b}{\tau}\left[\frac{e_k}{\tau}-e_m\cos\sigma+f_m\sin\sigma\right] \\
& +\frac{g}{\tau}\left[\frac{1}{\tau}f_k-e_m\sin\sigma-f_m\cos\sigma\right] \\
& +\frac{1}{2\tau^2}\left[b^{sh}e_k+g^{sh}f_k\right].
\end{aligned} \tag{B.5}$$

We can see that in the no-transformer case, i.e., $\tau = 1$ and $\sigma = 0$, with $g^{sh} = 0$, (B.4) and (B.5) match the expansion (1.19) for I_{km}, as desired. Further,

$$\begin{aligned}
p_{km} &= \operatorname{Re} V_k I_{km}^* \\
&= e_k\left\{\frac{g}{\tau}\left[\frac{e_k}{\tau}-e_m\cos\sigma+f_m\sin\sigma\right]-\frac{b}{\tau}\left[\frac{1}{\tau}f_k-e_m\sin\sigma-f_m\cos\sigma\right]\right\} \\
&+f_k\left\{\frac{b}{\tau}\left[\frac{e_k}{\tau}-e_m\cos\sigma+f_m\sin\sigma+\right]+\frac{g}{\tau}\left[\frac{1}{\tau}f_k-e_m\sin\sigma-f_m\cos\sigma\right]\right\} \\
&+\frac{g^{sh}}{2\tau^2}(e_k^2+f_k^2).
\end{aligned} \tag{B.6}$$

This can be rewritten as

$$\begin{aligned}
p_{km} = {} & \frac{1}{\tau}\left[\frac{e_k}{\tau}-e_m\cos\sigma\right](g\,,\,b)\binom{e_k}{f_k}+\frac{1}{\tau}\left[\frac{1}{\tau}f_k-f_m\cos\sigma\right](-b\,,\,g)\binom{e_k}{f_k} \\
& +\frac{g^{sh}}{2\tau^2}(e_k^2+f_k^2) \\
& +\frac{ge_kf_m+be_ke_m+bf_kf_m-gf_ke_m}{\tau}\sin\sigma.
\end{aligned} \tag{B.7}$$

In the no-transformer case this expression evaluates to

$$(e_k - e_m)(g\,,\,b)\binom{e_k}{f_k} + (f_k - f_m)(-b\,,\,g)\binom{e_k}{f_k} + \frac{g^{sh}}{2}(e_k^2 + f_k^2), \tag{B.8}$$

which is the same as (1.19) when $g^{sh} = 0$. Note that in (B.7) the third term vanishes when there is no transformer, and the second term vanishes when there is no shunt susceptance. Next we will compute an expression for p_{mk}. In the transformer case the line is not symmetric and we first need to compute I_{mk}. We have

$$I_{mk} = -\frac{1}{\tau e^{j\sigma}} y V_k + \left(y + \frac{y^{sh}}{2} \right) V_m \tag{B.9}$$

$$\begin{aligned}
&= -\frac{1}{\tau}(\cos\sigma - j\sin\sigma)(g + jb)(e_k + jf_k) \\
&\quad + \left(g + g^{sh}/2 + j(b + b^{sh}/2)\right)(e_m + jf_m) \\
&= -\frac{1}{\tau}(g + jb)[e_k\cos\sigma + f_k\sin\sigma + j(-e_k\sin\sigma + f_k\cos\sigma)] \\
&\quad + (g + g^{sh}/2)e_m - (b + b^{sh}/2)f_m + j\left[(b + b^{sh}/2)e_m + (g + g^{sh}/2)f_m\right].
\end{aligned} \tag{B.10}$$

Therefore,

$$\begin{aligned}
\operatorname{Re} I_{mk} &= -\frac{g}{\tau}[e_k\cos\sigma + f_k\sin\sigma] + \frac{b}{\tau}[-e_k\sin\sigma + f_k\cos\sigma] \\
&\quad + (g + g^{sh}/2)e_m - (b + b^{sh}/2)f_m,
\end{aligned} \tag{B.11}$$

$$\begin{aligned}
\operatorname{Im} I_{mk} &= -\frac{g}{\tau}[-e_k\sin\sigma + f_k\cos\sigma] - \frac{b}{\tau}[e_k\cos\sigma + f_k\sin\sigma] \\
&\quad + (b + b^{sh}/2)e_m + (g + g^{sh}/2)f_m.
\end{aligned} \tag{B.12}$$

Thus,

$$\begin{aligned}
p_{mk} &= \operatorname{Re} V_m I_{mk}^* \\
&= e_m\left\{-\frac{g}{\tau}[e_k\cos\sigma + f_k\sin\sigma] + \frac{b}{\tau}[-e_k\sin\sigma + f_k\cos\sigma]\right\} \\
&\quad + f_m\left\{-\frac{g}{\tau}[-e_k\sin\sigma + f_k\cos\sigma] - \frac{b}{\tau}[e_k\cos\sigma + f_k\sin\sigma]\right\} \\
&\quad + (g + g^{sh}/2)(e_m^2 + f_m^2) \\
&= \left[e_m - \frac{1}{\tau}e_k\cos\sigma\right](g\,,\,b)\binom{e_m}{f_m} + \left[f_m - \frac{1}{\tau}f_k\cos\sigma\right](-b\,,\,g)\binom{e_m}{f_m} \\
&\quad + \frac{g^{sh}}{2}(e_m^2 + f_m^2) \\
&\quad + \frac{-g e_m f_k - b f_m f_k - b e_m e_k + g f_m e_k}{\tau}\sin\sigma \\
&= \left[e_m - \frac{1}{\tau}e_k\cos\sigma - \frac{1}{\tau}f_k\sin\sigma\right](g\,,\,b)\binom{e_m}{f_m} \\
&\quad + \left[f_m - \frac{1}{\tau}f_k\cos\sigma + \frac{1}{\tau}e_k\sin\sigma\right](-b\,,\,g)\binom{e_m}{f_m} + \frac{g^{sh}}{2}(e_m^2 + f_m^2).
\end{aligned} \tag{B.13}$$

We now turn to the representation of p_{km} and p_{mk} using polar coordinates:

$$\begin{aligned}
S_{km} = V_k I_{km}^* &= V_k \left(\frac{1}{\tau^2} V_k^* - \frac{1}{\tau} e^{-j\sigma} V_m^* \right) y^* + \frac{1}{2\tau^2}(g^{sh} - j b^{sh}) V_k V_k^* \\
&= V_k \left(\frac{1}{\tau^2} V_k^* - \frac{1}{\tau} e^{-j\sigma} V_m^* \right)(g - jb) + \frac{1}{2\tau^2}(g^{sh} - j b^{sh}) V_k V_k^* \\
&= |V_k|^2 \frac{g}{\tau^2} - |V_k||V_m| \frac{g}{\tau} \cos(\theta_{km} - \sigma) - |V_k||V_m| \frac{b}{\tau} \sin(\theta_{km} - \sigma) + \frac{g^{sh}}{2\tau^2} |V_k|^2 \\
&\quad + j \left[-|V_k|^2 \frac{b}{\tau^2} + |V_k||V_m| \frac{b}{\tau} \cos(\theta_{km} - \sigma) - |V_k||V_m| \frac{g}{\tau} \sin(\theta_{km} - \sigma) - \frac{b^{sh}}{2\tau^2} |V_k|^2 \right].
\end{aligned} \tag{B.14}$$

Similarly,

$$\begin{aligned}
S_{mk} &= |V_m|^2 g - |V_k||V_m| \frac{g}{\tau} \cos(\theta_{mk} + \sigma) - |V_k||V_m| \frac{b}{\tau} \sin(\theta_{mk} + \sigma) + \frac{g^{sh}}{2} |V_m|^2 \\
&\quad + j \left[-|V_m|^2 b + |V_k||V_m| \frac{b}{\tau} \cos(\theta_{mk} + \sigma) - |V_k||V_m| \frac{g}{\tau} \sin(\theta_{mk} + \sigma) - \frac{b^{sh}}{2} |V_m|^2 \right].
\end{aligned} \tag{B.15}$$

Hence the active power loss equals

$$\begin{aligned}
&p_{km} + p_{mk} = \\
&\left(\frac{|V_k|^2}{\tau^2} + |V_m|^2 \right) g - 2g \frac{|V_k|}{\tau} |V_m| \cos(\theta_{km} - \sigma) + \frac{g^{sh}}{2\tau^2} |V_k|^2 + \frac{g^{sh}}{2} |V_m|^2.
\end{aligned} \tag{B.16}$$

We can directly derive this equation by considering point k_1 in Figure B.1. The power injection into the line, at k_1, is equal to p_{km} (i.e., it equals $\frac{V_k}{N}(N * I_{km})^* = V_k I_{km}^* = p_{km}$). Moreover, by construction, the voltage magnitude at k_1 equals $|V_k|/\tau$ and the phase angle difference from k_1 to m equals $\theta_{km} - \sigma$. We can now recover (B.16) from (1.12), with the last two terms accounting for shunts.

Bibliography

[1] IEEE Power Engineering Society Power System Dynamic Performance Committee. Causes of the 2003 major grid blackouts in North America and Europe, and recommended means to improve system dynamic performance. In *IEEE Power Engineering Society General Meeting*, 2004. (Cited on pp. 137, 139)

[2] S. Ahmed. *Strategic planning under uncertainty: Stochastic programming approaches*. PhD thesis, University of Illinois at Urbana-Champaign, 2000. (Cited on p. 250)

[3] R.K. Ahuja, T.L. Magnanti, and J.B. Orlin. *Network flows: Theory, algorithms, and applications*. Prentice-Hall, 1993. (Cited on pp. 15, 17, 106, 110, 128)

[4] W. Ai and S. Zhang. Strong duality for the CDT subproblem: A necessary and sufficient condition. *SIAM Journal on Optimization*, 19:1735–1756, 2009. (Cited on p. 36)

[5] F. Al-Khayyal and J. Falk. Jointly constrained biconvex programming. *Mathematics of Operations Research*, 8:273–286, 1983. (Cited on p. 97)

[6] F. Alvarado and S.S. Oren. Transmission system operation and interconnection. *U.S. Department of Energy national transmission grid study papers—issue papers*, 2002. `http://certs.lbl.gov`. (Cited on p. 24)

[7] G. Andersson. *Dynamics and control of electric power systems*. Power Systems Laboratory, ETH Zürich, 2003. (Cited on p. 26)

[8] G. Andersson. *Modelling and analysis of electric power systems*. Power Systems Laboratory, ETH Zürich, 2004. (Cited on pp. 1, 4, 26, 27, 29, 145, 179, 182, 212)

[9] M. Anghel, K.A. Werley, and A.E. Motter. Stochastic model for power grid dynamics. *Proceedings of the 40th Hawaii International Conference on System Sciences*, 2007. (Cited on pp. 152, 153, 154, 155, 156, 157, 158, 161, 162, 177)

[10] C. Audet and J.E. Dennis Jr. Mesh adaptive direct search algorithms for constrained optimization. *SIAM Journal on Optimization*, 17:188–217, 2006. (Cited on p. 267)

[11] E. Balas and R. Jeroslow. Canonical cuts on the unit hypercube. *SIAM Journal on Applied Mathematics*, 23:61–69, 1972. (Cited on p. 82)

[12] Ross Baldick. *DC power flow in rectangular coordinates*. Manuscript, University of Texas, 2013. (Cited on p. 52)

[13] J.F. Bard and J.T. Moore. A branch and bound algorithm for the bilevel programming problem. *SIAM Journal on Scientific and Statistical Computing*, 11:281–292, 1990. (Cited on p. 76)

[14] J. Bard and J. Moore. An algorithm for the discrete bilevel programming problem. *Naval Research Logistics*, 39:419–435, 1991. (Cited on p. 76)

[15] A.I. Barvinok. Feasibility testing for systems of real quadratic equations. *Discrete Computational Geometry*, 10:1–13, 1993. (Cited on pp. 38, 39)

[16] A.I. Barvinok. Problems of distance geometry and convex properties of quadratic maps. *Discrete Computational Geometry*, 13:189–202, 1995. (Cited on p. 40)

[17] A.I. Barvinok. A remark on the rank of positive semidefinite matrices subject to affine constraints. *Discrete Computational Geometry*, 25:23–31, 2001. (Cited on p. 40)

[18] A.I. Barvinok. Personal communication, 2014. (Cited on p. 39)

[19] A. Beck and Y.C. Eldar. Strong duality in nonconvex quadratic optimization with two quadratic constraints. *SIAM Journal on Optimization*, 17:844–860, 2006. (Cited on p. 36)

[20] C. Beltrán and L.M. Pardo. On Smale's 17th problem: a probabilistic positive answer. *Foundations of Computational Mathematics*, 8(1):1–43, 2008. (Cited on p. 32)

[21] C. Beltrán and L.M. Pardo. On Smale's 17th problem: Average polynomial time to compute affine and projective solutions. *Journal of the American Mathematical Society*, 22:363–385, 2009. (Cited on p. 32)

[22] J.F. Benders. Partitioning procedures for solving mixed-variables programming problems. *Numerische Mathematik*, 4(1):238–252, 1962. (Cited on p. 78)

[23] S.J. Benson, Y. Ye, and X. Zhang. Solving large-scale sparse semidefinite programs for combinatorial optimization. *SIAM Journal on Optimization*, 10:443–461, 2000. (Cited on p. 47)

[24] R. Bent, D. Bienstock, and M. Chertkov. Synchronization-aware and algorithm-efficient chance constrained optimal power flow. In *Proceedings 2013 IREP Symposium-Bulk Power System Dynamics and Control*, 2013. (Cited on pp. 48, 49, 50)

[25] A.R. Bergen and V. Vittal. *Power systems analysis*. Prentice-Hall, 1999. (Cited on pp. 1, 12, 26, 27, 141, 179, 182, 212)

[26] D. Bertsimas, E. Litvinov, X.A. Sun, J. Zhao, and T. Zheng. Adaptive robust optimization for the security constrained unit commitment problem. *IEEE Transactions on Power Systems*, 28(1):52–63, February 2013. (Cited on p. 57)

[27] D. Bienstock. Histogram models for robust portfolio optimization. *Journal of Computational Finance*, 11:1–64, 2007. (Cited on p. 94)

[28] D. Bienstock. Optimal adaptive control of cascading power grid failures. Preprint, arXiv:1012.4025, 2010. (Cited on pp. 158, 159, 167, 170)

[29] D. Bienstock. Adaptive online control of cascading blackouts. In *IEEE Power Engineering Society General Meeting*, 2011. (Cited on pp. 158, 159, 167, 170)

[30] D. Bienstock. Optimal control of cascading power grid failures. In 50*th IEEE Conference on Decision and Control and European Control Conference*, pages 2166–2173, 2011. (Cited on pp. 158, 159, 162, 167, 170, 171, 234, 242, 245, 263)

[31] D. Bienstock. A note on polynomial solvability of the CDT problem. Preprint, arXiv:1406.6429, 2013. (Cited on p. 40)

[32] D. Bienstock, A. Bernstein, D. Hay, M. Uzunoglu, and G. Zussman. Sensitivity analysis of power grid vulnerability to large-scale cascading failures. In *ACM Performance Evaluation Review, Special issue of papers from ACM Greenmetrics*'12, volume 40, pages 33–37, 2012. (Cited on p. 162)

[33] D. Bienstock, A. Bernstein, D. Hay, M. Uzunoglu, and G. Zussman. Power grid vulnerability to geographically correlated failures—analysis and control implications. In *Proceedings of INFOCOM* '14, 2014. (Cited on p. 162)

[34] D. Bienstock, J. Blanchet, and J. Li. Stochastic models and control for electrical power line temperature. In *Proceedings of* 51*st Annual Allerton Conference*, pages 1344–1348, 2013. (Cited on pp. 155, 177)

[35] D. Bienstock, M. Chertkov, and S. Harnett. Chance-constrained optimal power flow: Risk-aware network control under uncertainty. *SIAM Review*, 56:461–495, 2014. (Cited on p. 134)

[36] D. Bienstock and N. Diaz. Blocking small cuts in a network, and related problems. *SIAM Journal on Computing*, 22:482–499, 1993. (Cited on p. 76)

[37] D. Bienstock and S. Mattia. Using mixed-integer programming to solve power grid blackout problems. *Discrete Optimization*, 4:115–141, 2007. (Cited on pp. 126, 151, 264)

[38] D. Bienstock and B. McClosky. Tightening simple mixed-integer sets with guaranteed bounds. *Mathematical Programming*, 133:337–363, 2012. (Cited on p. 88)

[39] D. Bienstock and A. Michalka. Polynomial solvability of variants of the trust-region subproblem. In *Proceedings of the* 25*th Annual ACM-SIAM Symposium on Discrete Algorithms (SODA* 2014), pages 380–390, 2014. (Cited on p. 38)

[40] D. Bienstock and G. Muñoz. LP approximations to mixed-integer polynomial optimization problems. Preprint, arXiv:1501.00288, 2014. (Cited on p. 47)

[41] D. Bienstock and G. Muñoz. On linear relaxations of OPF problems. Preprint, arXiv:1411.1120, 2014. (Cited on p. 275)

[42] D. Bienstock and A. Verma. The N-k problem in power grids: New models, formulations, and numerical experiments. *SIAM Journal on Optimization*, 20:2352–2380, 2010. (Cited on pp. 72, 81, 99, 100, 101, 103, 104, 105, 115, 116, 117, 118, 119, 120, 121, 123, 129, 150)

[43] J.R. Birge and F. Louveaux. *Introduction to stochastic programming*. Springer Series in Operations Research and Financial Engineering. Springer, 2011. (Cited on p. 126)

[44] L. Blum, M. Shub, and S. Smale. On a theory of computation and complexity over the real numbers: NP-completeness, recursive functions and universal machines. *Bulletin of the American Mathematical Society (N.S.)*, 21(1):1–46, 1989. (Cited on p. 39)

[45] A. Bose. Computer simulation of power systems for operator training. In *Energy Conversion Engineering Conference,* 1989, *IECEC*-89, pages 165–169, vol. 1, 1989. (Cited on pp. 172, 234)

[46] S. Bose, D.F. Gayme, S. Low, and M.K. Chandy. Optimal power flow over tree networks. In *Proceedings of the* 49*th Annual Allerton Conference*, pages 1342–1348, 2011. (Cited on p. 47)

[47] S. Boyd and L. Vandenberghe. *Convex optimization*. Cambridge University Press, 2004. (Cited on pp. 35, 43, 51, 97, 108)

[48] S. Boyd and L. Vandenberghe. Additional exercises for *Convex Optimization*, Section 16.6. `http://www.stanford.edu/~boyd/cvxbook/bv_cvxbook_extra_exercises.pdf`, 2012. (Cited on pp. 48, 49)

[49] D. Braess. Über ein Paradox der Verkerhsplannung. *Unternehmenstorchung*, 12:258–268, 1968. (Cited on pp. 74, 151, 187)

[50] M. Burer and P. Wolfe. An algorithm for quadratic programming. *Naval Research Logistics Quarterly*, 3:95–110, 1956. (Cited on p. 122)

[51] S. Burer and K.M. Anstreicher. Second-order-cone constraints for extended trust-region subproblems. *SIAM Journal on Optimization*, 23:432–451, 2013. (Cited on pp. 37, 40)

[52] S. Burer and A.N. Letchford. On nonconvex quadratic programming with box constraints. *SIAM Journal on Optimization*, 20:1073–1089, 2009. (Cited on p. 34)

[53] S. Burer and A.N. Letchford. Non-convex mixed-integer nonlinear programming. *Surveys in Operations Research and Management Science*, 17:97–106, 2012. (Cited on p. 34)

[54] S. Burer and B. Yang. The trust region subproblem with non-intersecting linear constraints. *Mathematical Programming*, 149(1-2):253–264, 2015. (Cited on pp. 37, 40)

[55] M.B. Cain, R.P. O'Neill, and A. Castillo. *History of optimal power flow and formulations*. FERC staff technical paper, 2012. (Cited on p. 33)

[56] M.B. Cain, R.P. O'Neill, and A. Castillo. *The computational testing of AC optimal power flow using the current voltage formulations*. FERC staff technical paper, 2012. (Cited on pp. 34, 53, 54, 55, 56, 181)

[57] M.B. Cain, R.P. O'Neill, and A. Castillo. *The IV formulation and linearizations of the AC optimal power flow problem*. FERC staff technical paper, 2012. (Cited on p. 34)

[58] J. Carpentier. Contribution à l'étude du dispatching économique. *Bulletin de la Société Française des Électriciens*, 8:431–447, 1962. (Cited on p. 33)

[59] B.A. Carreras, V.E. Lynch, I. Dobson, and D.E. Newman. Critical points and transitions in an electric power transmission model for cascading failure blackouts. *Chaos*, 12(4):985–994, 2002. (Cited on pp. 144, 145, 146, 148, 149, 150, 151, 152, 156)

[60] B.A. Carreras, V.E. Lynch, I. Dobson, and D.E. Newman. Complex dynamics of blackouts in power transmission systems. *Chaos*, 14(3):1–33, 2004. (Cited on pp. 144, 145, 146, 148, 150, 151, 152, 178)

[61] B.A. Carreras, V.E. Lynch, M.L. Sachtjen, I. Dobson, and D.E. Newman. Modeling blackout dynamics in power transmission networks with simple structure. In *Hawaii International Conference on System Sciences*, 2001. (Cited on pp. 144, 145, 146, 148, 150, 151, 152, 178)

[62] CERC. Report on the grid disturbance on 30th July 2012 and grid disturbance on 31st July 2012. `http://www.cercind.gov.in`, 2012. (Cited on pp. 139, 140)

[63] C. Chen, A. Atamtürk, and S. Oren. Bound tightening for the alternating current optimal power flow problem. *IEEE Transactions on Power Systems*, to appear, 2015. (Cited on p. 47)

[64] H.-D. Chiang, A.J. Flueck, K.S. Shah, and N. Balu. CPFLOW: A practical tool for tracing power system steady-state stationary behavior due to load and generation variations. *IEEE Transactions on Power Engineering*, pages 623–634, 1995. (Cited on p. 233)

[65] H.-D. Chiang and F.F. Wu. Stability of nonlinear systems described by second-order vector differential equations. *IEEE Transactions on Circuits and Systems*, CAS-35:703–710, 1988. (Cited on p. 233)

[66] H.-D. Chiang, F.F. Wu, and P.P. Varaiya. Foundations of the potential energy boundary surface method for power system transient stability analysis. *IEEE Transactions on Circuits and Systems*, CAS-35:160–173, 1988. (Cited on p. 233)

[67] F.R.K. Chung. *Spectral graph theory*. CMBS Lecture Notes 92, American Mathematical Society, Providence, RI, 1997. (Cited on pp. 7, 9)

[68] F.R.K. Chung, R.L. Graham, and R.M. Wilson. Quasi-random graphs. *Combinatorica*, 9:345–362, 1987. (Cited on p. 166)

[69] CIGRE. Mathematical model for evaluation of conductor temperature in the steady (or quasi-steady) state (normal operation). *ELECTRA*, 44:109–115, 1992. (Cited on p. 180)

[70] C. Coffrin and P. Van Hentenryck. A linear-programming approximation of AC power flows. *INFORMS Journal on Computing*, pages 718–734, 2014. (Cited on pp. 51, 52)

[71] Federal Energy Regulatory Commission and North American Electric Reliability Corporation. Arizona-Southern California Outages on September 8, 2011. `http://www.ferc.gov/legal/staff-reports/04-27-2012-ferc-nerc-report.pdf`, 2012. (Cited on pp. 139, 141)

[72] A.R. Conn, K. Scheinberg, and L.N. Vicente. *Introduction to derivative-free optimization*. MPS-SIAM Book Series on Optimization, SIAM, Philadelphia, 2009. (Cited on pp. 114, 272)

[73] A.R. Conn, N.I.M. Gould, and P.L. Toint, editors. *Trust-region methods*. MPS/SIAM Series on Optimization, SIAM, Philadelphia, 2000. (Cited on pp. 36, 37, 114)

[74] ARKI Consulting and Development. CONOPT. `http://www.conopt.com`. (Cited on pp. 54, 114)

[75] K.J. Cormican, D.P. Morton, and R.K. Wood. Stochastic network interdiction. *Operations Research*, 46:184–197, 1998. (Cited on p. 127)

[76] F. Cucker and P. Bürgisser. Solving polynomial equations in smoothed polynomial time and a near solution to Smale's 17th problem. *Proceedings of the 42nd ACM Symposium on Theory of Computing*, pages 503–512, 2010. (Cited on p. 32)

[77] F. Cucker and P. Bürgisser. *Condition: The geometry of numerical algorithms*. Springer, 2013. (Cited on p. 272)

[78] S. Dash, O. Günlük, and A. Lodi. On the MIR closure of polyhedra. In *Integer Programming and Combinatorial Optimization*, pages 337–351. Springer, Berlin, Heidelberg, 2007. (Cited on p. 94)

[79] C.M. Davis and T. J. Overbye. Multiple element contingency screening. *IEEE Transactions in Power Systems*, 26:1294–1301, 2011. (Cited on p. 61)

[80] C.L. DeMarco. A new method of constructing Lyapunov functions for power systems. In *IEEE International Symposium on Circuits and Systems*, 1988, pages 905–908, 1988. (Cited on pp. 144, 233)

[81] C.L. DeMarco. A phase transition model for cascading network failure. *IEEE Control Systems Magazine*, 21:40–51, 2001. (Cited on pp. 144, 163, 233)

[82] S. DeNegre. *Interdiction and discrete bilevel linear programming*. PhD thesis, Lehigh University, 2011. (Cited on p. 76)

[83] B. Kocuk, S.S. Dey, and X.A. Sun. Inexactness of SDP relaxation and valid inequalities for optimal power flow. *IEEE Transactions on Power Systems*, to appear, 2015. (Cited on p. 47)

[84] S. Le Digabel. Algorithm 909: NOMAD: Nonlinear optimization with the MADS algorithm. *ACM Transactions on Mathematical Software*, 37:44:1–44:15, 2011. (Cited on p. 267)

[85] I. Dobson. Observations on the geometry of saddle node bifurcation and voltage collapse in electrical power systems. *IEEE Transactions on Circuits and Systems*, 39:240–243, 1992. (Cited on p. 12)

[86] I. Dobson, B.A. Carreras, V.E. Lynch, and D.E. Newman. An initial model for complex dynamics in electric power system blackouts. In *Hawaii International Conference on System Sciences*, 2001. (Cited on pp. 144, 145, 146, 148, 150, 151, 152, 178)

[87] I. Dobson, B.A. Carreras, V.E. Lynch, and D.E. Newman. Complex systems analysis of series of blackouts: cascading failures, critical points, and self-organization. *Chaos*, 17:1–13, 2007. (Cited on p. 165)

[88] I. Dobson, B.A. Carreras, and D.E. Newman. Complex dynamics of blackouts in power transmission systems. *Probability in the Engineering and Informational Sciences*, 19:15–332, 2005. (Cited on pp. 163, 164, 166)

[89] I. Dobson and H.-D. Chiang. Towards a theory of voltage collapse in electric power systems. *System and Control Letters*, 13:253–263, 1989. (Cited on p. 233)

[90] V. Donde, V. López, B. Lesieutre, A. Pinar, and J. Meza. Severe multiple contingency screening in electric power systems. *IEEE Transactions on Power Systems*, 23:406–417, 2008. (Cited on p. 106)

[91] J. Dupacová and R. Wets. Asymptotic behavior of statistical estimators and of optimal solutions of stochastic optimization problems. *The Annals of Statistics*, 16:1517–1549, 1988. (Cited on p. 130)

[92] M.J. Eppstein and P.D.H. Hines. A "random chemistry" algorithm for identifying collections of multiple contingencies that initiate cascading failure. *IEEE Transactions on Power Systems*, 27:1698–1705, 2012. (Cited on pp. 125, 160, 161, 235)

[93] EPRI. *Transmission reliability evaluation for large-scale systems (TRELSS)*. Technical Report, 2006. (Cited on p. 163)

[94] P. Erdös and A. Rényi. On random graphs. I. *Publicationes Mathematicae Debrecen*, 6:290–297, 1959. (Cited on pp. 166, 201)

[95] P. Erdös and J. Spencer. *Probabilistic methods in combinatorics*. Akadémiai Kiadó, Budapest, 1974. (Cited on p. 166)

[96] EUROSTAG. Software for the simulation of power system dynamics. `http://www.eurostag.be`. (Cited on p. 163)

[97] Neng Fan, R. Chen, and J. Watson. N-1-1 contingency-constrained optimal power flow by interdiction methods. In *Power and Energy Society General Meeting*, 2012 IEEE, pages 1–6, July 2012. (Cited on p. 57)

[98] E.B. Fisher, R.P. O'Neill, and M.C. Ferris. Optimal Transmission Switching. *IEEE Transactions on Power Systems*, 23:1346–1355, 2008. (Cited on pp. 263, 264, 267)

[99] R. Fitzmaurice, A. Keane, and M. O'Malley. Effect of short-term risk-aversive dispatch on a complex system model for power systems. *IEEE Transactions on Power Systems*, 26:460–469, 2011. (Cited on p. 144)

[100] R. Fletcher, S. Leyffer, and P. Toint. A brief history of filter methods. *SIAG/Optimization Views-and-News*, 18:2–12, 2007. (Cited on p. 272)

[101] U.S.-Canada Power System Outage Task Force. Report on the August 14, 2003 blackout in the United States and Canada: causes and recommendations. `http://energy.gov/sites/prod/files/oeprod/DocumentsandMedia/BlackoutFinal-Web.pdf`, (2004). (Cited on pp. 137, 138, 140, 153, 154, 155)

[102] B.A. Foster and D.M. Ryan. An integer programming approach to the vehicle scheduling problem. *Operational Research Quarterly* (1970–1977), 27:367–384, 1976. (Cited on p. 82)

[103] M. Fukuda, M. Kojima, K. Murota, and K. Nakata. Exploiting sparsity in semidefinite programming via matrix completion I: General framework. *SIAM Journal on Optimization*, 11:647–674, 2000. (Cited on p. 47)

[104] M.R. Garey and D.S. Johnson. *Computers and intractability: A guide to the theory of NP-completeness*. W. H. Freeman & Co., New York, 1990. (Cited on p. 265)

[105] F. Glover. Improved linear integer programming formulations of nonlinear integer problems. *Management Science*, 22:455–460, 1975. (Cited on pp. 93, 94)

[106] J.D. Glover, M.S. Sarma, and T.J. Overbye. *Power system analysis and design*. CENGAGE Learning, 2012. (Cited on pp. 1, 10, 26, 27, 179, 182, 212, 213, 234)

[107] D. Goldfarb and F. Alizadeh. Second-order cone programming. *Mathematical Programming*, 95:3–51, 2003. (Cited on p. 35)

[108] D. Goldfarb and G. Iyengar. Robust portfolio selection problems. *Mathematics of Operational Research*, 28:1–38, 2003. (Cited on p. 38)

[109] G.H. Golub and C.F. Van Loan. *Matrix computations* (3rd ed.). Johns Hopkins University Press, Baltimore, MD, 1996. (Cited on p. 62)

[110] N.I.M. Gould, S. Lucidi, M. Roma, and P.L. Toint. Solving the trust-region subproblem using the Lanczos method. *SIAM Journal on Optimization*, 9:504–525, 1999. (Cited on p. 36)

[111] D. Grigoriev, D.V. Pasechnik, and E. de Klerk. Quadratic optimization subject to a fixed number of quadratic constraints is polynomial-time. *Lecture slides*, 2002. (Cited on p. 39)

[112] V. Goel and I.E. Grossmann. A class of stochastic programs with decision dependent uncertainty. *Mathematical Programming*, 108:355–394, 2006. (Cited on p. 250)

[113] T. Güler and G. Gross. Detection of island formation and identification of causal factors under multiple line outages. *IEEE Transactions on Power Systems*, 22:505–513, 2007. (Cited on p. 61)

[114] A. Gupte, S. Ahmed, M.S. Cheon, and S. Dey. Solving mixed integer bilinear problems using MILP formulations. *SIAM Journal on Optimization*, 23:721–744, 2013. (Cited on p. 94)

[115] GUROBI. Gurobi Optimizer. `http://www.gurobi.com/`. (Cited on p. 54)

[116] J. Han and A. Papavasiliou. The impacts of transmission topology control on the European electricity network. *IEEE Transactions on Power Systems*, forthcoming, 2015. (Cited on p. 263)

[117] R.C. Hardiman, M.T. Kumbale, and Y.V. Makarov. An advanced tool for analyzing multiple cascading failures. In *Eighth International Conference on Probability Methods Applied to Power Systems*, Ames, Iowa, 2004. (Cited on p. 163)

[118] K.W. Hedman, M.C. Ferris, R.P. O'Neill, E.B. Fisher, and S.S. Oren. Co-optimization of generation unit commitment and transmission switching with N-1 reliability. *IEEE Transactions on Power Systems*, 25:1052–1063, 2010. (Cited on p. 263)

[119] K.W. Hedman, R.P. O'Neill, E.B. Fisher, and S.S. Oren. Optimal transmission switching—Sensitivity analysis and extensions. *IEEE Transactions on Power Systems*, 23:1469–1479, 2008. (Cited on pp. 263, 264, 267)

[120] K.W. Hedman, S.S. Oren, and R.P. O'Neill. A review of transmission switching and network topology optimization. *Proceedings of the* 2011 *IEEE PES*, pages 1–7, 2011. (Cited on p. 263)

[121] H. Held and D.L. Woodruff. Heuristics for multi-stage interdiction of stochastic networks. *Journal of Heuristics*, 11:483–500, 2005. (Cited on p. 250)

[122] H.L. Hijazi, C. Coffrin, and P. Van Hentenryck. *Convex quadratic relaxations for mixed-integer nonlinear programs in power systems*. `http://www.optimization-online.org/DB_HTML/2013/09/4057.html`, 2013. (Cited on p. 51)

[123] N. G. Hingorani and L. Gyugyi. *Understanding FACTS: concepts and technology of flexible AC transmission systems*. IEEE, 2000. (Cited on p. 7)

[124] I.A. Hiskens and R.J. Davy. Exploring the power flow solution space boundary. *IEEE Transactions on Power Systems*, 16:389–95, 2001. (Cited on pp. 12, 53)

[125] IBM. ILOG CPLEX Optimizer. `http://www-01.ibm.com/software/integration/optimization/cplex-optimizer/index.html`. (Cited on pp. 54, 97)

[126] IEEE. IEEE standard for calculating the current-temperature of rare overhead conductors. *Std* 738-2006 *(Revision of IEEE Std* 738-1993*)*, pages 1–59, January 2007. (Cited on pp. 153, 177)

[127] M. Ilić, E. Allen, J. Chapman, C. King, J. Lang, and E. Litvinov. Preventing future blackouts by means of enhanced electric power systems control: From complexity to order. *Proceedings of IEEE*, 93:1920–1941, 2005. (Cited on p. 230)

[128] M. Ilić and J. Zaborski. *Dynamics and control of large electric power systems*. Wiley and Sons, 2000. (Cited on pp. 7, 26)

[129] E. Israeli and R.K. Wood. Shortest path network interdiction. *Networks*, 40:97–111, 2002. (Cited on p. 76)

[130] R.A. Jabr. Optimal power flow using an extended conic quadratic formulation. *IEEE Transactions on Power Systems*, 23:1000–1008, 2008. (Cited on p. 42)

[131] U. Janjarassuk and J.T. Linderoth. Reformulation and sampling to solve a stochastic network interdiction problem. *Networks*, 52:120–132, 2008. (Cited on pp. 76, 127, 130, 131, 133, 134)

[132] T. Jonsbraten. *Optimization models for petroleum field exploitation*. PhD thesis, Norwegian School of Economics and Business Administration, 1998. (Cited on p. 250)

[133] H. Kerivin and A.R. Mahjoub. Design of survivable networks: A survey. *Networks*, 46:1–21, 2005. (Cited on p. 60)

[134] T. Kim, S.J. Wright, D. Bienstock, and S. Harnett. Vulnerability analysis of power systems. Preprint, arXiv:1503.02360, 2015. (Cited on pp. 121, 122, 123)

[135] B. Kirby and E. Hirst. Reliability management and oversight. *U.S. Department of Energy National transmission grid study papers—Issue papers*, 2002. `http://certs.lbl.gov`. (Cited on p. 24)

[136] B.J. Kirby. Spinning reserve from responsive loads. `http:web.ornl.gov/sci/ees/etsd/pes/pubs/ORNLTM200319.pdf`, 2003. (Cited on p. 172)

[137] B.J. Kirby, J. Dyer, C. Martinez, R.A. Shoureshi, R. Guttromson, and J. Dangle. Frequency control concerns in the North American electric power system. *Oak Ridge National Laboratory*, ORNL/TM-2003/41, 2002. (Cited on p. 22)

[138] A.J. Kleywegt, A. Shapiro, and T. Homem-de-Mello. The sample average approximation method for stochastic discrete optimization. *SIAM Journal on Optimization*, 12:479–502, 2001. (Cited on pp. 127, 131)

[139] A. Klos and J. Wojcicka. Physical aspects of the nonuniqueness of load flow solutions. *Electrical Power and Energy Systems*, 13:268–276, 1991. (Cited on p. 12)

[140] T. Koch, T. Achterberg, E. Andersen, O. Bastert, T. Berthold, R.E. Bixby, E. Danna, G. Gamrath, A.M. Gleixner, S. Heinz, A. Lodi, H. Mittelmann, T. Ralphs, D. Salvagnin, D.E. Steffy, and K. Wolter. MIPLIB 2010. *Mathematical Programming Computation*, 3:103–163, 2011. (Cited on p. 60)

[141] B. Kocuk, H. Jeon, S. Dey, J. Linderoth, J. Luedtke, and X. Sun. A cycle-based formulation and valid inequalities for DC power transmission problems with switching. Preprint, arXiv:1412.6245, 2014. (Cited on p. 264)

[142] B. Bouzaiene-Ayari, L. Al-Kanj, and W. B. Powell. A probability model for grid faults using incomplete information. IEEE Transactions on Smart Grid (to appear), 2015. (Cited on p. 247)

[143] J.B. Lasserre. An explicit exact SDP relaxation for nonlinear 0-1 programs. In K. Aardal and A. M. H. Gerards, editors, *Integer programming and combinatorial optimization*, number 2081 in Lecture Notes in Computer Science, pages 293–303, Springer, 2001. (Cited on p. 41)

[144] J.B. Lasserre. Global optimization with polynomials and the problem of moments. *SIAM Journal on Optimization*, 11:796–817, 2001. (Cited on p. 41)

[145] M. Laurent. Sums of squares, moment matrices and polynomial optimization. In M. Putinar and S. Sullivant, editors, *Emerging applications of algebraic geometry*, number 149 in IMA Volumes in Mathematics and Its Applications, pages 157–270, Springer, 2009. (Cited on p. 41)

[146] J. Lavaei and S. Low. Zero duality gap in optimal power flow problem. *IEEE Transactions on Power Systems*, 27:92–107, 2012. (Cited on pp. 42, 44, 45, 46, 47, 48)

[147] K. Lehmann, A. Grastien, and P. Van Hentenryck. The complexity of DC-switching problems. Preprint, arXiv:1411.4369, 2014. (Cited on p. 264)

[148] B.C. Lesieutre, D.K. Molzahn, A.R. Borden, and C.L. DeMarco. Examining the limits of the application of semidefinite programming to power flow problems. In *Proceedings of the 49th Annual Allerton Conference on Communication, Control, and Computing*, pages 1492–1499, 2011. (Cited on pp. 46, 47)

[149] H. Liao. *Cascading failures in power grids: modeling, metrics and mitigation*. PhD thesis, Carnegie Mellon University, 2011. (Cited on p. 230)

[150] H. Liao and M. Ilić. Stress test model of cascading failures in power grids. *Proceedings of Bulk Power System Dynamics and Control* VII, Charleston, S.C., 2007. (Cited on p. 230)

[151] C. Lim and J.C. Smith. Algorithms for discrete and continuous multicommodity flow network interdiction problems. *IIE Transactions*, 39:15–26, 2007. (Cited on p. 76)

[152] J. Linderoth and S. Wright. Decomposition algorithms for stochastic programming on a computational grid. *Computational Optimization and Applications*, 24(2-3):207–250, 2003. (Cited on p. 126)

[153] M. Liu and G. Gross. Effectiveness of the distribution factor approximations used in congestion modeling. In *Proceedings of the 14th Power Systems Computation Conference*, Seville, Spain, pages 24–28, 2002. (Cited on p. 61)

[154] J. Luedtke, M. Namazifar, and J. Linderoth. Some results on the strength of relaxations of multilinear functions. *Mathematical Programming*, 136:325–351, 2012. (Cited on pp. 97, 98)

[155] Z.-Q. Luo, J.-S. Pang, and D. Ralph. *Mathematical programs with equilibrium constraints*. Cambridge University Press, 1996. (Cited on p. 76)

[156] J. Ma, Y. Makarov, and Z. Dong. Phasor measurement unit and its application in modern power systems. In *Emerging Techniques in Power System Analysis*, pages 147–184. Springer, Berlin, Heidelberg, 2010. (Cited on p. 29)

[157] R. Madani, G. Fazelnia, S. Sojoudi, and J. Lavaei. Finding low-rank solutions of sparse linear matrix inequalities using convex optimization. `http://www.ieor.berkeley.edu/~lavaei/LMI_Low_Rank.pdf`, 2014. (Cited on p. 47)

[158] R. Madani, G. Fazelnia, S. Sououdi, and J. Lavaei. Low-rank solutions of matrix inequalities with applications to polynomial optimization and matrix completion problems. *Proceedings of* 2014 *CDC*, 2014. (Cited on p. 47)

[159] W.-K. Mak, D.P. Morton, and R.K. Wood. Monte Carlo bounding techniques for determining solution quality in stochastic programs. *Operations Research Letters*, 24:47–56, 1999. (Cited on pp. 130, 131)

[160] G.P. McCormick. Computability of global solutions to factorable nonconvex programs: part I—convex underestimating problems. *Mathematical Programming*, 10:147–175, 1976. (Cited on p. 97)

[161] J.D. Meiss. *Differential dynamical systems*. Mathematical Modeling and Computation. SIAM, Philadelphia, 2007. (Cited on pp. 231, 232)

[162] D. Molzahn. *Application of semidefinite optimization techniques to problems in electric power systems*. PhD thesis, University of Wisconsin, 2013. (Cited on pp. 46, 47)

[163] D.K. Molzahn and I.A. Hiskens. Sparsity-exploiting moment-based relaxations of the optimal power flow problem. *IEEE Transactions on Power Systems*, 30(6):3168–3180, 2015. (Cited on pp. 42, 47)

[164] J.J. Moré and D.C. Sorensen. Computing a trust region step. *SIAM Journal on Scientific and Statistical Computing*, 4:553–572, 1983. (Cited on p. 36)

[165] D.P. Nedic, I. Dobson, D.S. Kirschen, B.A. Carreras, and V.E. Lynch. Criticality in a cascading failure blackout model. *Electrical Power & Energy Systems*, 28(9):627–633, 2006. (Cited on p. 163)

[166] NERC. *Balancing and frequency control*. `http://www.nerc.com/docs/oc/rs/NERCBalancingandFrequencyControl040520111.pdf`. (Cited on pp. 22, 25)

[167] NERC. *Frequency response*. `http://www.nerc.com/filez/standards/Frequency_Response.html`. (Cited on pp. 22, 27)

[168] NERC. *Mandatory standards subject to enforcement*. `http://www.nerc.net/standardsreports/standardssummary.aspx`. (Cited on p. 57)

[169] NERC. *Standard TPL-001-0.1*. `http://www.nerc.com/files/TPL-001-0_1.pdf`. (Cited on p. 57)

[170] J. Nocedal and S.J. Wright. *Numerical optimization*. Springer, Berlin, 2006. (Cited on p. 272)

[171] V.I. Norkin, G.Ch. Pflug, and A. Ruszczynski. A branch and bound method for stochastic global optimization. *Mathematical Programming*, 83:425–450, 1996. (Cited on p. 130)

[172] University of Washington. *Power systems test archive*. `http:www.ee.washington.edu/research/pstca/`. (Cited on pp. 47, 191)

[173] R.P. O'Neill, E.A. Krall, K.W. Hedman, and S.S. Oren. A model and approach to the challenge posed by optimal power systems planning. *Mathematical Programming*, 140:239–266, 2013. (Cited on p. 29)

[174] Ziena Optimization. *Knitro*. `http://www.ziena.com/knitro.htm`. (Cited on pp. 54, 114)

[175] J. Ostrowski, J. Wang, and C. Liu. Transmission switching with connectivity-ensuring constraints. *IEEE Transactions on Power Systems*, 29(6):2621–2627, 2014. (Cited on p. 263)

[176] T.J. Overbye. *Power system operations and visualization*. `http://www.pserc.wisc.edu/research/white_papers.aspx`. (Cited on pp. 1, 24)

[177] T.J. Overbye. A load flow calculation method for ill-conditioned power systems. *IEEE Transactions on Power Apparatus Systems*, PAS-100:1736–1743, 1981. (Cited on p. 12)

[178] T.J. Overbye. A power flow measure for unsolvable cases. *IEEE Transactions on Power Systems*, 9:1359–1365, 1994. (Cited on pp. 12, 13, 33, 55, 183)

[179] T.J. Overbye. Computation of a practical method to restore a power flow solvability. *IEEE Transactions on Power Systems*, 10:280–286, 1995. (Cited on pp. 12, 13, 33, 55, 183)

[180] G. Pataki. On the rank of extreme matrices in semidefinite programs and the multiplicity of optimal eigenvalues. *Mathematics of Operations Research*, 23:460–469, 1998. (Cited on p. 40)

[181] J.-P. Paul and K.R.W. Bell. A flexible and comprehensive approach to the assessment of large-scale power system security under uncertainty. *International Journal of Electrical Power and Energy Systems*, 26:265–272, 2004. (Cited on p. 163)

[182] R. Pfitzner, K. Turitsyn, and M. Chertkov. Controlled tripping of overheated lines mitigates power outages. Preprint, arXiv:1104.4558v2, 2011. (Cited on pp. 158, 159, 160, 162, 170, 171, 235)

[183] R. Pfitzner, K. Turitsyn, and M. Chertkov. Statistical classification of cascading failures in power grids. In *IEEE Power Engineering Society General Meeting*, 2011. (Cited on p. 158)

[184] G.Ch. Pflug. Online optimization of simulated Markovian processes. *Mathematics of Operations Research*, 15:381–395, 1990. (Cited on p. 250)

[185] D.T. Phan. Lagrangian duality and branch-and-bound algorithms for optimal power flow. *Operations Research*, 60:275–285, 2012. (Cited on p. 47)

[186] C.A. Phillips. The network inhibition problem. In *Proceedings of the 25th Annual ACM Symposium on Theory of Computing*, ACM Press, New York, pages 776–685, 1993. (Cited on pp. 106, 107)

[187] A. Pinar, J. Meza, V. Donde, and B. Lesieutre. Optimization strategies for the vulnerability analysis of the electric power grid. *SIAM Journal on Optimization*, 20:1786–1810, 2010. (Cited on pp. 106, 107, 108, 109, 111)

[188] I. Pólik and T. Terlaky. A survey of the S-lemma. *SIAM Review*, 49:371–418, 2007. (Cited on pp. 37, 38, 40)

[189] PowerWorld Corporation. PowerWorld Simulator. `http://www.powerworld.com`. (Cited on p. 55)

[190] C. Raack, A.M.C.A. Koster, S. Orlowski, and R. Wessäly. On cut-based inequalities for capacitated network design polyhedra. *Networks*, 57(2):141–156, 2011. (Cited on p. 60)

[191] Y.G. Rebours and D.S. Kirschen. *What is spinning reserve?* The University of Manchester, 2005. (Cited on pp. 28, 172)

[192] Y.G. Rebours, D.S. Kirschen, M. Trotignon, and S. Rossignol. A survey of frequency and voltage control ancillary services. Part I: Technical features. *IEEE Transactions on Power Systems*, 22:350–357, 2007. (Cited on pp. 24, 27, 172)

[193] Y.G. Rebours, D.S. Kirschen, M. Trotignon, and S. Rossignol. A survey of frequency and voltage control ancillary services. Part II: Economic features. *IEEE Transactions on Power Systems*, 22:358–366, 2007. (Cited on p. 172)

[194] F. Rendl and H. Wolkowicz. A semidefinite framework for trust region subproblems with applications to large scale minimization. *Mathematical Programming*, 77:273–299, 1997. (Cited on pp. 36, 37)

[195] J. Renegar. On the efficiency of Newton's method in approximating all zeros of a system of complex polynomials. *Mathematics of Operations Research*, 12(1):121–148, 1987. (Cited on p. 32)

[196] A.D. Rikun. A convex envelope formula for multilinear functions. *Journal of Global Optimization*, 10:425–437, 1997. (Cited on p. 97)

[197] R.T. Rockafellar and S. Uryasev. Optimization of conditional value-at-risk. *Journal of Risk*, 2:21–41, 2000. (Cited on p. 134)

[198] J. Salmeron, K. Wood, and R. Baldick. Analysis of electric grid security under terrorist threat. *IEEE Transactions on Power Systems*, 19:905–912, 2004. (Cited on p. 124)

[199] J. Salmeron, K. Wood, and R. Baldick. Worst-case interdiction analysis of large-scale electric power grids. *IEEE Transactions on Power Systems*, 24:96–104, 2009. (Cited on p. 124)

[200] San Diego Union-Tribune. Grid operators didn't share info during blackout. `http://www.utsandiego.com/news/2011/sep/22/5-authorities-handle-electric-grid-didnt-share-inf/`. (Cited on p. 139)

[201] A. Schecter and R.P. O'Neill. Exploration of the AC OPF feasible region for the standard IEEE test set. *FERC Technical Conference, Docket No. AD*10-12003, 2012. (Cited on pp. 53, 54)

[202] R. Sedgewick. *Algorithms in C: Fundamentals, data structures, sorting, searching, and graph algorithms*. Addison-Wesley Professional, 2001. (Cited on p. 117)

[203] G.B. Sheble and G.N. Fahd. Unit commitment literature synopsis. *IEEE Transactions on Power Systems*, 9:128–135, 1994. (Cited on p. 25)

[204] T. Shiina and J. R. Birge. Stochastic unit commitment problem. *International Transactions in Operational Research*, 11:19–32, 2004. (Cited on p. 126)

[205] M. Shub and S. Smale. Complexity of Bézout's theorem. I. Geometric aspects. *Journal of the American Mathematical Society*, 6(2):459–501, 1993. (Cited on p. 32)

[206] M. Shub and S. Smale. Complexity of Bézout's theorem V. Polynomial time. *Theoretical Computer Science*, 133:141–164, 1994. (Cited on p. 32)

[207] C. Singh and I. Hiskens. Direct assessment of protection operation and non-viable transients. *IEEE Transactions on Power Systems*, 16:427–434, 2001. (Cited on p. 233)

[208] R.M. Van Slyke and R. Wets. *L-shaped linear programs with applications to optimal control and stochastic programming*. *SIAM Journal on Applied Mathematics*, 17:638–663, 1969. (Cited on p. 131)

[209] S. Smale. Differentiable dynamical systems. *Bulletin of the American Mathematical Society*, 73:747–817, 1967. (Cited on p. 231)

[210] S. Smale. Mathematical problems for the next century. *Mathematics: Frontiers and Perspectives, American Mathematical Society*, pages 271–294, 2000. (Cited on p. 32)

[211] Stanford Business Software. *MINOS*. `http://www.sbsi-sol-optimize.com/asp/sol_product_minos.htm`. (Cited on pp. 54, 114)

[212] Stanford Business Software. *SNOPT*. `http://www.sbsi-sol-optimize.com/asp/sol_product_snopt.htm`. (Cited on pp. 54, 114)

[213] S. Sojoudi and J. Lavaei. Physics of power networks makes hard optimization problems easy to solve. In *IEEE Power Engineering Society General Meeting*, 2012. (Cited on p. 47)

[214] R.J. Stern and H. Wolkowicz. Indefinite trust region subproblems and nonsymmetric eigenvalue perturbations. *SIAM Journal on Optimimization*, 5:286–313, 1995. (Cited on p. 38)

[215] M. Stoer. *Design of survivable networks*. Lecture notes in mathematics. Springer-Verlag, Berlin, New York, 1992. (Cited on p. 60)

[216] J. Sturm. Using SeDuMi 1.02, a MATLAB toolbox for optimization over symmetric cones. *Optimization Methods and Software*, 11:625–653, 1999. (Cited on p. 47)

[217] J. Sturm and S. Zhang. On cones of nonnegative quadratic functions. *Mathematics of Operations Research*, 28:246–267, 2003. (Cited on pp. 37, 38)

[218] B. Sturmfels. *Solving systems of polynomial equations*. Department of Mathematics, University of California at Berkeley, 2002. (Cited on p. 32)

[219] M. Tawarmalani and N.V. Sahinidis. A polyhedral branch-and-cut approach to global optimization. *Mathematical Programming*, 103:225–249, 2005. (Cited on p. 82)

[220] The New York Times. Much of east coast phone service is disrupted by Jersey cable break. `http://www.nytimes.com/1988/11/19/us/much-of-east-coast-phone-service-is-disrupted-by-jersey-cable-break.html`. (Cited on p. 60)

[221] K.S. Turitsyn and P.A. Kaplunovich. Fast algorithm for N-2 contingency problem. Preprint, arXiv:1211.0728, 2012. (Cited on pp. 61, 64, 65)

[222] R.J. Vanderbei. LOQO: An interior point code for quadratic programming. *Optimization Methods and Software*, 11:451–484, 1998. (Cited on pp. 113, 114)

[223] R.J. Vanderbei and D.F. Shanno. An interior-point algorithm for nonconvex nonlinear programming. *Computational Optimization and Applications*, 13:231–252, 1999. (Cited on p. 113)

[224] M.V. Venkatasubramanian, R. Goddard, Y. Li, and Y. Chen. *Coordination of transmission line transfer capabilities*. Power Systems Engineering Research Center Publication 02-28, 2002. (Cited on p. 176)

[225] A. Wächter and L.T. Biegler. On the implementation of a primal-dual interior point filter line search algorithm for large-scale nonlinear programming. *Mathematical Programming*, 106:25–57, 2006. (Cited on pp. 34, 54, 113, 114)

[226] H. Wan, J.D. McCalley, and V. Vittal. Increasing thermal rating by risk analysis. *IEEE Transactions on Power Systems*, 14:815–828, 1999. (Cited on p. 177)

[227] Wikipedia. 2011 Southwest blackout. `http://en.wikipedia.org/wiki/2011_Southwest_blackout` Reviewed February 2012. (Cited on p. 139)

[228] Wikipedia. Operating reserve. `http://en.wikipedia.org/wiki/Operating_reserve`. (Cited on p. 172)

[229] R.K. Wood. Deterministic network interdiction. *Mathematical and Computer Modelling*, 17:1–18, 1993. (Cited on pp. 76, 107, 127)

[230] H. Xiao and E.M. Yeh. Cascading link failure in the power grid: A percolation-based analysis. In *Proceedings of the IEEE International Workshop on Smart Grid Communications*, 2011. (Cited on pp. 165, 166, 204)

[231] V.A. Yakubovich. S-procedure in nonlinear control theory. *Vestnik Leningrad University*, 1:62–777, 1971. (Cited on p. 37)

[232] Y. Ye. A new complexity result on minimization of a quadratic function with a sphere constraint. In C. Floudas and P. Pardalos, editors, *Recent advances in global optimization*. Princeton University Press, 1992. (Cited on p. 37)

[233] Y. Ye and S. Zhang. New results on quadratic minimization. *SIAM Journal on Optimization*, 14:245–267, 2003. (Cited on p. 37)

[234] R.D. Zimmerman, C.E. Murillo-Sánchez, and D. Gan. *MATPOWER steady-state operations, planning and analysis tools for power systems research and education*. `http://www.pserc.cornell.edu/matpower/`. (Cited on pp. 9, 11, 34, 53, 123, 125, 191)

[235] R.D. Zimmerman, C.E. Murillo-Sánchez, and D. Gan. MATPOWER, a MATLAB power system simulation package. *IEEE Transactions on Power Systems*, 26(1):12–19, 2011. (Cited on p. 191)

[236] M. Zuckerberg. *A set theoretic approach to lifting procedures for* 0, 1 *integer programming*. PhD thesis, Columbia University, 2004. (Cited on p. 88)

Index